Open Standards and the Digital Age

How did openness become a foundational value for the networks of the twenty-first century? *Open Standards and the Digital Age* answers this question through an interdisciplinary history of information networks that pays close attention to the politics of standardization. For much of the twentieth century, information networks such as the monopoly Bell System and the American military's Arpanet were closed systems subject to centralized control. In the 1970s and 1980s, however, engineers in the United States and Europe experimented with design strategies to create new digital networks. In the process, they embraced discourses of "openness" to describe their ideological commitments to entrepreneurship, technological innovation, and participatory democracy. The rhetoric of openness has flourished – for example, in movements for open government, open source software, and open access publishing – but such rhetoric also obscures the ways the Internet and other "open" systems still depend heavily on hierarchical forms of control.

Andrew L. Russell is an Assistant Professor of History and the Director of the Program in Science and Technology Studies in the College of Arts and Letters at the Stevens Institute of Technology in Hoboken, New Jersey. Before arriving at Stevens, he was a Postdoctoral Fellow in Duke University's John Hope Franklin Humanities Institute. Russell earned his PhD from Johns Hopkins University (2007), MA from the University of Colorado at Boulder (2003), and BA from Vassar College (1996). His work has been published in journals such as *IEEE Annals of the History of Computing*, *Enterprise & Society*, and *Information & Culture*. Russell has been awarded fellowships from the Charles Babbage Institute, the Association for Computing Machinery, the Institute for Electrical and Electronics Engineers, and the American Society for Information Science and Technology.

CAMBRIDGE STUDIES IN THE EMERGENCE OF GLOBAL ENTERPRISE

The world economy has experienced a series of globalizations in the past two centuries, and each has been shaped by business enterprises, by their national political contexts, and by new sets of international institutions. Cambridge Studies in the Emergence of Global Enterprise focuses on those business firms that have given the global economy many of its most salient characteristics, particularly regarding how they have fostered new technology, new corporate cultures, new networks of communication, and new strategies and structures designed to meet global competition. All the while, they have accommodated to changes in national and international regulations, environmental standards, and cultural norms. This is a history that needs to be understood because we all have a stake in the performance and problems of global enterprise.

We are especially pleased to have Andrew L. Russell's *Open Standards and the Digital Age: History, Ideology, and Networks* in our series. The subject of open standards in the digital world of the Third Industrial Revolution is of importance to both scholars and those who frame public policy. There is a deep and broad movement calling for greater transparency and openness in all fields of modern science and technology. Yet, we have very few solidly researched studies of how, exactly, these demands play out, how the institutions they create evolve, and how the leaders of these organizations frame and re-frame their proposals. We have virtually no well-researched studies that place these developments in a comparative, historical framework. Russell's book achieves this goal, reaching back to the nineteenth-century roots of telecommunications and sweeping forward from telegraph to the telephone, from the monopoly Bell System through the digital transformation of the late twentieth century. In a world now dependent on electronic modes of communication and calculation, the power to shape standards is a crucial aspect of our economic, political, social, and cultural development.

Editors

Louis Galambos, The Johns Hopkins University
Geoffrey Jones, Harvard Business School

OTHER BOOKS IN THE SERIES

Teresa da Silva Lopes, *Global Brands: The Evolution of Multinationals in Alcoholic Beverages*

Christof Dejung and Niels P. Petersson, *The Foundations of Worldwide Economic Integration: Power, Institutions, and Global Markets, 1850–1930*

William J. Hausman, *Global Electrification: Multinational Enterprise and International Finance in the History of Light and Power, 1878–2007*

Christopher Kobrak, *Banking on Global Markets: Deutsche Bank and the United States, 1870 to the Present*

Christopher Kobrak, *National Cultures and International Competition: The Experience of Schering AC, 1851–1950*

Christopher D. McKenna, *The World's Newest Profession: Management Consulting in the Twentieth Century*

Johann Peter Murmann, *Knowledge and Competitive Advantage: The Coevolution of Firms, Technology, and National Institutions*

Neil Rollings, *British Business in the Formative Years of European Integration, 1945–1973*

Open Standards and the Digital Age

History, Ideology, and Networks

ANDREW L. RUSSELL

Stevens Institute of Technology

CAMBRIDGE
UNIVERSITY PRESS

32 Avenue of the Americas, New York NY 10013-2473, USA

Cambridge University Press is part of the University of Cambridge.

It furthers the University's mission by disseminating knowledge in the pursuit of
education, learning, and research at the highest international levels of excellence.

www.cambridge.org
Information on this title: www.cambridge.org/9781107612044

© Andrew L. Russell 2014

This publication is in copyright. Subject to statutory exception
and to the provisions of relevant collective licensing agreements,
no reproduction of any part may take place without the written
permission of Cambridge University Press.

First published 2014

Printed in the United States of America

A catalog record for this publication is available from the British Library.

Library of Congress Cataloging in Publication data
Russell, Andrew L., 1975–
Open standards and the digital age : history, ideology, and networks / Andrew L. Russell.
    pages   cm. – (Cambridge studies in the emergence of global enterprise)
Includes bibliographical references and index.
ISBN 978-1-107-03919-3 (hardback) – ISBN 978-1-107-61204-4 (paperback)
    1. Standardization – United States – History.   2. Information technology – Standards – United
States – History.   3. Telecommunication – Standards – United States – History.   I. Title.
T59.2.U6R87   2014
602'.18–dc23   2013038663

ISBN 978-1-107-03919-3 Hardback
ISBN 978-1-107-61204-4 Paperback

Cambridge University Press has no responsibility for the persistence or accuracy of URLs
for external or third-party Internet Web sites referred to in this publication and does not
guarantee that any content on such Web sites is, or will remain, accurate or appropriate.

To my father and mother, Lawrence Keith Russell and Carol Pereicich Russell.

Contents

Tables and Figures

Tables

Figures

Acknowledgments

An astonishing number of people and institutions – teachers, friends, schools, and foundations – have helped me research and develop the ideas that I have tried to articulate in this book. I could not possibly list all of my intellectual debts here, but I am delighted to acknowledge the most substantial of them.

First, I am pleased to thank my colleagues at the University of Colorado at Boulder, Johns Hopkins University, Duke University, Stevens Institute of Technology, and in the professional communities that cluster around the Society for the History of Technology, the Business History Conference, and the Special Interest Group for Computers, Information, and Society. In particular, Bill Leslie and Louis Galambos patiently helped me pull meaning out of the acronym-laden snippets that I collected during my dissertation research. Conversations and correspondence with the following people also have improved my work immeasurably: Jane Anderson, Hyungsub Choi, John Day, Michael Aaron Dennis, Nathan Ensmenger, Margaret B. W. Graham, Thomas Haigh, Dale Hatfield, Sheldon Hochheiser, Richard R. John, Susan D. Jones, Robert H. Kargon, Sharon Kingsland, Harry Marks, Thomas J. Misa, Craig N. Murphy, Eric Nystrom, Mark Pittenger, James A. Schafer, Nick Smith, Daniel P. Todes, Steven W. Usselman, Philip J. Weiser, JoAnne Yates, and countless students in my classes at Stevens Institute of Technology. Three of my colleagues – James E. McClellan III, Valérie Schafer, and Lee Jared Vinsel – read multiple chapters of this manuscript as it was in progress, and their encouragement and gentle criticisms helped me at crucial moments. James Pelkey generously provided access to unique materials and interviews that he had collected, and gave me a flash of inspiration as I was revising the manuscript.

Despite euphoric claims that everything worth finding can now be found on the Internet, librarians and archivists still provide essential wisdom and support to help researchers find things that are not yet (or never will be) one or two clicks away. I wish I knew the names of all of the library staff at Colorado, Johns Hopkins, Duke, and the Herbert Hoover Presidential Library who handled my odd requests and late returns gracefully and charitably – you all have my deep appreciation. I would also like to thank Linda Beninghove, Adam Winger, and the inimitable Interlibrary Loan team at the S. C. Williams Library at Stevens Institute of Technology; William D. Caughlin at the AT&T Archives and History Center in San Antonio, Texas; George Kupczak at the AT&T Archives and History in Warren, New Jersey; Marc Weber at the Computer History Museum in Mountain View, California; and Katie Charlet, Arvid Nelson, and Stephanie Crowe at the Charles Babbage Institute for the History of Information Technology at the University of Minnesota, Minneapolis.

Archival research and historical scholarship are costly and time consuming, and the returns that come from investments in them do not generate quick or predictable profits (at least in the economic sense of the term). I am therefore very happy to acknowledge sources of financial support that have helped me complete this book: the Adelle and Erwin Tomash Fellowship in the History of Information Processing and the Arthur L. Norberg Travel Fund, both administered by the Charles Babbage Institute at the University of Minnesota; the Association for Computing Machinery History Committee; the American Society for Information Science & Technology History Fund; the John Hope Franklin Humanities Institute at Duke University; the IEEE Life Members Committee; the IEEE Computer Society History Committee; the College of Arts & Sciences and the Interdisciplinary Telecommunications Program at the University of Colorado; the Department of the History of Science and Technology at Johns Hopkins; and, most recently, the College of Arts & Letters at Stevens and its dedicated Dean Lisa Dolling.

I would like to thank Lewis Bateman and his team of editorial assistants at Cambridge University Press for their skill and patience as we turned my manuscript submission into a book. The anonymous reviewers commissioned by Cambridge University Press provided new insights and did their best to help me improve the manuscript. An earlier version of some material in Chapter 8 appeared as "'Rough Consensus and Running Code' and the Internet-OSI Standards War," *IEEE Annals of the History of Computing* 28 (2006): 48–61, and I am thankful for permission to reuse it.

Although the topic of standardization has the potential to "induce narcolepsy" (as an editor warned me long ago), I would like to thank the friends, family, and strangers who confirmed my instinct that such a mundane topic actually taps into deep human emotions and societal concerns. My sense of the meaning of standards and standardization – and not only in our present digital age – has been sharpened through endless hours of casual conversation about computers, cars, music, soccer, shower curtains, and so on. To those who have taken part in those conversations (you know who you are!): thank you.

Finally, I am lucky to have terrific friends and family who make my life richer and better in the moments when I have been contemplating this book and in the moments when I have set it aside. I am profoundly grateful for the friendship of Brian Shaw and Seth David Halvorson, and for the love of my wife, Lesley Bernier Russell. I dedicate this book to my parents, Larry and Carol Russell. The admiration, gratitude, and love I feel for them leave me at a loss for words.

Acronyms

ACM	Association for Computing Machinery
AESC	American Engineering Standards Committee
AFNOR	Association Française de Normalisation
AIEE	American Institute of Electrical Engineers
AIME	American Institute of Mining Engineers
ANSI	American National Standards Institute
AREMWA	American Railroad Engineering and Maintenance of Way Association
ARPA	Advanced Research Projects Agency
ARPANET	Advanced Research Projects Agency Network
ASA	American Standards Association
ASCE	American Society of Civil Engineers
ASCII	American Standard Code for Information Interchange
ASME	American Society of Mechanical Engineers
ASTM	American Society for Testing Materials
AT&T	American Telephone and Telegraph
BBN	Bolt, Beranek and Newman
BESC	British Engineering Standards Committee
BSI	British Standards Institute
CCITT	International Telegraph and Telephone Consultative Committee
CIDR	Classless Inter-Domain Routing
CII	Compagnie Internationale pour l'Informatique
CLNP	ConnectionLess Network Protocol
CNET	Centre National d'Etudes des Telecommunications
CNRI	Corporation for National Research Initiatives
DARPA	Defense Advanced Research Projects Agency
DISY	Distributed Systems

ECMA	European Computing Machinery Association
ECSA	Exchange Carriers Standards Association
FCC	Federal Communications Commission
FORTRAN	Formula Translating System
GEC	General Engineering Circular
GOSIP	Government Open Systems Interconnection Profile
HDNA	Honeywell Distributed Network Architecture
HDSA	Honeywell Distributed Systems Architecture
IAB	Internet Advisory Board (1984–1986)
IAB	Internet Activities Board (1986–1992)
IAB	Internet Architecture Board (1992–present)
IBM	International Business Machines
ICC	International Computation Centre
ICCB	Internet Configuration Control Board
IEC	International Electrotechnical Commission
IEEE	Institute of Electrical and Electronic Engineers
IESG	Internet Engineering Steering Group
IETF	Internet Engineering Task Force
IFIP	International Federation for Information Processing
IMP	Interface Message Processor
INARC	Internet Architecture Task Force
INWG	International Network Working Group
IP	Internet Protocol
IPRs	Intellectual Property Rights
IPTO	Information Processing Techniques Office
IRIA	Institut de Recherche en Informatique et en Automatique
ISO	International Organization for Standardization
ITU	International Telecommunications Union
MAP	Manufacturing Automation Protocol
MCI	Microwave Communications, Inc.
MIT	Massachusetts Institute of Technology
NCP	Network Control Program
NPL	National Physical Laboratory
NSF	National Science Foundation
NTEA	National Telephone Exchange Association
NWG	Network Working Group
OSI	Open Systems Interconnection
OSIC	Open Systems Interconnection Committee
PRnet	Packet Radio Network
PTT	Post, Telegraph, and Telephone
RCA	Radio Corporation of America
RFC	Request for Comments
ROAD	Routing and Addressing
SATnet	Satellite Radio Network

SC	Subcommittee
SNA	System Network Architecture
SPARC	Standards Planning and Requirements Committee
TC	Technical Committee
TCP	Transmission Control Program (before 1977)
TCP	Transmission Control Protocol (after 1977)
TOP	Technical Office Protocol
UCLA	University of California, Los Angeles
UNESCO	United Nations Educational, Scientific, and Cultural Organization
W3C	World Wide Web Consortium
WG	Working Group

I

Introduction

The architects of the twenty-first-century digital age proclaim that *openness* is their foundational value. Their work is exemplified in movements that embrace open science, open access publishing, open source software, open innovation business strategies, open education, and so on.[1] President Barack Obama's "Open Government Initiative," announced on his first day in office in 2009, captured the collective spirit of these efforts: "We will work together to ensure the public trust and establish a system of transparency, public participation, and collaboration. Openness will strengthen our democracy and promote efficiency and effectiveness in government."[2]

The ideology of openness, as it is articulated and practiced in the early twenty-first century, would have surprised the critic and historian of technology Lewis Mumford. In his 1964 essay, "Authoritarian and Democratic Technics," he wondered: "Why has our age surrendered so easily to the controllers, the manipulators, the conditioners of an authoritarian technics?"[3] The advocates of openness, however, have not surrendered so easily. Instead, they sense the dawn of a radical, almost utopian transformation in which power hierarchies are flattened, secret activities are made transparent, individuals are

[1] Paul A. David, "Understanding the Emergence of 'Open Science' Institutions: Functionalist Economics in Historical Context," *Industrial and Corporate Change* 13 (2004): 571–589; Steven Weber, *The Success of Open Source* (Boston: Harvard University Press, 2004); Peter Suber, *Open Access* (Cambridge, MA: The MIT Press, 2012); Henry Chesbrough, *Open Innovation: The New Imperative for Creating and Profiting from Technology* (Boston: Harvard Business Review Press, 2003); Michael A. Peters and Rodrigo G. Britez, eds., *Open Education and Education for Openness* (Rotterdam, The Netherlands: Sense Publishers, 2008).

[2] Barack Obama, "Transparency and Openness in Government," The White House, http://www.whitehouse.gov/open (accessed August 19, 2010).

[3] Lewis Mumford, "Authoritarian and Democratic Technics," *Technology & Culture* 5 (1964): 1–8. See also David E. Nye, "Shaping Communication Networks: Telegraph, Telephone, Computer," *Social Research* 64 (1997): 1067–1091.

empowered, and knowledge itself is democratized. They welcome innovation, thrive on diversity and change, insist on the sanctity of autonomy and freedom, and have an unqualified disdain for authority, restrictions, and gatekeepers.

The technological foundations that sustain this vision are digital: the Internet, the Web, and cellular telephone networks. These technologies together create a communication infrastructure that has the potential to evade the ability of established authorities to control, censor, or ignore. For individuals, "open" is shorthand for transparent, welcoming, participatory, and entrepreneurial; for society at large, "open" signifies a vast increase in the flow of goods and information through a global, market-oriented system of exchange. In the most general sense, it conveys independence from the threats of arbitrary power and centralized control. It is a discourse that speakers of the English language have found easy to adopt, given the common use of expressions such as "open-minded," "open to the public," and "open for business." "Openness" thus describes a marriage of technology and ideology and a fusion of technology, democracy, and entrepreneurial capitalism.

Although readers may believe that openness is inextricably linked to digital technologies born in the late twentieth and early twenty-first centuries, the purpose of this book is to explain the historical forces – creations and legacies of a pre-digital age – that forged our twenty-first-century open world. Some of these historical forces already have been the subject of a large number of academic and popular studies. Whether they describe the "rise of the network society," the advent of the "flat world," or the dawn of the "information age," many of these studies tend to celebrate American-style high-tech globalization. But the praise is not universal: critics rightly note that the world is not flat, that power asymmetries persist between various nodes in the network society, and that traditional bureaucracies such as national governments and industrial corporations have in many ways gained in strength and not withered away.[4]

Rather than join the chorus as a critic or champion of globalization in the twenty-first century, I pursue a historical objective in *Open Standards and the*

---

[4] Manuel Castells, *The Rise of the Network Society* (Cambridge, MA: Blackwell Publishers, 1996); Thomas L. Friedman, *The World Is Flat: A Brief History of the Twenty-First Century* (New York: Farrar, Straus and Giroux, 2005); Louis Galambos, "Recasting the Organizational Synthesis: Structure and Process in the Twentieth and Twenty-First Centuries," *Business History Review* 79 (2005): 1–37; Louis Galambos, *The Creative Society – And the Price Americans Paid for It* (New York: Cambridge University Press, 2012); Alfred E. Eckes and Thomas W. Zeiler, *Globalization and the American Century* (New York: Cambridge University Press, 2003); John Gray, *False Dawn: The Delusions of Global Capitalism* (New York: The New Press, 2000); Phillipe Legrain, *Open World: The Truth About Globalization* (Chicago: Ivan R. Dee, 2004); Jack Goldsmith and Tim Wu, *Who Controls the Internet? Illusions of a Borderless World* (New York: Oxford University Press, 2006); Evgeny Morozov, *The Net Delusion: The Dark Side of Internet Freedom* (New York: Public Affairs, 2011); Stanley Fish, "Anonymity and the Dark Side of the Internet," January 3, 2011, http://opinionator.blogs.nytimes.com/2011/01/03/anonymity-and-the-dark-side-of-the-internet (accessed January 17, 2012); Nathan Ensmenger, "The Digital Construction of Technology," *Technology & Culture* 53 (2012): 753–776.

Digital Age: to explain how this state of affairs came into being. There already exists a conventional view, a theory of causation that permeates the literature: new technologies drive social change. A typical expression of this view may be found in Andrew Shapiro's 1999 book, *The Control Revolution*, whose argument was aptly summarized by its subtitle: "How the Internet Is Putting Individuals in Charge and Changing the World We Know." Another example may be found in the distinctive prose of journalist Thomas Friedman, who referred to new information technologies such as personal computing, Internet telephony, and wireless devices as "steroids" that are "amplifying and turbo-charging all the other flatteners."[5]

The consequences of all of this turbocharged flattening, of course, depend on one's point of view: when Google complained in 2010 that the Chinese government was censoring search results from google.cn, the Chinese newspaper *Global Times* defended China's right to protect itself from American "information imperialism." Google's high-minded defense of the freedom of expression was, the *Global Times* declared, a ruse – a "disguised attempt to impose its values on other cultures in the name of democracy."[6] The inherent contradictions and tensions bundled within terms such as "openness" and "transparency" have been further exposed by activists such as Chelsea Manning, Aaron Swartz, and Edward Snowden who put powerful institutions in uncomfortable positions by publicizing data that were intended to be secret. In other words, openness (and its ally, transparency) is easy to promote in rhetoric but more complicated to adhere to in practice.

One comes away from the popular accounts of high-tech globalization with an oversimplified, linear, and somewhat deterministic view of the relationship between technology and society: for better and for worse, the Internet and digital technologies have thrust an unprecedented era of openness on us. A more sophisticated and nuanced interpretation has, somehow, failed to find its way into the mainstream, even though the foundations for such an interpretation have been laid over the past fifty years by communities of historians, sociologists, and political scientists who study science and technology. Recent scholars, many of them contributing or responding to the theoretical framework of the "social construction of technology," emphasize the ways that technological artifacts, networks, and systems can embody and advance the cultural and political values of their makers. This literature, taken as a whole, has sharpened our understanding of the ways that technologies and societies are mutually constitutive – or, as historian Thomas Hughes summarized, how technological systems are "both socially constructed and society shaping."[7]

---

[5] Friedman, *The World Is Flat*, 187.
[6] *Global Times*, "The Real Stake in the 'Free Flow of Information,'" January 22, 2010, http://opinion.globaltimes.cn/editorial/2010–01/500324.html (accessed January 17, 2012).
[7] Thomas P. Hughes, "The Evolution of Large Technological Systems," in Wiebe E. Bijker, Thomas P. Hughes, and Trevor Pinch, eds., *The Social Construction of Technological Systems: New Directions in the Sociology and History of Technology* (Cambridge, MA: The MIT Press,

To comprehend the culture and technology of the twenty-first-century open world, we need to reject monocausal technological determinism, build on the work of social constructivists, and look closely at the "system builders" as well as at the social and institutional forces that shaped and constrained their actions. We should gain a better understanding of technology, but at the same time we must also come to terms with the organizational, economic, political, and cultural forces that do not exist independently of technology. Hence the central questions of this book: Who designed the digital foundations of open systems? How did they act upon their commitments to transparency, decentralization, and innovation? How are open systems different from previous systems and previous regimes of control, and how are they similar?

Two conceptual tools are useful to build historical explanations for these questions. The first, *ideology*, refers to a comprehensive set of ideas about the past, present, and future orderings of politics and society. I am aware that Marxist scholars and literary theorists (such as Raymond Williams) tend to use the term pejoratively to refer to illusions, false consciousness, and class conflict, but I do not use the term in that way. Instead, I follow historians such as Bernard Bailyn and John Kasson who use the term in a more neutral and descriptive sense to refer simply to worldviews or systems of meaning making. A definition in the *Oxford English Dictionary* also is helpful:

4. A systematic scheme of ideas, usu. relating to politics or society, or to the conduct of a class or group, and regarded as justifying actions, esp. one that is held implicitly or adopted as a whole and maintained regardless of the course of events.[8]

In the chapters that follow we will see many examples in which engineers, managers, and regulators made their ideological beliefs explicit in speeches, published articles, and private meetings. We will also see many other cases in which ideological beliefs were, as the *OED*'s definition suggests, held

---

1989), 51; Langdon Winner, "Do Artifacts Have Politics?" *Daedelus* 109 (1980): 121–136; W. Bernard Carlson, "The Telephone as Political Instrument: Gardiner Hubbard and the Political Construction of the Telephone, 1875–1880," in Michael Thad Allen and Gabrielle Hecht, eds., *Technologies of Power: Essays in Honor of Thomas Parke Hughes and Agatha Chipley Hughes* (Cambridge, MA: The MIT Press, 2001); Lawrence Lessig, *Code and Other Laws of Cyberspace* (New York: Basic Books, 1999); Janet Abbate, *Inventing the Internet* (Cambridge, MA: The MIT Press, 1999).

8 Raymond Williams, *Keywords: A Vocabulary of Culture and Society* (New York: Oxford University Press, 2003), 153–157; Clifford Geertz, "Ideology as a Cultural System," in *The Interpretation of Cultures* (New York: Basic Books, 1973); Hayden White, "Method and Ideology in Intellectual History: The Case of Henry Adams," in Dominick LaCapra and Steven L. Kaplan, eds., *Modern European Intellectual History: Reappraisals and New Perspectives* (Ithaca, NY: Cornell University Press, 1982); *Oxford English Dictionary*, 3rd ed., s.v. "Ideology, n." http://dictionary.oed.com (accessed January 17, 2012); Bernard Bailyn, *The Ideological Origins of the American Revolution* (Cambridge, MA: The Belknap Press, 1992), viii; John F. Kasson, *Civilizing the Machine: Technology and Republican Values in America, 1776–1900* (New York: Hill and Wang, 1999); Frank Ninkovich, "Ideology, the Open Door, and Foreign Policy," *Diplomatic History* 6 (1982): 185–208.

implicitly – that is, unstated yet always to be found as justifications for action. In all cases, the technologies discussed in this book were the subject of ideological beliefs and claims about the proper control of information and communication networks.[9]

The second conceptual tool, *critique*, names the process through which ideological beliefs may be put into practice. My starting point here is Michel Foucault's insight that "critique only exists in relation to something other than itself" – that is, it is always a response to external conditions. Critique, Foucault continued, is "the art of not being governed quite so much ... not to be governed thusly, like that, in this way."[10] Critique is therefore a strategic response that "must be an instrument for those who fight, resist, and no longer want what is. It must be used in processes of conflict, confrontation, and resistance attempts.... It is a challenge to the status quo."[11]

The philosopher Gerald Raunig, picking up on Foucault's notion of critique as a response to existing conditions, suggested that it might also be the foundation of a more productive act: "At the same time," he argued in 2008, "critique also means re-composition [and] invention."[12] In this view, critique is much more than a process of criticizing, belittling, or tearing down; it is also a process of creating, promoting, and building anew. Accordingly, the concept of critique in cultural theory can come to share common ground with Joseph Schumpeter's notion of "creative destruction" in economic theory: both concepts emphasize that creativity and innovation do not occur in a vacuum, but rather respond in an active and at times aggressive way to that which already exists.[13]

Throughout this book, I extend this line of inquiry to study acts of critique (and, in many cases, creative destruction) that were advanced by telephone and computer engineers. In some cases, these engineers offered *explicit* critiques – clear remarks on existing political, market, and technical issues – in private meetings, conference presentations, and written publications. In other cases, their critiques and challenge to the status quo were *implicit* in their creation of new artifacts, new software, and new institutions – collective acts of recomposition and invention that were not always accompanied by explicit

---

[9] For an insightful discussion of various meanings of "control," see Gustav Sjoblom, "Control in the History of Computing: Making an Ambiguous Concept Useful," *IEEE Annals of the History of Computing* 33 (2011): 88–90.

[10] Michel Foucault, "What Is Critique?" in David Ingram, ed., *The Political* (Malden, MA: Blackwell, 1978), 192–193, 208.

[11] Michel Foucault quoted in Gerald Raunig, "What Is Critique? Suspension and Recomposition in Textual and Social Machines," *Transversal* (2008), http://eipcp.net/transversal/0808/raunig/en (accessed January 17, 2012). See also Terry Eagleton, *The Function of Criticism* (New York: Verso, 2006).

[12] Raunig, "What Is Critique?"

[13] Joseph A. Schumpeter, *Capitalism, Socialism, and Democracy* (New York: Harper & Row, 1976 [1942]).

claims of superiority. When we interpret engineering practice as acts of implicit and explicit critique, we will be in a better position to understand how new standards and new networks emerge as components of broader visions that respond to the past and present and that seek to redistribute power and control in the future. The "open systems" created in the late twentieth century, and the "open standards" described by the title of this book, thus constitute critiques and rejections of ideologies of centralized control.[14]

Making Open Systems

By the time engineers started referring to their computer and information networks as open systems in the late twentieth century, the concept of openness had become loaded with particular political, economic, technical, and cultural meanings. How did the term accumulate such symbolic weight? What did the language of openness mean to the people who deployed it, and what did it critique?

To answer these questions in the American context we need to begin by recalling the American colonies in the late eighteenth century, when American colonists split from Britain and created a new political system and a new information system at the same historical moment. The new political system rejected the British monarchy in favor of an American republic; the new information system rejected the British notion that kept knowledge, and therefore power, in the hands of the Crown and its allies. Both changes shared a basic ideological orientation: new liberal and local practices replaced the old restrictive and remote traditions.

At first, the information system in the American colonies – that is, the institutions, rules, and physical means of disseminating information – replicated European precedents set by absolutist rulers in England and France and the authorities in the Vatican who controlled the production and dissemination of printed materials through regimes of licensing, stamp taxes, state-backed guilds, and monopoly control of postal networks. In the American colonies, change came to this restrictive ancien régime as part of a broader colonial reaction against British rule. The major turning point came with the Stamp Act, passed by Parliament in February 1765 and scheduled to take effect at the beginning of November of the same year. For a majority of colonists, the amount of the tax did not threaten economic hardship; nevertheless, many reacted furiously to the notion that Parliament could tax them without their consent.[15] They summarized their principles in a slogan that every American schoolchild memorizes: No taxation without representation.[16]

---

[14] Raunig, "What is Critique?" See also Helen Nissenbaum, "How Computer Systems Embody Values," *Computer* (March 2001): 118–120.

[15] Edmund S. Morgan, ed., *Prologue to Revolution: Sources and Documents on the Stamp Act Crisis, 1764–1766* (Chapel Hill: University of North Carolina Press, 1959), 45–49.

[16] Robert Darnton, "An Early Information Society: News and Media in Eighteenth-Century Paris," *American Historical Review* 105 (2000): 1–35; Starr, *Creation of the Media*, 28–39; Brown,

Some opposition to the Stamp Act cited a natural bond between unfettered communication and political liberty, such as a famous series of 1765 newspaper articles written by John Adams.[17] Adams warned his readers of the political dangers of the Stamp Act ("a design is formed to strip us in great measure of the means of knowledge," he wrote), but he also warned of the Act's *material* dangers. In keeping with British traditions but running against American expectations, the Act required a wide range of colonial documents – including commercial and legal forms such as ship inventories, court records, contracts, and wills; public documents such as newspapers and pamphlets; and recreational goods such as almanacs, calendars, and playing cards – to be printed on paper that was manufactured with a stamp impressed in it. Such stamped paper could be purchased only from designated agents who, in turn, received it from printers in London. Understood in material terms, the Act was an attempt to favor state-sponsored monopolies, establish a new supply-and-distribution hierarchy, and mandate new standard stationary upon a resistant network of lawyers, printers, merchants, citizens, consumers, and users.[18]

The Stamp Act, "as if designed to inflame the most articulate" (in Paul Starr's apt summary), ultimately inspired colonists such as Patrick Henry and Samuel Adams to create the first intercolonial networks of resistance, including the Sons of Liberty and Committees of Correspondence, against British authority.[19] After American independence was declared and defended, literate Americans shared the expectation that neither government officials nor religious leaders or social elites had the power or the right to dictate what could or could not be purchased and read.[20] To be a good citizen, an American needed to be literate and knowledgeable. "Print had come to be seen as indispensable to political life" as early as 1765, observed the literary historian Michael Warner, "and could appear to men such as [John] Adams to be the primary

---

Knowledge is Power, 41 ("restrictive character of European information systems"); and Stephen Botein, "'Meer Mechanics' and an Open Press: the Business and Political Strategies of Colonial American Printers," *Perspectives in American History* 9 (1975): 127–225.

[17] Bernard Bailyn, *The Ideological Origins of the American Revolution* (Harvard University Press, 1967); Edmund S. Morgan and Helen Morgan, *The Stamp Act Crisis: Prologue to Revolution* (Chapel Hill: University of North Carolina Press, 1995/1953); Starr, *Creation of the Media*, 65–67; John Adams, "A Dissertation on the Canon and Feudal Law," Charles F. Adams, ed., *The Works of John Adams, III* (Boston: Little and Brown, 1851), 447–464.

[18] For a survey of surviving "America" stamps, see Adolph Koeppel, *The Stamps that Caused the American Revolution: The Stamps of the 1765 Stamp Act for America* (Manhasset, NY: Town of North Hempstead American Revolution Bicentennial Commission, 1976).

[19] Starr, Creation of the Media, 65.

[20] Of course, not all Americans shared the same literary tastes, nor did they have equal access to books, newspapers, and pamphlets. The information marketplace certainly was more dynamic in the northeast, but slaves, along with the rural and urban poor in the north and especially in the south, did not share in the information abundance of the early republican era. Brown, *Knowledge is Power*, 197–244 and 268–296.

agent of world emancipation."[21] But most printers were not concerned with world emancipation; they were a pragmatic group who were more interested in their personal economic circumstances. They wanted to sell more newspapers and books. As a result, the republican print ideology depended on – and was a construction of – a commercial vision of an open print marketplace. In historian Marcus McCorison's summary, the "traditional aim of the newspaperman was to profess political, economic, and social subservience to no faction and to keep his press freely open."[22] The historian Richard Brown notes that these commercial values of printers meshed well with American expectations of free speech and open communication: "Free speech was not always and everywhere honored, especially in the slave states, but free speech and open access to America's information system was the normal expectation."[23]

American politicians began to use the discourse of openness conspicuously at the turn of the twentieth century, in the context of American foreign policy in China. The "Open Door" policy took shape through a series of notes that U.S. Secretary of State John Hay sent to imperial administrations in Britain, France, Germany, Italy, Japan, and Russia. Hay was concerned that imperial competition over Chinese ports might disrupt opportunities for American commercial interests. To prevent China from being carved into imperial spheres of influence, Hay advanced an informal policy that would, in principle, respect Chinese territorial and administrative integrity.

Historians interpret the Open Door as an expression of the expansionist urges of the American economic system that was built on an implicit critique of European-style imperialism. It was, in one summary, an effort on the part of American diplomats and industrialists to make "the best use of mutual distrust inherent in the European principle." Open Door ideology called for a system of international relations that favored fair trade, peaceful cultural expansion, minimal government intrusion, and the avoidance of political and especially military entanglements. Although the new American principles ostensibly called for moralistic nonintervention in foreign affairs, historians such as William Appleman Williams have argued that the Open Door in practice marked the emergence of a new style of intervention where economic imperialism replaced the older practices of territorial colonialism, thus allowing Americans to "win the victories without the wars." Open Door ideology, in Williams's influential reading, became "the keystone of twentieth-century American diplomacy." By insisting on a "world open to American ideas and influence," Open Door ideology created the conditions where Americans could pursue "informal empire"

---

[21] Michael Warner, *The Letters of the Republic: Publication and the Public Sphere in Eighteenth-Century America* (Boston: Harvard University Press, 1990), 32. See also Brown, *Knowledge is Power*, 287; and Brown, *The Strength of a People*, 49–118.

[22] Marcus A. McCorison, "Forward," Bernard Bailyn and John B. Hench, eds., *The Press & the American Revolution* (Boston: Northeastern University Press, 1981), 2.

[23] Brown, "Early American Origins of the Information Age," 52.

in the Pacific while proclaiming the high-minded and progressive principles of national self-determination and international peace.[24]

In the middle decades of the twentieth century, prominent philosophers, scientists, and theorists in the United States and Europe embraced "open" metaphors in their work. Unlike the Open Door policy's explicit confrontation with international diplomacy in the early twentieth century, the mid-century experts who wrote about open societies, open economies, and open systems retreated from geopolitical realities and instead applied the discourses of openness to their theoretical, structural, and systematic investigations.

For example, the philosophers Henri Bergson and Karl Popper posed stark dichotomies between social customs that they called "open" and "closed." Bergson drew a distinction in *The Two Sources of Morality and Religion* (1932) between "closed morality" and "open morality," where the former concept described an exclusionary and static society that promoted social cohesion and carried a deep preoccupation with strict internal obedience and war against all enemies. "Open morality," on the other hand, had a higher regard for creativity and progress; it was, according to Bergson, inclusive, universal, and peaceful. Popper's more famous 1945 book, *The Open Society and Its Enemies*, sketched an equally stark contrast. Popper argued that knowledge is provisional and fallible – not absolute – and therefore societies must accept new and diverse points of view. "Open societies" thus required moral foundations of humanitarianism, equality, and political freedom. One common thread between Bergson and Popper was their clear identification of openness with a wide array of progressive, peaceful, and inclusive values. Together they built a philosophical case that led to an inevitable (if understated) conclusion: open morality and open societies were superior alternatives to fascist and communist oppression.[25]

Experts in economics, sociology, and mathematics also adopted open metaphors in the 1930s and 1940s and, even more than the philosophers, took refuge from political and military conflict in the specialized language of their own disciplines. In 1939, the British Keynesian economist George Shackle published an article titled "The Multiplier in Closed and Open Systems," a commentary

---

[24] Ninkovich, "Ideology, the Open Door, and Foreign Policy"; Yoneyuki Sugita, "The Rise of an American Principle in China: A Reinterpretation of the First Open Door Notes Toward China," in Richard Jensen, Jon Davidann and Yoneyuki Sugita, eds. *Trans-Pacific Relations: America, Europe, and Asia in the Twentieth Century* (Westport, Connecticut: Praeger Press, 2003), 3–20; William Appleman Williams, "Open Door Interpretation," in *Encyclopedia of American Foreign Policy: Volume 2*, Alexander Deconde, ed. (New York: Charles Scribner's Sons, 1978), 703–710; William Appleman Williams, *The Tragedy of American Diplomacy*, (New York: WW Norton & Company, 1972 [1959]), 45–57; Jerry Israel, *Progressivism and the Open Door: America and China, 1905–1921* (Pittsburgh: University of Pittsburgh Press, 1971); Andrew J. Bacevich, *American Empire: The Realities and Consequences of U.S. Diplomacy* (Cambridge: Harvard University Press, 2002).

[25] Henri Bergson, *The Two Sources of Morality and Religion* (New York: Doubleday Anchor, 1935 [1932]); Karl Popper, *The Open Society and its Enemies* (London: Routledge, 1945).

on the theoretical uncertainties inherent in export and import values in an "open economy."[26] The language of open systems also appeared in the work of the sociologist Talcott Parsons, who in 1943 described the "Kinship System of the Contemporary United States" as an "open, multilineal, conjugal system," one in which individuals choose their marriage partners rather than having marriages arranged on their behalf. In 1945, the term "open systems" appeared again in a different context – this time in the journal *Philosophy of Science*. Arturo Rosenblueth and Norbert Wiener published an article, "The Role of Models in Science," where they contrasted theoretical models that they called "closed box" and "open box." The distinction between these two types of models came from the number of fixed finite variables that each system had: fewer in closed boxes, many more in open boxes. "All scientific problems," they explained, "begin as closed-box problems, i.e., only a few of the significant variables are recognized. Scientific progress consists in a progressive opening of those boxes."[27]

Amid these provocative yet scattered ideas of the 1930s and 1940s, the Austrian biologist Ludwig von Bertalanffy began to build a general account of the properties of all open systems. In his 1950 article, "The Theory of Open Systems in Physics and Biology," von Bertalanffy drew on insights from biology, thermodynamics, and evolutionary theory to propose a basic definition: "A system is closed if no material enters or leaves it; it is open if there is import and export and, therefore, change of the components." Later that year, von Bertalanffy argued for the existence of a set of *"general system laws* which apply to any system of a certain type, irrespective of the particular properties of the system or elements involved." These laws formed the basis of his proposal for a "General System Theory," a direct critique of the "mechanistic world-view" that had "found its ideal in the Laplacean spirit." In place of this static worldview, von Bertalanffy proposed a "unification of science" around the "central problem" of "dynamic interaction" whose "general principles are to be defined by System Theory."[28]

[26] G.L.S. Shackle, "The Multiplier in Closed and Open Systems," *Oxford Economic Papers* 2 (May 1939): 135–144. Shackle did not comment in his article on his choice of the term "open system," but in a later interview he recalled the many "famous" and "thrilling" visitors who he met as a graduate student in F.A. Hayek's seminar in the mid-1930s – including Karl Popper on his first lecture in England in 1936. Soon after publishing his article, Shackle joined Churchill's staff of economic advisors for the duration of World War II. "An Interview with G.L.S. Shackle," *The Austrian Economics Newsletter* (1983), available from http://mises.org/journals/aen/aen4_1_1. asp (accessed January 27, 2012).

[27] Talcott Parsons, "The Kinship System of the Contemporary United States," *American Anthropologist*, New Series, 45 (1943): 22–38; Arturo Rosenblueth and Norbert Wiener, "The Role of Models in Science," *Philosophy of Science* 12 (1945): 316–321.

[28] Ludwig von Bertalanffy, "The Theory of Open Systems in Physics and Biology," *Science* New Series 111 (1950): 23–29; Ludwig von Bertalanffy, "An Outline of General System Theory," *British Journal for the Philosophy of Science* 1 (1950): 139–164.

The explanatory power of systems theory in general – and open systems in particular – continued to accumulate comment and support in the 1950s, 1960s, and 1970s. According to the sociologist W. Richard Scott, the new open system models of this era were part of a "broad intellectual movement that, beginning in the mid-1950s, swept through the sciences, affected established fields such as biology, and created new fields such as cybernetics." Open-system concepts were especially influential in the new field of organization studies, where the transition from "closed to open-system models" marked "a watershed event that forever altered our view of organizations."[29] The defining questions of the vibrant new discipline came from scholars, such as Herbert Simon, Richard Cyert, James Marsh, and Oliver Williamson, for whom systems thinking necessarily entailed an opening outward: How does a system interact with its environment? In human systems, how do people negotiate the boundaries between an organization and its environment in the presence of persistent uncertainty?[30]

In general terms, the intellectuals (such as Popper, Shackle, Wiener, and von Bertalanffy) who built the theoretical foundations of open systems from the 1940s to the 1970s avoided explicit commentary on the centralization of political and cultural power in the fascist and communist states of Europe, the humanitarian catastrophes of World War II, and the anxiety of the early Cold War. Yet it would be a mistake to consider their work as nonideological. As much as they avoided commenting directly on current affairs, their ideological commitments to individualism, democracy, and market economies were as strong as they were implicit. Their theories of open systems reflected their shared view of the problems that confronted them: societies could not exist in isolation, progress could not result from repressive political regimes, and the imperatives of moral development, political freedom, scientific progress, and personal autonomy required peoples (and nation-states) to engage with their broader environments in a peaceful way. The midcentury open system theorists were, in short, implicitly defending liberalism and Western modernity by defining and justifying its scientific and structural foundations.

[29] W. Richard Scott, "Ecosystems and the Structuring of Organizations," in Larry V. Hedges and Barbara L. Schneider, *The Social Organization of Schooling* (New York: Russell Sage Foundation, 2005), 37.

[30] John G. Maurer, ed., *Readings in Organization Theory: Open-Systems Approaches* (New York: Random House, 1971); Louis R. Pondy and Ian I. Mitroff, "Beyond Open System Models of Organization," *Readings in Organizational Behavior* 1 (1979): 3–39 ("for the last decade, thinking and research in the field of organization theory have been dominated by a point of view labeled as open system models"); Denise M. Rousseau, "Assessment of Technology in Organizations: Closed versus Open Systems Approaches," *The Academy of Management Review* 4 (1979): 531–542; W. Richard Scott, *Organizations: Rational, Natural, and Open Systems* (Englewood Cliffs, NJ: Prentice-Hall, 1981); and Oliver E. Williamson, "Review of *Organizations: Rational, Natural, and Open Systems* by W. Richard Scott," *Journal of Economic Literature* 20 (1982): 585–586.

Some of the implicit ideological assumptions of open system theorists became explicit in the 1960s and 1970s as engineers began to use the language of open systems to justify new strategies for building technological systems. In their work we can see the deeper meaning of the open concept shifting a second time: if the first shift took place in midcentury as theorists ignored the explicit diplomatic and industrial aspects of the American Open Door policy, the second shift took place as engineers adopted the language of open systems to advocate increases in efficiency and consumer welfare that could be realized through competition and standardization. They endeavored to build – in a literal sense – an open world.

The first instance I can find in which engineers explicitly embraced the principles of open systems occurred in the movement to create modular standards for materials used in the construction and building industries. The modular concept, first published in the early 1930s, promised to eliminate waste, create room for multiple suppliers, preserve competition, and generate industrywide economies of scale: if architects, manufacturers, and builders adhered to modular standards, they could save money and reduce waste in all phases of the construction process, including design, manufacturing, and on-site assembly. In a 1966 article in the industry journal *Modular Quarterly*, the Canadian building engineers Neil Hutcheon and S. R. Kent used the term "open system" to refer to this fantasy of a competitive and standardized industry. Hutcheon and Kent defined open systems as those that "envisage the construction of various kinds and sizes of building from selected factory-made components conforming to the system." The "requirement for complete standardization and interchangeability," they noted, faced substantial technical and organizational obstacles because the "integration of standard units or components becomes progressively difficult as the components are increased in size and complexity." Indeed, the lessons of industrial history were discouraging: "The development of mass production of other manufactured goods has rarely proceeded on an open system basis."[31]

Although their article was only a tentative exploration of the development of industrial open systems, Hutcheon and Kent anticipated the fundamental traits of subsequent open system projects: the belief that standardization and competition would save money and increase efficiency, the difficulties of standardization in the face of technical and social complexity, and the historical tendency for closed systems to dominate industrial production. They could not have known that these were the precise characteristics of the international computer networking industry of the 1950s, 1960s, and 1970s. Since the late 1950s, computer researchers in Europe and the United States had been taking

---

[31] N. B. Hutcheon and S. R. Kent, "Influence of Size, Function, and Design on the Standardization of Components," *Modular Quarterly* (1966): 20–21; Andrew L. Russell, "Modularity: An Interdisciplinary History of an Ordering Concept," *Information & Culture: A Journal of History* 47 (2012): 257–287.

small steps to define common standards for their immature industry, including standards for programming languages, terminology, and the transmission of digital data. This task was unbelievably complex. Computer hardware was expensive and inflexible, software was experimental and unstable, and a wide variety of professionals and organizations were trying to steer the standardization process in different directions.

The problem of coordination was especially acute in the effort to create standards for computer networks: the technologies were unstable, and there were few incentives for competitors to cooperate. The most pressing problem, then, was not only the absence of mature technical standards – it was the absence of an organization that had the authority to create standards that would be obeyed "across diverse computer lines and international boundaries."[32] Moreover, the two most likely candidates for international standardization were closed systems: IBM's System Network Architecture, and the X.25 data transmission standard created by a group of national telephone monopolies under the auspices of the International Telegraph and Telephone Consultative Committee (CCITT). Both IBM and the telecom monopolies in the CCITT were trying to position themselves for the imminent convergence between the computing and communication industries. Neither one wanted to cede control.

The threat of IBM domination was felt most acutely by firms in the computer industry and by national governments in Europe.[33] IBM, after all, had been the undisputed leader of the international computing industry for years. With the 1974 release of its System Network Architecture, industry observers thought that IBM would begin to extend its dominance to the nascent data communication industry. No single vendor or institution had enough power to confront IBM and beat it in the marketplace; the best bet for the sizable contingent of anti-IBM manufacturers and users, therefore, was to join forces, devise new collaborative mechanisms, and build a technological alternative before it was too late – that is, before IBM's System Network Architecture became the global de facto standard.

In 1977, a small group of British computer scientists proposed a new project to create "network standards needed for open working." One of these computer scientists, Jack Houldsworth, defined "open working" not in technical language, but rather in layperson's terms: "The ability of the user or program of any computer to communicate with the user or the program of any other." He posed this vision in contrast to "the traditional computing scene" that was "dominated by installations that were planned and implemented as self-contained, 'closed' systems with little regard for the possibility of their

---

[32] Bachman, "Domestic and International Standards Activities for Distributed Systems," September 28, 1978, SPARC/DISY 138, Box 18, Folder 12, Charles W. Bachman Papers (CBI 125), Charles Babbage Institute, University of Minnesota [hereafter "Bachman Papers"].

[33] Simon Nora and Alain Minc, *The Computerization of Society: A Report to the President of France* (Cambridge, MA: The MIT Press, 1981).

interworking with each other." The conclusion, in Houldsworth's eyes, was obvious: open network standards "will be of significant economic value to computer manufacturers and users."[34]

The British proposal was embraced by the International Organization for Standardization (ISO), which created in August 1977 a new subcommittee on Open Systems Interconnection (OSI) to "investigate the need for standardization in the area of open systems."[35] One year later, the subcommittee chairman, the American database expert Charles W. Bachman, synthesized what he had learned in a brief essay. Bachman reinforced Houldsworth's interpretation of an open system as something that enabled "open working": "The goal of Open System Interconnection is to define standard interfaces and standard protocols which would permit any person, any terminal, any physical process, and any computer, which used these standards, to exchange information." But Bachman pushed his definition one step further, in a way that emphasized democratic collaboration. He continued, "The adjective 'open' means to imply that all participants come to the system as equal partners." By opening the *process* of standardization to all interested parties, Bachman recognized that his committee would be entering a competitive and turbulent domain where "the political problems will challenge the most astute statesman."[36]

The OSI project ultimately failed to create consensus international standards for open systems. As I describe in Chapter 7, it collapsed under its own weight, a victim of an excess of democracy in its design and standardization processes. But where OSI standards faltered, the discourse of open systems continued to be mobilized in the 1990s – particularly by those who opposed IBM and wished to decentralize the locus of control over computing. Most users, as computer scientist Martin Libicki summarized, "defined open as any system that would get them out from under IBM's thumb." Carl Cargill, a standards strategist for Sun Microsystems, summarized the ideology of open systems in 1996: "The concept of 'Open Systems' has become a convenient icon to express all that is good about computing and the promise that computing can hold. It, too, has undergone significant shifts in its meaning, but it has always been held up as the ideal to which all computing should subscribe."[37]

Open systems thus lived on as an icon or ideal that, at its root, was always a critique of centralized control. The open system ideals did more than criticize IBM or the telephone monopolies; they also proposed new methods of

---

[34] Jack Houldsworth, "Standards for Open Network Operation," *Computer Communications* 1 (1978): 5–12; Bachman, "Domestic and International Standards Activities for Distributed Systems."

[35] "Resolution 11. Establishing Subcommittee 16 – Open System Interconnection," May 11, 1977, Bachman Papers, Box 19, Folder 1.

[36] Bachman, "Domestic and International Standards Activities for Distributed Systems."

[37] Martin Libicki, *Standards: The Rough Road to the Common Byte* (Cambridge, MA: Harvard Program on Information Resources Policy, 1994), 11; Carl Cargill, *Open Systems Standardization: A Business Approach* (Upper Saddle River, NJ: Paladin Consulting, 1997), 70.

governance, a newly democratized means to produce industrywide standards, and a new and more decentralized industry structure for the data communications industry. The open systems ideology that was developed in computing between the 1970s and the 1990s embodied several assumptions articulated in previous open system visions in diplomacy, economics, philosophy, and engineering.[38] These assumptions included

- an economic commitment to global markets;
- a moral support of international and multicultural ties;
- a political opposition to centralized power – either in governments or in monopolies – that threatened individual autonomy;
- a belief that technical professionals could achieve these economic, moral, and political aspirations through cooperation and standardization.

In the 1980s and 1990s, the newly invigorated ideology of open systems continued to be mobilized as a critique and rallying cry within the world of computing. The computer manufacturer Sun Microsystems, for example, built an open systems advertising strategy that promised to liberate customers from being painted into a corner and locked in to proprietary products that were unreliable and expensive to maintain. In 1998, a group of programmers led by Eric Raymond coined the term "open source" to gain support from engineers who sought alternatives to proprietary software but who perceived Richard Stallman's crusade for "free software" as too radical. More recently, "open libraries" has emerged as a mantra for orienting the libraries of the twenty-first century around the virtues of open access, open data sets, open source software, and open government practices.[39]

At the same time, powerful incumbents such as IBM and American Express have learned to mobilize the rhetoric and technologies of openness to reify their positions as market leaders and advance their own proprietary ambitions. Although the exuberant rhetoric of openness often signifies a favorable set of political, cultural, economic, and technical values, open systems are neither inevitable nor, in some cases, even preferable. Manufacturers – particularly those in dominant positions – use closed systems to maintain their position. Regulators and consumers also have found that closed systems can be more efficient, more secure, and more user friendly. Scientists and theorists recognize that closed systems (or "closed-box" problems, to borrow Wiener's phrase) are easier to manage and to solve. The same virtue – simplicity – also convinced Hutcheon and Kent that closed building systems could be more efficient, and

[38] T. A. Critchley and K. C. Batty, *Open Systems: The Reality* (New York: Prentice Hall, 1993), 10–15; David M. Piscitello and A. Lyman Chapin, *Open Systems Networking: TCP/IP and OSI* (Reading, MA: Addison-Wesley, 1993), 5.

[39] Weber, *Success of Open Source*; Christopher M. Kelty, *Two Bits: The Cultural Significance of Free Software* (Durham, NC: Duke University Press, 2008); Matthew Nelson, ed., "Open Libraries: More Than Just Open Books or Open Doors," *Oregon Library Association Quarterly* 16 (2010).

that it "may be more economical at times to design on a non-standard basis." Economists argue that the single-minded pursuit of openness can impede standardization and damage innovation, and legal scholars have shown how closed and hierarchical information systems can be more secure, more efficient, and more convenient for users.[40]

Each of these examples remind us that the benefits of openness are by no means indisputable, and that substantial distance exists between the rhetoric of open systems and the realities of industrial development. System builders are always engaged in ideological and discursive work, not merely technical work. Their practice is simultaneously ideological and technological, abstract and concrete, discursive and material. They face a persistent problem: How can they exercise control in a way that can achieve the reconciliation of diversity into a unified standard? To answer this question, we need more precision in our understanding of the production and control of interfaces between components, systems, and their environments. In other words, we need to distinguish between different types of standards and look more closely at where they come from and how they govern.

Open Standards

Standardization is a social process by which humans come to take things for granted. Through standardization, inventions become commonplace, novelties become mundane, and the local becomes universal. It is a *historical* and therefore contested process whose success depends upon the obfuscation of its founding conflicts and contingencies. Successful standards, if they are noticed at all, simply appear as authoritative, objective, uncontroversial, and natural. Standards are, as other scholars have noted, "recipes for reality" whose black boxes are rarely opened and whose subjectivity and contingency are rarely revealed.[41]

[40] Hutcheon and Kent, "Influence of Size, Function, and Design"; Jonathan Zittrain, *The Future of The Internet – And How to Stop It* (New Haven, CT: Yale University Press, 2008); Joseph Farrell and Philip J. Weiser, "Modularity, Vertical Integration, and Open Access Policies: Towards a Convergence of Antitrust and Regulation in the Internet Age," *Harvard Journal of Law and Technology* 17 (2003): 85–134; W. Russell Neuman, Lee McKnight, and Richard Jay Solomon, *The Gordian Knot: Political Gridlock on the Information Highway* (Cambridge, MA: The MIT Press, 1998), 94–112.

[41] For richly theorized discussions of standards and standardization, see W. A. Shewart, "Nature and Origin of Standards of Quality," *Bell System Technical Journal* 37 (1958): 1–22; Paul A. David and Shane Greenstein, "The Economics of Compatibility Standards: An Introduction to Recent Research," *Economics of Innovation and New Technology* 1 (1990): 3–41; U.S. Congress Office of Technology Assessment, *Global Standards: Building Blocks for the Future* (Washington, DC: U.S. Government Printing Office, 1992); Paul A. David, "Standardization Policies for Network Technologies: The Flux between Freedom and Order Revisited," in Richard Hawkins, Robin Mansell, and Jim Skea, eds., *Standards, Innovation and Competitiveness* (Aldershot: Edward Elgar, 1995); Cargill, *Open Systems Standardization*; Susanne K. Schmidt and Raymund Werle, *Coordinating Technology: Studies in the International Standardization of Telecommunications*

The distinction between a standard and standardization is instructive. The term "standard" often refers to customs or norms, but it also has more specific meanings that refer to documented practices or to a particular type of product: a technical specification for the composition, interfaces, or characteristics of a given material, such as the quality of a steel rail, the size and angle of a screw thread, or a common measure of electrical resistance. In all cases, the standards I examine in this book refer to agreements between producers and consumers. As we will see, the process by which these actors come to agreements is complex and deserves close scrutiny. "Standardization" is the term that describes the process of making standards – a process that entails a wide range of practices and ideas with distinct political, economic, and cultural dimensions. Specific industrial standards embody the dominant values and assumptions that emerge during the process of standardization. They are simultaneously social constructions and material realities.

As standards emerge from contested contexts, they immediately function as a means of control within the political and economic order. In the United States, for example, the scope of industrial standardization grew as an integral part of broader discourses of rational planning and attempts to exert systematic control over the complexities of technological modernity. One distinctive characteristic of the American system of standardization – a characteristic that it borrowed from international meetings of engineers in the nineteenth century – is its development of organizational capabilities for setting standards that do not depend on the authority of the federal government.[42]

Two fundamental questions guide my effort to conceptualize standardization as a social and historical phenomenon: First, what do standards accomplish? Second, who makes standards? In response to the first question, theorists of standardization agree that most standards fall into three general categories: performance, measurement, and compatibility.

Performance: These standards specify ways to perform certain tasks or carry out certain designs. Performance standards seek to ensure a minimum level of quality by specifying

---

(Cambridge, MA: The MIT Press, 1998); Geoffrey Bowker and Susan Leigh Star, *Sorting Things Out: Classification and Its Consequences* (Cambridge, MA: The MIT Press, 1999); James Sumner and Graeme J. N. Gooday, "Introduction: Does Standardization Make Things Standard?" *History of Technology* 28 (2008): 1–13; Stefan Timmermans and Steven Epstein, "A World of Standards but not a Standard World: Toward a Sociology of Standards and Standardization," *Annual Review of Sociology* 36 (2010): 69–89; Lawrence Busch, *Standards: Recipes for Reality* (Cambridge, MA: The MIT Press, 2011).

[42] David F. Noble, *America by Design: Science, Technology, and the Rise of Corporate Capitalism* (New York: Oxford University Press, 1977); Edwin T. Layton, *The Revolt of the Engineers: Social Responsibility and the American Engineering Profession* (Cleveland, OH: Press of Case Western Reserve University, 1971); Marina Moskowitz, *Standard of Living: The Measure of the Middle Class in Modern America* (Baltimore; The Johns Hopkins University Press, 2004); Andrew L. Russell, "Industrial Legislatures: The American System of Standardization," in *International Standardization as a Strategic Tool* (Geneva: International Electrotechnical Commission, 2006).

either a *process*, such as the ISO 9000 Quality Management Principles, or a *result*, such as a safe and accident-free workplace.

Measurement: These standards specify an objective and quantifiable unit of measurement such as a meter, gallon, or ohm. Measurement standards make it possible to compare physical qualities such as length, volume, or electrical current.

Compatibility: These standards define interfaces between discrete objects. Compatibility standards generate efficiencies and economies of scale in the production process, and promote interoperability between complementary products. Interfaces between various components of computer hardware provide many familiar examples of compatibility standards, such as Universal Serial Bus and Ethernet ports.

Of these three categories, my primary focus in here is the third category: the creation of compatibility standards for communication networks. Scholars who have studied the interconnection of various components of communication networks often use three concepts – *infrastructure*, *platforms*, and *network effects* – to describe the important role of standards that enable compatibility and interoperability. The utility of these concepts lies in their connotations of stability, support, and external benefits that standards can generate. They emphasize the potential for a heap of standardized components to combine into a cohesive and flexible network that can, in turn, sustain more complex social and economic activity.[43]

The second fundamental question posed above – who makes standards? – pushes us to conceptualize standardization as a power-laden process. On whose authority does something become "standard"? How are different types of standards created and enforced? In response to these questions, theorists identify three varieties of standards: *de facto*, *de jure*, and *voluntary consensus*.

De facto standards arise from common usage or market acceptance. Individual people or firms often generate these standards, which spread either through the efforts of a sponsor or in a more organic way. Examples include Microsoft's word-processing software and the desktop metaphor in computer interfaces.

De jure standards are mandated by regulators at the local, state, federal, and/or international level. Governments commonly test *conformance* with mandated standards, and can legally (and at times severely) punish noncompliance. Examples include the Environmental Protection Agency's rules for fuel economy and the

---

[43] National Telecommunications and Information Administration, *The National Information Infrastructure: An Agenda for Action* (1993), http://www.ibiblio.org/nii (accessed January 4, 2012); Michael L. Katz and Carl Shapiro, "Systems Competition and Network Effects," *The Journal of Economic Perspectives* 8 (1994): 93–115; Richard R. John, "Recasting the Information Infrastructure for the Industrial Age," in Alfred D. Chandler, Jr. and James Cortada, eds., *A Nation Transformed by Information: How Information Has Shaped the United States from Colonial Times to the Present* (New York: Oxford University Press, 2000); Philip J. Weiser, "Law and Information Platforms," *Journal of Telecommunications and High Technology Law* 1 (2002): 1–35; Steven W. Usselman, "Public Policies, Private Platforms: Antitrust and American Computing," in Richard Coopey, ed., *Information Technology Policy: An International History* (New York: Oxford University Press, 2004).

Federal Communications Commission's Part 68 rules that govern telephone terminal equipment.

Voluntary consensus standards are specified within a range of private institutions, including engineering societies, trade associations, accredited standards-setting organizations, and industry consortia. "Consensus" refers to the collaborative process used to develop these standards; "voluntary" indicates that nobody is legally compelled to use these standards. However, there can be strong economic incentives that encourage conformance, and many parties involved in developing these standards make a priori commitments to adopt them. Two examples of consensus standards are the TCP/IP networking protocols and the HTML language for the World Wide Web.

Of these three varieties of standards, the voluntary consensus process is the most complicated and least understood. A small but growing literature on the topic illustrates how consensus standardization is a fundamentally political process, one in which different stakeholders seek to exercise control over the direction and shape of technological change. Standards committees do not *make* technology, rather they make *agreements about* technology. Despite the appealing connotations of the term, "consensus methods" are neither seamless nor harmonious. Consensus standardization, by definition, requires exclusion in order to define the scope of inclusion. The process ideally creates more winners than losers, but it is certain that there will always be losers.[44]

For much of the history of communication and information networks, engineers created standards through the bureaucratic machinery of centralized institutions – industrial monopolies such as AT&T, government agencies in charge of postal, telegraph, and telephone service (PTT), or international regulatory bodies such as the International Telecommunications Union. These authorities created and enforced the most important standards related to signal transmission and translation. Consensus standards, on the other hand, were either non-existent or of marginal importance. This situation began to change between the 1920s and the 1970s as groups such as the British Engineering Standards Committee, the American Standards Association (ASA), the International Standards Organization, the International Federation for Information Processing, and the Institute of Electrical and Electrotechnical Engineers created standards for terminology, electrical interfaces, computer languages, and data communications. When widespread mistrust of IBM and the monopoly PTTs coalesced into the open systems movement in the late 1970s and 1980s, computer and networking engineers combined the core practices of consensus standardization with their open systems ideology to create a hybrid – *open standards* – that became the predominant mode of standardization for the digital infrastructure of the open world.

[44] Cargill, *Open Systems Standardization*, 52–67; JoAnne Yates and Craig N. Murphy, *The International Organization for Standardization (ISO): Global Governance through Voluntary Consensus* (New York: Routledge, 2008).

Despite ongoing efforts to publicize the "principles for the modern standards paradigm," definitions of "open standards" and contemporary notions of openness more generally vary greatly among the scholars and practitioners who examine and use the term.[45] Most definitions identify three categories of participants in the standards process – creators, implementers, and end users – and argue that more openness generally means a larger scope of inclusion for end users and a fourth constituency, the general public. Further, most definitions agree on the paramount importance of two features: well-defined *procedures* that guarantee public participation in the production of standards, and liberal terms of *access* that allow public use of standardized technologies. The second feature – access – has been the site of the most bitter debates in standards committees, particularly when they need to define a policy for handling the intellectual property rights (IPRs) of contributors. In general, many creators demand payments for their IPRs, but many users and implementers prefer that IPRs be licensed on a low-cost or royalty-free basis. In practice, standards committees resolve this impasse by defining licenses that specify reasonable and nondiscriminatory terms for using the standard. Licenses vary greatly, but most fall on a spectrum between exclusive licensing (which would be most lucrative for owners of IPRs) and fully open, royalty-free terms. The explosion of scholarly and professional literature devoted to this issue indicates that it is a source of deep concern and uncertainty for all constituents in the standards-setting process.

The advocates of open standards also emphasize the importance of well-defined procedures for creating standards that can incorporate the concerns of end users and the general public. They value regular procedures that enable interested parties to review a proposal, comment or vote on it, and, if necessary, lodge an appeal. All of these aspects of open standards are not new; they are direct descendants of the principles of consensus standardization that have their roots in international engineering conferences of the late nineteenth and early twentieth centuries and were first articulated in the United States by the ASA in the 1920s. In 1928, the ASA Constitution and Procedure declared that a "national standard implies a consensus of those substantially concerned with

---

[45] The following discussion draws on Carl Cargill, "Evolution and Revolution in Open Systems," *StandardView* 2 (1994): 3–13; Jonathan Band, "Competing Definitions of 'Openness' in the NII," in Brian Kahin and Janet Abbate, eds., *Standards Policy for Information Infrastructure* (Cambridge, MA: The MIT Press, 1995); Ken Krechmer, "Open Standards Requirements," *International Journal of IT Standards and Standardization Research* 4 (2006): 43–61; Elliott Maxwell "Open Standards, Open Source, and Open Innovation: Harnessing the Benefits of Openness," *Innovations: Technology, Governance, Globalization* 1 (2006): 119–176; Joel West, "Seeking Open Infrastructure: Contrasting Open Standards, Open Source, and Open Innovation," *First Monday* 12 (June 2007), http://firstmonday.org/ojs/index.php/fm/article/view/1913/1795 (accessed September 16, 2013); Andrew Updegrove, "ICT Standards Today: A System Under Stress," in Laura DeNardis, ed., *Opening Standards: The Global Politics of Interoperability* (Cambridge, MA: The MIT Press, 2011); and OpenStand, "The Modern Standards Paradigm – Five Key Principles," http://open-stand.org/principles (accessed September 5, 2012).

its scope and provisions." To achieve such a consensus, ASA rules required that all proposals for standards must come from committees that could, in turn, demonstrate a "balance of interests" between representatives from creators and implementers of standards. The ASA's clever consensus-building process – a superior alternative to the courts and to commissions, in the eyes of its many advocates – created the basis of widely adopted rules for creating, maintaining, and disputing standards. In other words, the ASA established standards for setting standards.[46]

A standards process that calls itself open will usually adopt clear rules for consensus and due process, but in recent years standards committees have democratized participation in deliberations by lowering barriers to entry or eliminating membership fees or requirements altogether. The most prominent example of radical openness is the Internet Engineering Task Force, a group that conducts much of its work on public email lists that anyone can join. In practice, discussions tend to focus on specialized topics that require advanced knowledge of computer networking to understand, thus having the effect of excluding substantive contributions from a vast majority of the general public.

In sum, it is possible to distill an *ideology of open standards* from the spectrum of open standards definitions that presently exist. First, open standards strive to honor the ideals of participatory democracy, including commitments to fairness, transparency, due process, and rights of appeal. Second, open standards embrace the ideal of a vibrant market economy that has negligible barriers to entry and liberal terms for using standardized technologies. Third, proponents of open standards share the implicit conviction – which, in some cases, manifests itself as a religious zeal – that their work is just and that the forces of technological and social progress are on their side.[47] They share a common foe – namely, anyone who advocates closed or centralized control over the production and use of standards.

The purpose of *Open Standards and the Digital Age* is to show how the ideology of open standards came into being, why it came into being in a specific time and place, and why it resonated so deeply with people who worried – and continue to worry – about the concentration of power in a global society. My story is, in effect, the convergence of two stories: the history of communication networks and the history of a collaborative form of consensus standardization. The two stories do not share the same origins; indeed, they converged fully only in the last decades of the twentieth century, when consensus standards were recast as open standards.

I begin my account of their convergence in Chapter 2 by describing how American expectations of open and unfettered communication were challenged

[46] *American Engineering Standards Committee Year Book* (New York: American Engineering Standards Committee, 1928), 68.
[47] Kelty, *Two Bits*, 64–94.

in the late nineteenth century by Western Union's monopoly control of the telegraphs. The public debate about ownership and control over American telegraphy foreshadows a key theme in this book in which Americans struggled to reconcile their fear of government control, their trust in a market economy, and their expectations of open communication.

In Chapters 2 and 3, I examine the institutions that American engineers created between 1880 and 1930 to bring some order to the American system of standardization. In many sectors of the industrial economy, engineers believed that a lack of common standards created inefficiency and waste. Before 1900, de facto standards were the norm: the federal government refused to use its authority to back specific standards, and no organizations existed that could harmonize industrial practice. In the late 1800s, engineers in the electrical, chemical, and railroad industries began to experiment with professional committees to set industry standards; by the early 1900s, their experiments had generated the organizational capabilities for creating and sustaining a consensus around standard industrial practices. I emphasize that the practices of consensus standardization were invented with a great deal of effort and controversy and with a great deal of help from the American political culture of the 1920s that stressed the progressive potential of engineering and the value of associations among Americans working in the public and private sectors.

In the United States, the most important institution for the development of consensus standards was the American Engineering Standards Committee (AESC), formed in 1918. The AESC, inspired by earlier developments in the United States and in Britain, grew steadily in its first ten years before it was reconstituted as the American Standards Association in 1928. As they matured in a wide variety of industrial practices in the twentieth century, consensus standards bodies shared common ideological assumptions: a belief in the rational methods of professional engineering, a progressive view of the power of science and technology to improve society, a conviction that cooperation among engineers within "industrial legislatures" was superior to adversarial conflicts in existing judicial and political institutions, and, especially in the United States, an assumption that the private sector should cooperate and should lead.

In Chapter 4, I return to the history of communication networks by considering the growth of the American telephone industry and the formation of the monopoly Bell System and the standards strategies of executives in its parent company, AT&T. Whereas many critics characterized AT&T's managerial style as one of centralized control, my examination of standardization in the Bell System shows that AT&T's control was not as monolithic as its critics suggested. AT&T executives such as Bancroft Gherardi, AT&T vice president and chief engineer from 1920 to 1938, struggled mightily to reconcile the diversity of local practices with his desire to achieve systemwide efficiencies through standardization. Gherardi faced significant difficulties within the vast organizational structure of the Bell System, but he faced even greater problems with standardization projects that spanned its boundaries and thus evaded his managerial control. To help him solve these problems, Gherardi turned to the industry standards bodies – including the young American Standards Association, for which he served as president in 1932 and 1933.

During the New Deal era that spanned from the 1930s to the 1970s, a variety of critiques of centralized control disrupted the viability of AT&T's monopoly. By the 1960s, critics of AT&T found ready allies in critics of IBM's dominance of the computer industry. I examine this loose alliance among entrepreneurs, regulators, engineers, and counterculture computer "hackers" in Chapter 5. By the end of the 1970s, critiques of IBM and AT&T, despite their divergent origins and targets, converged on a common coordinating mechanism for standardization: a loose network of technical committees that operated beyond the reach of any dominant company. It was these critiques, rather than any inherent characteristics of digital technology, that structured the realm of ideological and practical possibilities for the convergence between telecommunications and computing technologies.

Chapters 6, 7, and 8 explain how the two major threads in my story – the history of communication networks and the history of consensus standardization – came together in the 1970s, 1980s, and 1990s. This history typically is cast as the "history of the Internet," a story in which the TCP/IP Internet emerged in the 1970s and 1980s by virtue of its support by the American government, its simple design, and its enthusiastic user base. This version of history is not incorrect, but it is certainly incomplete to the extent that "Internet history" excludes the political, technical, and cultural diversity that falls within the broader rubric that I use: the history of networking.[48] These chapters also demonstrate that the creation of standards for computer networks is not an exclusively American story. Rather, the principles of openness and consensus have international origins and global consequences.

In Chapter 6, I describe a series of international collaborations in the 1970s among American, British, and French computer researchers who tried, and ultimately failed, to agree on a single design for packet-switched computer networks. Instead, the future of computer networking appeared to be on the verge of a battle between IBM's proprietary System Network Architecture and public data networks based on the X.25 standard produced by the International Telecommunications Union. These foes proved to be too powerful for the packet-switched researchers. In 1976, the packet-switching research community splintered into two groups: one inspired by the French computer scientists Louis Pouzin and Hubert Zimmermann, and the other funded by the American Department of Defense and led by Vinton Cerf and Robert Kahn.

The former group – minus Pouzin, who had been punished by the French government for his iconoclasm – continued its work under the auspices of

---

[48] For a complementary view of networking history, see Martin Campbell-Kelly and Daniel D. Garcia-Swartz, "The History of the Internet: The Missing Narratives," *Journal of Information Technology* 28 (2013): 18–33.

ISO's Subcommittee SC16, Open Systems Interconnection (OSI). In Chapter 7, I describe their attempt to design a seven-layer computer network architecture through the democratic mechanisms of international standardization. Openness was OSI's founding justification, noblest aim, and fatal flaw. The history of OSI – a subject that has so far eluded the attention of historians – has some troubling lessons for those who champion inclusivity, openness, and multistakeholder governance. Indeed, these were the very values that opened up OSI's process to strategies of delay and disruption and to internal conflicts that proved impossible to resolve.

In Chapter 8, I explain how the Internet's leaders in the 1980s built on the Arpanet's tradition of autocratic control over network architecture and standards. Advocates of the Internet standards process bragged about their institutional bypass of the "official" standards process and celebrated a new political philosophy coined by MIT computer scientist David D. Clark: "We reject: kings, presidents and voting. We believe in: rough consensus and running code." Internet engineers embraced their informal and unconventional identity. By the late 1990s, the Internet's commercial success convinced outsiders that their critique of the formal standards process was a harbinger of a new mode of global standardization. The Internet's advocates recast it as an unqualified success story and a model of open and democratized innovation; in the process, they obscured or simply ignored the Internet engineering community's rejection of the basic features of democracy such as membership and voting rights. I argue that the Internet's emergence was both a cause and a consequence of a broader shift in the organization of international standards that rejected the formal democratic procedures of the American National Standards Institute, ISO, and CCITT. The new regime of the 1980s was a disorganized network of ad hoc alliances and standards consortia that only needed to establish "rough consensus" among a limited membership before publishing technical documents that were self-designated open standards.

The origins of the twenty-first-century open world cannot be understood simply by following a linear history of communication technologies from the printing press to the telegraph, telephone, and Internet. By expanding the narrative to include the invention of consensus standardization, *Open Standards and the Digital Age* demonstrates how information networks were critiques that were consciously designed to project new ideas about technology and control into the future. The creation of the digital infrastructure of the open world reflects the faith in market capitalism, reverence for technological expertise, and skepticism toward concentrated power that were defining characteristics of the late twentieth century. This history ends in the early twenty-first century, not with an open utopia but rather with digitized and reinvigorated critiques of centralized control. These critiques indicate the persistence of a bundle of contradictions and unresolved questions over power and control in a digitized open world.

2

Ideological Origins of Open Standards I: Telegraph and Engineering Standards, 1860s–1900s

> Americans of all ages, all conditions, and all dispositions constantly form associations. They have not only commercial and manufacturing companies, in which all take part, but associations of a thousand other kinds, religious, moral, serious, futile, general or restricted, enormous or diminutive.... Wherever at the head of some new undertaking you see the government in France, or a man of rank in England, in the United States you will be sure to find an association.
>
> – Alexis de Tocqueville, *Democracy in America*, 1840[1]

Had he been able to visit America in 1900, Alexis de Tocqueville would have seen that Americans continued to "constantly form associations." Americans continued to form commercial and manufacturing companies, as Tocqueville witnessed in the 1830s. Moreover, representatives of those companies formed additional associations to pursue common interests and ambitions. Through these combinations of associations, Americans developed technical standards to harmonize crucial aspects of industrial production.

Such voluntary industrial and corporate associations are by and large missing from histories of the late nineteenth-century American industrial economy. Histories of this era of American capitalism tend to focus on two general types of activity: disorganized and cutthroat competition in markets, and hierarchical command structures that developed within large corporations.[2] Only

---

[1] Alexis de Tocqueville, *Democracy in America: Part the Second, The Social Influence of Democracy* (New York: J & HG Langley, 1840), 114.

[2] Alfred D. Chandler, Jr., *The Visible Hand: The Managerial Revolution in American Business* (Cambridge, MA: Belknap Press, 1977); Philip Scranton, *Endless Novelty: Specialty Production and American Industrialization, 1865–1925* (Princeton, NJ: Princeton University Press, 1997); Naomi R. Lamoreaux, Daniel M. G. Raff, and Peter Temin, "Beyond Markets and Hierarchies: Toward a New Synthesis of American Business History," *American Historical Review* 108 (2003): 404–433; Richard White, *Railroaded: The Transcontinentals and the Making of Modern America* (New York: W. W. Norton, 2011).

recently have historians paid more attention to a third type of activity, which they conceptualize as hybrids of markets (which facilitate singular transactions) and hierarchies (which provide permanent structures for repeated transactions). William Cronon's discussion in *Nature's Metropolis* of the Chicago Board of Trade provides perhaps the best-known example of this hybrid form of coordination mechanism in the mid-nineteenth century. Cronon explains how the Board of Trade's grading system decoupled grain ownership from its market price by creating standards of quality for different types of wheat. It thus functioned as an institution that created a level playing field for all grain merchants and prevented the possibility of control by a few powerful ones.[3]

The Chicago Board of Trade is but one example of a broader and underappreciated type of institution that developed standards in an effort to bring order to market transactions and industrial development. Between the 1860s and 1900s, industrial engineers created a wide variety of standard-setting committees that, in turn, created and disseminated guidelines and designs for an even wider variety of industrial tools, processes, and products. These committees were important in their own era because they maintained fluidity in the dynamic industrial economy of the late nineteenth century: by defining standards, they facilitated the existence of multiple sources of supply and thus provided a means for small- and medium-sized industrial firms to avoid the specter of monopoly. As subsequent chapters in this book will show, these committees were also important in later eras because they created institutional and ideological precedents that computer and telecommunications engineers used in the latter decades of the twentieth century as they, too, sought to generate network effects and a competitive market structure.

The history of industrial standards in the late nineteenth century, and my specific claim that the production of these standards fits within a conceptual middle ground between markets and hierarchies, might strike some readers as something of a detour from the gradual evolution of communication technologies of the digital age. After all, many standards for American telephone and telegraph networks in the nineteenth century were established within the corporate hierarchies of AT&T and Western Union, and the self-conscious movement for open systems and open standards did not begin until the 1970s. But the key principles and formative practices of open standards – due process, consensus, and a balance of interests – were not inventions of the computer age; rather, their roots stretch back to the late nineteenth century, when engineers first experimented with specialized committees to set industrywide standards. In order to understand the technological and ideological history of the twenty-first-century digital age, it is necessary to disrupt the familiar linear narrative of communication networks (telegraph to telephone to Internet) and explore how the key principles and formative practices of industrial standardization

---

[3] William Cronon, *Nature's Metropolis: Chicago and the Great West* (New York: W. W. Norton, 1991), 104–142.

emerged from a variety of American industrial practices in the late nineteenth century.

Accordingly, Chapters 2 and 3 describe the growth of industrial standards committees between the 1880s and 1920s – the institutional and ideological foundations of the open digital age. This chapter studies standards in telegraph networks and professional engineering societies and trade associations in the late nineteenth century; Chapter 3 examines the movement to create "American standards" in the 1910s and 1920s. Together, they describe the formation of institutions designed to reconcile conflicting technological practices and create orderly methods for the creation of American standards. Tocqueville would not have been surprised that American engineers and businessmen created their own private associations and did not depend on government or on "men of rank."

In these chapters I emphasize two aspects of American industrial standard-setting committees. First, these groups experimented with an organizational form – neither market nor hierarchy – that they believed would reconcile conflicts among diverse people, ideas, practices, and technologies. Standards committees were, in a general sense, part of a broader organizational response to the disorder that industrial capitalism brought to American society. In standards committees, power did not flow along the hierarchical lines that we are accustomed to seeing in corporate or state bureaucracies; rather, power was distributed among a community of engineers, including representatives of firms that both competed and cooperated with one another. My analysis therefore departs from a Chandlerian interpretation of American business history that emphasizes a small number of large firms. It explains instead the growth of institutions that could prevent large firms from automatically dictating standards to small firms. Associations that set standards could therefore facilitate cooperation among a large number of small firms and, indeed, among firms of all sizes.[4]

A second point of emphasis in these two chapters is the rhetorical and ideological strategies that American engineers used to promote and legitimize their new approach of cooperative standardization. Near the end of the twentieth century, the voluntary consensus style of standardization that emerged from these experiments would be recast as "open standards." The engineers who developed private standard-setting committees in the late nineteenth and early twentieth centuries, however, did not use the rhetorical appeal of "openness" to promote their work. Rather, they tended to use moralistic terms such as "consensus" and "cooperation" that set their approach apart from prevailing conceptions of American industry as the realm of ruthless and untrustworthy capitalists and robber barons.

---

[4] Louis Galambos, "Technology, Political Economy, and Professionalization: Central Themes of the Organizational Synthesis," *Business History Review* 57 (1983): 471–493; Robert H. Wiebe, *The Search for Order, 1877–1920* (New York: Hill and Wang, 1967).

Western Union: Open Expectations Meet Monopoly Capitalism

American mistrust of centralized power has remained constant even as the objects of skepticism have changed over time. Throughout the 1700s, American colonists viewed government officials (the British Crown and Parliament) as censors and oppressors, imagining instead the colonists themselves, merchants, and printers to be the true guardians of virtue and liberty. By the late 1800s, however, colonial suspicions and hopes had been inverted: many Americans had come to view self-interested industrialists as oppressors and increasingly turned to government officials to preserve liberty.

Frank Parsons, a Progressive educator, lawyer, reformer, and critic of Western Union, captured this inversion of colonial paranoia in an 1896 article. He argued that the "essence of royalty and aristocracy is not in the title but in the overgrown power which one man possesses over his fellows. The board of directors of the Western Union is as truly a body of aristocrats as the lords and dukes of England." The American Congress, Parsons continued, had blundered by granting excessive "privileges to private individuals," thus establishing "a far more powerful and therefore more dangerous aristocracy than any that could possibly be created by the mere bestowal of titles of nobility."[5]

What was it about American telegraphy that prompted Parsons to compare the directors of Western Union to British aristocrats? Telegraphy had brought economic and social benefits to Americans, who could use it to access information and thus participate in the vibrant market economy. At the same time, however, telegraphy facilitated concentrations of wealth and power within larger, richer, and more tightly integrated firms. And, for critics such as Parsons, Western Union was the epitome to an abusive Gilded Age monopoly.[6]

How did telegraphy come to embody these contradictions? Telegraph networks, like all technologies, were products of the society that created and used them. In this sense, the inconsistencies and self-contradictions of telegraphy registered both the growing pains of a nation at war with itself as well as the depths of its struggle to reconcile the ideals of democracy with the forces of industrial capitalism. The proper role for the American state was, characteristically, open to interpretation and dispute: Should it stimulate industrial growth directly through funds for "internal improvements," protect citizens from the sharp edges of industrial capitalism, or simply get out of the way and

---

[5] Frank Parsons, "The Telegraph Monopoly, Part VII," *The Arena* 16 (1896): 193.

[6] JoAnne Yates, "The Telegraph's Effect on Nineteenth Century Markets and Firms," *Business and Economic History* 15 (1986), 149–164; Richard D. DuBoff, "The Telegraph in Nineteenth-Century America: Technology and Monopoly," *Comparative Studies in Society and History* 26 (1984): 571–574; Chandler, *Visible Hand*, 188–205; David Hochfelder, "The Communications Revolution and Popular Culture," in William L. Barney, ed., *A Companion to 19th-Century America* (Malden, MA: Blackwell Publishers, 2001), 308–313; James W. Carey, "Technology and Ideology: The Case of the Telegraph," in *Communication as Culture: Essays on Media and Society* (New York: Routledge, 1992), 201–230; Frank Parsons, "The Telegraph Monopoly, Part VI," *The Arena* 16 (1896): 70–84.

let self-interested capitalists express their creative energies? Because it raised these questions but never answered them fully, American telegraphy was both a microcosm and a driver of dominant themes in nineteenth-century American political economy.[7]

The question of federal patronage looms large in the early history of the telegraph in the United States. The inventor Samuel Morse failed to attract federal sponsorship for his telegraph experiments on a trip to Washington, D.C., in 1838, but he continued to seek federal support. In 1843, Congress appropriated $30,000 to build a line from Washington to Baltimore; on May 25, 1844, Morse paid tribute to his patron by organizing the first public demonstration of the telegraph within the Supreme Court chamber in the Capitol building. Morse's technological and symbolic achievement, however, failed to convince Congress to purchase his patent rights as a basis for the construction of public telegraph networks. Consequently, Morse searched for private investors and, in 1845, began to license the rights to build and operate networks through his agent, Amos Kendall. Kendall, building on his experience as postmaster general from 1835 to 1840, began by planning for lines along busy postal routes. Despite his best efforts, authority over financing, construction, and operations soon splintered among a vast number of investors, speculators, and promoters. In the absence of a coordinated system architecture, an era of "methodless enthusiasm" ensued.[8]

During the next decade, the telegraph industry entered what one historian characterized as a period of "wasteful" competition, featuring "unbridled, reckless construction programs" and "neither consistent nor rational approaches to rate making" between rival companies.[9] By the early 1850s, three or more competing telegraph lines blanketed major cities in the Northeast and Midwest, resulting in a maze of overhead cables along urban streets and confusion at terminal stations and in telegram dispatch and delivery. Amid the frenzy of competition, entrepreneurs and financiers never seriously considered a more cooperative approach, with the possible exception of Kendall's establishment of the American Telegraph Confederation in 1853. Kendall hoped that the confederation could harmonize the activities of the various companies in the industry and thus increase the quality and reliability of messages sent

[7] Richard R. John, *Network Nation: Inventing American Telecommunications* (Boston: Harvard University Press, 2010); Harry L. Watson, *Liberty and Power: The Politics of Jacksonian America* (New York: Hill and Wang, 1990); Rebecca Edwards, *New Spirits: Americans in the Gilded Age, 1865–1905* (New York: Oxford University Press, 2005).

[8] John, *Network Nation*, 24–114; Robert Luther Thompson, *Wiring a Continent: The History of the Telegraph Industry in the United States, 1832–1866* (Princeton, NJ: Princeton University Press, 1947), 37–96; Paul Starr, *The Creation of the Media: Political Origins of Modern Communications* (New York: Basic Books, 2004), 162–165; Frank G. Carpenter, "Henry Clay on Nationalizing the Telegraph," *The North American Review* 154 (1892): 380–382.

[9] Lester G. Lindley, *The Constitution Faces Technology: The Relationship of the National Government to the Telegraph, 1866–1884* (New York: Arno Press, 1975), 3–10.

across the wire. Kendall was handicapped, however, by Morse's insistence that he deal only with companies that licensed Morse's patents. Because a majority of telegraph companies depended on other patents – or viewed Morse's claims as absurdly broad and therefore illegitimate – Kendall's confederation was doomed from the start. It possessed neither the organizational capabilities nor the authority needed to stabilize the industry. Voluntary cooperation was not enough.[10]

Telegraph executives saw consolidation – rather than competition or cooperation – as the only strategy that could overcome fragmentation, ensure stability, and pay dividends to stockholders. A flurry of mergers and acquisitions, many of them hostile, concentrated the industry into regional monopolies in the 1850s. Network operators began to realize the economies of scale and scope that resulted from the standardization of both the human (dispatch and delivery) and technical (equipment and transmission) aspects of the industry – a dynamic that twentieth-century economists eventually would describe as "network effects." The consolidations of the 1850s – particularly the Treaty of Six Nations that divided territory between the six leading telegraph firms – provided observers with the first industrial-era examples of the peculiar economics of network industries that would soon become articulated in the legal discourse of natural monopolies.[11]

Western Union was the main beneficiary of consolidation, thanks in large part to its expansion during the Civil War and the skilled labor that emerged from the military telegraph corps. Western Union, bolstered by these wartime profits as well as investments in equipment and personnel, bought out its remaining rivals in 1866 and became the first American industrial monopoly.[12] The firm's triumph in the market attracted congressional scrutiny and public mistrust. Congress passed the Telegraph Act of 1866 in which it, first, granted any company the right to build new telegraph lines along postal roads and public lands and, second, reserved the right to purchase the system outright and administer telegraph networks. Despite numerous attempts to follow through on the act's second provision – one historian counted more than seventy bills introduced between 1866 and 1890 to reform or nationalize the

---

[10] Thompson, *Wiring a Continent*, 260–264, 441–442; Richard B. DuBoff, "Business Demand and the Development of the Telegraph in the United States, 1844–1869," *Business History Review* LIV (1980): 462–465; John, *Network Nation*, 88–89.

[11] Thompson, *Wiring a Continent*, 259–330; Richard B. DuBoff, "The Rise of Communications Regulation: The Telegraph Industry, 1844–1880," *Journal of Communication* 34 (1984): 52–59; John, *Network Nation*, 157–158, 194–198.

[12] Thompson, *Wiring a Continent*, 383–426; Richard R. John, "Recasting the Information Infrastructure for the Industrial Age," in Alfred D. Chandler and James Cortada, eds., *A Nation Transformed by Information* (New York: Oxford University Press, 2000), 75–77; John, *Network Nation*, 101–103; DuBoff, "Business Demand and the Development of the Telegraph"; DuBoff, "Rise of Communications Regulation," 61–62.

industry – Congress never exercised its authority, significant competition never materialized, and Western Union reinforced its dominant position.[13]

Morse himself was one of the earliest and most insistent advocates of congressional action. Morse's views, as historian Richard John has shown in his eloquent book *Network Nation*, illustrate a direct link between the suspicion of government power that was so prominent in the revolutionary era and the suspicion of power concentrated in private hands that became increasingly common in the nineteenth century. Francis O. J. Smith, the Maine congressman who helped finance and publicize Morse's experiments, argued that only federal control could ensure that telegraphy would be "open to all" and "monopolized by none." Morse, Smith, and their associates were able to convince merchants, journalists, and railroad operators to support their campaign for federal control but were nevertheless unable to convince Congress to buy the telegraph patents and create a federal telegraph network. Such reluctance is not surprising, given the economic context of the Panic of 1837 and the era's widespread skepticism of public works and internal improvements.[14]

America's elites embraced Morse's technical invention but did not heed his political advice for what to do with it. Thanks to the shrewd management of Western Union presidents Hiram Sibley and William Orton, the company's monopoly grew stronger and its effects more widely felt. In the 1860s, 1870s, and 1880s, public debates over Western Union's monopoly consistently returned to what, if anything, the federal government should do. Defenders of Western Union, led by Orton and his successor, Norvin Green, argued that it was neither necessary nor expedient for the government to own or operate a commercial telegraph service. Green, writing in the *North American Review* in 1883, anchored his argument in the Constitution, which, in his reading, did not permit the federal government to own or operate telegraph networks. Moreover, Green argued that the American system was superior in terms of quality, speed, and cost when compared to the European telegraph networks that were operated by national governments. Green also warned that government control would concentrate too much power into a government department and thus create opportunities for abuse. For Green and other defenders of the Western Union monopoly, the lessons of European monarchical oppression and the American Revolution remained relevant even 100 years after the

---

[13] Senator John Sherman of Ohio, remembered more for his authorship of the Sherman Antitrust Act of 1890, was the architect and champion of the Telegraph Act of 1866. John, *Network Nation*, 114–140, 170–194; Lindley, *The Constitution Faces Technology*, 41–76; Daniel J. Czitrom, *Media and the American Mind: From Morse to McLuhan* (Chapel Hill: University of North Carolina Press, 1983), 27; David Hochfelder, "A Comparison of the Postal Telegraph Movement in Great Britain and the United States, 1866–1900," *Enterprise and Society* 1 (2000): 746–748.

[14] John, *Network Nation*, 24–64.

British surrender at Yorktown: government could not be trusted with control over the means of communication.[15]

Gardiner Hubbard, a Boston lawyer and one of Western Union's fiercest critics, published a rebuttal to Green that cast the legacy of the American Revolution in a different light. He noted that the Continental Congress recognized, in 1775, the need to establish "the speedy and secure conveyance of intelligence from one end of the continent to the other." Hubbard argued that subsequent legislative initiatives such as the creation of the Post Office Department, appropriations for telegraph experiments, and the Telegraph Act of 1866 followed through on this revolutionary expectation. Each of these examples supported the legal rationale for Congress to operate telegraph networks – thus refuting Green's argument.

Hubbard did not stop, however, with his legalistic analysis; he also made an economic and moral argument based on international comparisons. Western Union's high rates, he alleged, prevented Americans from realizing the full potential of the telegraph. On this count, the example of European governments was to be followed, not scorned: "Abroad the telegraph is used principally by the people for social correspondence; here by business men for business purposes." Hubbard lobbied Congress for many years to create a "postal telegraph" that would provide service for social correspondents while leaving intact Western Union's arrangements with business customers. Congress, constrained by the ideological climate of postbellum America, lacked the will to enact his plan.[16]

Hubbard, understandably discouraged from his failed promotion of the postal telegraph, sought out other venues to actualize his critique of Western Union's power. We will return to one such effort, Hubbard's financial support for the inventions of his son-in-law, Alexander Graham Bell, in Chapter 4. Hubbard's vision for the Bell Telephone Company was not merely a business plan, it was also a commitment to serving a group of customers – middle-class Americans – that Western Union had ignored. Western Union's corporate orientation, Hubbard insisted, deprived middle-class Americans of vital information about news and market conditions and enriched its owners at the expense of the American democracy. His eventual support for Bell and the telephone system thus was both an ideological and economic critique of Western Union.

[15] Norvin Green, "The Government and the Telegraph," *North American Review* 137 (1883): 422–434; Richard R. John, "The Politics of Innovation," *Daedelus* 127 (1998): 198–200.

[16] Gardiner G. Hubbard, "Government Control of the Telegraph," *North American Review* 137 (1883): 521–535. Telegraph historians estimate that business and press use of the telegraph accounted for between 70 and 95% of all traffic. DuBoff, "Business Demand and the Development of the Telegraph," 465–470; John, "Recasting the Information Infrastructure," 81–82; W. Bernard Carlson, "The Telephone as Political Instrument: Gardiner Hubbard and the Formation of the Middle Class in America," in Michael Allen and Gabrielle Hecht, eds., *Technologies of Power: Essays in Honor of Thomas Parke Hughes and Agatha Chipley Hughes* (Cambridge, MA: The MIT Press, 2001), 28–34. On Hubbard's lobbying efforts in the 1860s and 1870s, see Lindley, *The Constitution Faces Technology*, 177–213; and John, *Network Nation*, 126–133.

Over the long term, his sponsorship of technological competition through the creation of a new system – rather than his promotion of political interference within Western Union's existing system – was more successful.[17]

The debate between Green and Hubbard, like the telegraph reform debate more generally, indicates that Americans in the late nineteenth century still were sensitive to the animating concerns of the American Revolution of the late eighteenth century. Although Hubbard and Green interpreted the Revolution in different ways, both men grounded their views in what, at a certain level of abstraction, was a shared understanding of the American revolutionary experience: the ability to transmit intelligence was a fundamental right, and any powerful institution that opposed open communication was to be viewed with the gravest of suspicions. Another critic of Western Union, Boston University law professor Frank Parsons, also built his critique upon the lessons of the Revolution. Like Hubbard, Parsons feared that telegraphy would facilitate dangerous concentrations of power. Writing in 1896, Parsons summoned the spirit of the "patriots of '76" when he declared that "private monopoly involves the power of taxation *without representation* and for *private purposes* ... a double infringement of freedom."[18]

Parsons's rhetoric brings us to an important point: in the late 1870s, 100 years after the Stamp Act and the Revolutionary War, Americans retained their skepticism toward powerful institutions even as they were revising their views of what a powerful institution might be. Whereas government and private companies with government-granted privileges were the objects of suspicion in the late 1700s, Americans in the late 1800s grew increasingly wary of private companies that had enhanced their power through market competition and turned into abusive industrial monopolies. In the eyes of its critics, the Western Union monopoly provided a clear illustration of a patent holder and system operator that pushed its legitimate rights too far, thus preempting fair competition and killing entrepreneurial opportunity.

By the late nineteenth century, a variety of Progressive reformers such as Parsons became convinced that government was the only institution with the power to keep abusive monopolies in check. It only needed to summon the political support and administrative capacities to act. Historian Lester Lindley captured the circular logic of the problem: "The inability to propose workable alternatives to the telegraph monopoly points to a dilemma nineteenth century Americans faced. The prevailing concept of economic life was free and open competition with minimal government interference. Yet it was clear that competition was the evil which had prompted telegraph managers to seek order in their industry."[19]

---

[17] John, "Recasting the Information Infrastructure," 87–93; Carlson, "Telephone as Political Instrument," 38–45; John, *Network Nation*, 158–164.
[18] Frank Parsons, "The Telegraph Monopoly, Part VIII," *The Arena* 16 (1896), 363–364 (emphasis in original).
[19] Lindley, *The Constitution Faces Technology*, 2, 252–262. Lindley continued, "From this dilemma some type of government regulation – not government sponsored competition or

Critics perceived plenty of warning signs that pointed to the danger that Western Union could pose to traditional American civic ideals. The alliance forged by an 1867 agreement between Western Union and the Associated Press, for example, was a blockbuster deal between two major capitalist institutions. The agreement stated that the Associated Press would use Western Union lines exclusively, with the promise of priority treatment at discounted rates. Western Union, in return, secured a promise that the Associated Press would refrain from printing stories that supported either Western Union's competition or a government-backed postal telegraph. Not only did the arrangement make it practically impossible for serious competitors – to either company – to arise, it also concentrated control over the transmission of vital public information, including political opinions and market prices and conditions, into the hands of a very few power-hungry men.[20]

Hubbard's plan for a postal telegraph went unrealized, but he and other reformers contributed to a sea change in the political history of American network industries. Their unrelenting criticism reminded Americans that even though every other industrialized nation created a public bureaucracy to own and operate telegraph networks, the stalemate between American capitalists and reformers left control over telegraphy, somewhat perilously, at the awkward crossroads of industrial self-interest and government regulation.[21]

The antimonopoly critiques from reformers such as Gardiner Hubbard and Frank Parsons were typical expressions of concerns Americans felt as their society struggled with the consequences of industrial capitalism. Citizens who at one point felt secure in their "island communities"[22] grew uneasy as they recognized the power of capitalists in New York, Boston, and Chicago over daily affairs throughout the nation. The demands of stockholders trumped the concerns of citizens. Obvious alternatives, such as government regulation or outright ownership of telegraphy, were rendered impossible by both partisanship in Congress and the predominant ideology of private control.

Hubbard, Parsons, and other critics of monopoly such as Senator John Sherman slowly shaped an ideological setting and political discourse in which varieties of regulation created a middle ground between unfettered capitalism and outright nationalization. Americans learned that monopolies could be more efficient, but they could also impose severe economic, political, and social costs. Over time, officials in Congress and the executive branch developed the

ownership – appeared to be the only escape. However, the dilemma went unsolved for nearly half a century." Lindley, *The Constitution Faces Technology*, 260–261.
[20] Hubbard, "Government Control," 529–533; Hochfelder, "Communications Revolution," 311–315; Czitrom, *Media and the American Mind*, 21–29; Menaham Blondheim, *News over the Wires: The Telegraph and the Flow of Public Information in America, 1844–1897* (Cambridge, MA: Harvard University Press, 1994).
[21] For international comparisons and transnational context, see Starr, *Creation of the Media*, 153–189; and Hochfelder, "Postal Telegraph Movement in Great Britain and the United States."
[22] Wiebe, *The Search for Order*, xiii.

political will, organizational capabilities, and legal precedents – such as the Interstate Commerce Act (1887), the Sherman Antitrust Act (1890), and the Mann-Elkins Act (1910) – to extend the power of the federal government and restrain monopolistic abuse.

In the meantime, a wide variety of Americans and Europeans connected to the telegraph industry – executives, engineers, scientists, operators, and messengers – began to experiment with new ways to stabilize and simplify their work through standardization. Ownership was only one form of control, and a relatively abstract one at that. The complexities of daily practice in the telegraph business provided opportunities for other modes of control to arise. By establishing new institutions and practices to set standards, engineers and managers created alternatives to centralized control by a single institution – either Western Union or Congress – as well as security from the "methodless enthusiasm" of the unfettered market.[23]

Telegraphy Standards: Beyond Markets and Hierarchies

What standards were necessary to create functional telegraph systems? And what standards did telegraphy, in turn, inspire or catalyze? In the late nineteenth century, standards emerged in all aspects of the industry. In the early competitive phase of the industry there was little standardization, but its gradual emergence illustrates the wide variety of ways that techniques, practices, and customs became "standard," including de facto adoption, ad hoc inter-industry agreements, and international scientific committees. Once formed, standards were neither fixed nor permanent; rather, specifications and practices that arose as local practices could become de facto standards and, over time, gain widespread approval as consensus industry standards or even as government-mandated de jure standards. Varieties of standards in three realms of telegraphy can clarify this abstract discussion: Morse code, labor, and electrical measurement.

Morse Code

Beginning in the 1830s, Morse and his colleague Alfred Vail developed a system of encoding letters and numbers into dots and dashes. Their invention, which came to be known as Morse code, both facilitated the more efficient use of telegraph capacity and reduced the cost of using time on the wire. Instead of spelling each message out in Morse code, operators began to use Morse as a basis for more concise and "secret" codes published in codebooks. Some private codes existed simply as a way of keeping costs down. For example,

---

[23] Hochfelder, "Communications Revolution," 312; David Hochfelder, "A Comparison of the Postal Telegraph Movement in Great Britain and the United States, 1866–1900," *Enterprise & Society* 1 (2000): 757–758; DuBoff, "Rise of Communications Regulation," 63–64; Duboff, "Telegraph in Nineteenth-Century America," 572–573.

on the Austin Social Telegraph Line in the 1870s and 1880s, "Kps" meant "compliments," the numeral 3 stood for "What is the correct time?" and the numeral 88 was short for "Love and kisses." Beyond these social codes, business users and diplomats developed ciphers in an effort to maintain secrecy and confidentiality. In all cases, Morse code was an intermediate step: the real act of translation often occurred not through the fingers of telegraph operators, but in the hands and minds of private citizens who sent and received telegrams.[24]

Most telegraph clerks used Morse code as telegraph networks grew in the 1840s and 1850s, but they used it in different ways and to mean different things. There was no official standardized version of the code – mostly because no organization existed with the authority to make a code "official." Private codes worked well in the American context but created confusion in Europe where different countries had varying conventions and rules for coded messages. Morse's original code soon split (or "forked," as software engineers would say today) into two versions: American Morse (which had become the de facto standard used by the nascent railroad industry and by Western Union) and a revised version of Morse code created by a German, Friedrich Clemens Gerke, in 1848. To clarify and simplify matters, representatives from twenty European nations gathered in Paris in 1865 at the first meeting of the International Telegraph Union. The convention adopted Gerke's code as the standard International Morse, but Americans continued to use their own version, sometimes called Railroad Morse or American Morse, for several decades.[25]

Today we think of Morse code as a universal means for condensing and communicating information, but even this brief description shows how telegraphic communication occurred through a variety of overlapping standard – and nonstandard – conventions. It may be more accurate to think of Morse codes as small components of a rapidly expanding communications infrastructure, and to think of operators as agents who learned to use and manipulate components within the constraints of powerful institutions, local conditions, and their individual imaginations.

Labor

From the 1860s to the turn of the century, labor costs – wages for operators, clerks, and messengers – accounted for more than half of the capital that

---

[24] Tom Standage, *The Victorian Internet: The Remarkable Story of the Telegraph and the Nineteenth Century's On-Line Pioneers* (New York: Walker & Co., 1998), 109–126; Steven M. Bellovin, "Compression, Correction, Confidentiality, and Comprehension: A Look at Telegraph Codes," March 25, 2009, available from https://www.cs.columbia.edu/~smb/papers/codebooks.pdf (accessed October 21, 2012).

[25] George A. Codding and Anthony M. Rutkowski, *The International Telecommunication Union in a Changing World* (Dedham, MA: Artech House, 1982), 5.

Western Union used to operate telegraph networks.[26] Western Union, eager to increase profits through efficient management, experimented with a number of strategies to create "unity of administration" and "the certainty and uniformity of mechanism," which in many cases implied the standardization of work in all aspects of the business.[27]

Some aspects of telegraph administration proved easier to standardize than others. Western Union quickly established a structure for collecting and reporting information generated locally on an everyday basis into a form that executives and investors could understand. In the 1860s and 1870s, Western Union executives created a managerial hierarchy to account for expenditures and message transmission across four regional divisions (Eastern, Southern, Central, and Pacific). In doing so, Western Union mimicked innovations in organizational structure, pioneered by the railroads, to subdivide the telegraph business into routine tasks that could be accomplished by discrete levels of clerks and middle managers.[28]

Telegraph managers struggled to apply the logic of standardization and routinization to the work of telegraph operators. Errors in transmission could be both embarrassing and costly, and company rules and regulations increased pressure on operators to maintain the speed and volume of messages across the wire. As a result, operators endured stagnant wages and dirty, cramped, and sometimes dangerous office conditions. They also found ways to resist. Although Morse code mechanized and constrained the basic transmission of letters and words, a skilled telegraph operator could express his or her personality within the dots and dashes of the daily routine. Workers in distant offices – who would rarely if ever meet face-to-face – would strike up partnerships. Some of these partnerships blossomed into friendships (there are tales of the occasional romance as well) that helped operators endure long working hours and often uncomfortable working conditions. Judging from the many records and memoirs of telegraph operators, even the smallest flash of individuality could bring enough joy to distract operators – even momentarily – from their dull and repetitive work. Operators also resorted to less subtle means to resist the dehumanizing aspects of their employment, most spectacularly in the failed Great Strike of 1883 organized with the help of the Knights of Labor.[29]

Western Union's paternalistic style of labor relations found further expression in its treatment of another class of telegraph worker, the messenger

---

[26] In one estimate based on samples from the 1870s, Western Union's workforce was approximately 20% clerks, 60% telegraph operators, and 20% messenger boys. Gregory John Downey, *Telegraph Messenger Boys: Labor, Technology, and Geography, 1850–1950* (New York: Routledge, 2002), 25–27.

[27] Anonymous Western Union managers quoted in Downey, *Telegraph Messenger Boys*, 25.

[28] Chandler, *Visible Hand*, 197–205.

[29] Edwin Gabler, *The American Telegrapher: A Social History, 1860–1900* (New Brunswick, NJ: Rutgers University Press, 1988), 44–56, 79–85, and 151–158; Standage, *Victorian Internet*, 129–144.

boys. Messengers, approximately 20 percent of all Western Union employees, performed one of the most important roles in telegraph service: interacting with customers who were sending and receiving telegrams. Telegraph managers expended a great deal of effort to routinize this crucial aspect of their business. They printed rulebooks and manuals, referred to messenger boys by numbers rather than their names, and issued uniforms that messengers grudgingly financed out of their own wages. Telegraph managers embraced the military and mechanical ideology at the heart of telegraph messenger management to discipline messenger boys into functional components of the technological system. Such disciplinary tactics, as historian Gregory Downey has shown, attempted to hide the haphazard and complex nature of the telegraph business by projecting a consistently clean and boyish image to customers and to the general public.[30]

This brief sketch of episodes in the standardization of telegraph labor illuminates a broader pattern in industrial labor relations: standardization often was welcomed by efficiency-seeking managers but resisted by laborers who hoped to preserve the subjective, autonomous, and enjoyable aspects of their work. By the early twentieth century, as historian Paul Israel has shown, even the tasks of invention – once the realm of inspired individuals in machine shops – moved to industrial laboratories and became subject to the bureaucratic routines of corporate strategists who were themselves inspired and emboldened by the new ideology of scientific management. As they rejected independence in favor of the corporate life, "inventor-engineers came to personify engineering values of standardization and efficiency."[31]

Electrical Measurement
The international and transnational dimensions of telegraphy – and of industrial and scientific standardization more generally – become evident when we consider the dilemmas of electrical transmission in European networks. Telegraphy was the catalyst for the establishment of precise units of electrical resistance and, ultimately, the international organizational capabilities needed to create and sustain these standards. In 1861, the Electrical Standards Committee formed by the British Association for the Advancement of Science was the first such group to take up this challenge. Its participants included the leading authorities of British telegraphy and electrical science, including James Clerk Maxwell, Fleeming Jenkin, James Joule, John William Strut (later Lord Rayleigh), J. J. Thomson, and William Thomson (later Lord Kelvin).

In order to identify the best methods for measuring electrical resistance, the committee invited contributions from a number of distinguished foreign

---

[30] Downey, *Telegraph Messenger Boys*, 61–125.
[31] Paul Israel, *From Machine Shop to Industrial Laboratory: Telegraphy and the Changing Context of American Invention, 1830–1920* (Baltimore: The Johns Hopkins University Press, 1992), 3–4 and 152–183 (quote at 152).

scientists, including Ernst Esselbach, Joseph Henry, Henry Rowland, Werner von Siemens, and Wilhelm Weber. The topic of standardization drew such an illustrious crowd not because it was a disinterested pursuit of scientific questions but rather because of the strategic importance of the primary technological application of electricity for cable telegraphy. After a period of testing and negotiation, the committee reached a consensus on a standard unit of resistance – known first as the "unit of 1862," then as the "BA unit" or "BA ohm" – that was acceptable both to the theoretical physicists and practical electricians who worked closely with telegraph networks.[32]

The British Academy's Electrical Standards Committee hosted the numerous rivalries and disagreements among the various constituencies – academics and practitioners, Brits and Germans – who sought to advance and profit from electrical technologies. These meetings in the early 1860s laid the technical and organizational foundations for more regular and formal international electrical standardization in the International Electrical Congresses that met in conjunction with World's Fairs, beginning in Paris in 1881. Subsequent congresses met in Paris (1889), Frankfurt (1891), Chicago (1893), Geneva (1896), Paris (1900), St. Louis (1904), and Turin (1911).[33]

At the 1904 Electrical Congress in St. Louis, leading figures in electrical science, including the British inventor and engineer Colonel R. E. B. Crompton, Swedish Nobel Prize winner Svante Arrhenius, Lord Kelvin, and the American inventor Elihu Thomson, created the International Electrotechnical Commission (IEC). Their goal was to establish a representative body that could bring the "cooperative spirit that animates electrical workers" into a formal and permanent organization.[34] As the IEC grew and matured in the early twentieth century, electrical scientists understood perfectly well that standards were not exclusively technical matters but rather technically oriented instances of

---

[32] On electrical standardization, see Graeme J. N. Gooday, *The Morals of Measurement: Accuracy, Irony, and Trust in Late Victorian Electrical Practice* (New York: Cambridge University Press, 2004); Bruce J. Hunt, "The Ohm Is Where the Art Is: British Telegraph Engineers and the Development of Electrical Standards," *Osiris* 9, 2nd Series, Instruments (1994): 48–63; Bruce J. Hunt, "Doing Science in a Global Empire: Cable Telegraphy and Electrical Physics in Victorian Britain," in Bernard Lightman, ed., *Victorian Science in Context* (Chicago: University of Chicago Press, 1997), 312–333; Larry Randles Lagerstrom, *Constructing Uniformity: The Standardization of International Electromagnetic Measures, 1860–1912* (PhD dissertation, University of California at Berkeley, 1992), 7–81; Joseph O'Connell, "Metrology: The Creation of Universality by the Circulation of Particulars," *Social Studies of Science* 23 (1993): 129–173; Michael Kershaw, "The International Electrical Units: A Failure in Standardisation?" *Studies in History and Philosophy of Science* 38 (2007): 108–131; and Simon Schaffer, "Rayleigh and the Establishment of Electrical Standards," *European Journal of Physics* 15 (1994): 277–285.

[33] Tim Büthe, "Engineering Uncontestedness? The Origin and Institutional Development of the International Electrotechnical Commission (IEC)," *Business & Politics* 12 (2010).

[34] William Goldsborough, address to the 1904 Electrical Congress, St. Louis, quoted in Jeanne Erdman, "The Appointment of a Representative Commission," *ANSI Reporter: A Commemorative Tribute* (2004): 6.

diplomacy, with a heavy dose of international prestige and commercial power on the line.[35]

Taken together, the wide variety of standards developed for and inspired by electrical telegraphy illustrate the fundamental irony of standardization in the late nineteenth century: there was no standard way to make standards. Where some standards (such as rules for telegraph operators and uniforms for messenger boys) emerged from the demands of managers at a private firm, other standards (such as agreements for Morse code and the ohm) emerged from negotiations in international committees.

These negotiations and the standards they generated marked the beginning of a new era in control over American communication networks. In the ancien régime of the late eighteenth century, governments controlled the means of communication – whether printing press, stamped paper, or postal routes. By end of the nineteenth century, this state of affairs had changed. No single institution could control all aspects of telegraphy, and no single institution could dictate standards unilaterally. There were, to be sure, powerful nodes in the American information infrastructure of the late 1800s. But anyone who participated in information networks – from Western Union executives to newspaper readers to telegraph messenger boys – would have recognized that their participation was subject to forces beyond their control. Western Union was a powerful monopoly, but it was not monolithic.[36]

The forces driving change in late nineteenth-century networks were not only financial; they were also political and, increasingly, technological. Owners, managers, and other agents of capital were no longer the undisputed authorities. State and federal regulators, as well as international bureaucrats, began to exert more control than they had in the formative decades of American telegraphy. Moreover, a new class of economic actors – scientists and engineers – were gaining power over the fundamental tasks of telegraph transmission and innovation. Where debates over the political economy of communication in the nineteenth century tended to boil down to the basic question of ownership – should telegraph networks be operated by a chaotic mix of competing firms, a private monopoly, a government monopoly, or something else? – the emergence of professional engineers at the turn of the twentieth century posed

---

[35] Despite optimistic speculation that telegraphy would herald an era of world peace, political leaders in England, France, Germany, Portugal, and elsewhere embraced telegraphy as a tool of imperial aggression. See for example Daniel Headrick, *The Invisible Weapon: Telecommunications and International Politics, 1851–1945* (New York: Oxford University Press, 1991), 3–115.

[36] On changing styles of news and financial reporting, see Carey, "Technology and Ideology," 210–222; Hochfelder, "Communications Revolution," 312–313; and Standage, *Victorian Internet*, 170–180. On "the pinpricks of small firms to more serious bouts with larger concerns or consortia," see Gabler, *American Telegrapher*, 40. On the complexities of international coordination, see Keith A. Nier and Andrew J. Butrica, "Telegraphy Becomes a World System: Paradox and Progress in Technology and Management," *Essays in Economic and Business History* 6 (1988): 211–226.

new questions. To what ends did they work: the private gain of their employers or the public benefits of science? To whom would they be loyal, and to whom would they appear trustworthy? How would they use their newfound social power?

As engineers and scientists grappled with these questions, they developed a professional self-awareness and became a "new subspecies of economic man" (to appropriate Alfred Chandler's description of corporate managers of the same era).[37] In the process, they joined the types of voluntary associations that Tocqueville described and thus contributed to the formation of private and associational modes of control in American social and economic life. Significantly, very few looked to the American federal government to oversee these new modes of control. Government would play a key (and as yet undefined) role in this experiment – not to take over private enterprise but rather, as historian Bernie Carlson summarized, to "intervene and create a 'space' for new organizations and institutions."[38] These developments were central to the history of standardization, a field where the most significant organizational innovations of the late nineteenth century came from the scientists and engineers who were fueling the growth of the technology-based industrial economy. Before returning to the history of communication networks in America, we need to understand the broader contexts and specific varieties of institutional arrangements that existed within the nineteenth-century American system of engineering and industrial standardization.

The "American System" and the Ideological Origins of Cooperation

For much of the nineteenth century, standardization in American industry occurred in an ad hoc manner within individual firms. The "American system of manufactures" that was born in the federal armories relied on the use of interchangeable parts to facilitate faster and more efficient production. As mechanical engineers moved from the armories to firms that made other products, including machine tools, farm equipment, sewing machines, and bicycles, they brought with them techniques and tools that could mechanize production, which they hoped would make manufacturing more efficient and profitable. Some firms did not embrace interchangeable parts but nevertheless developed their own standard practices in custom and batch production to make products such as locomotives, furniture, and jewelry. On the whole, there was a great deal of variety in American industrial standardization in the early and mid-nineteenth century. Many firms set their own standards, but standardization did not necessarily imply large-scale production.[39]

---

[37] Chandler, *Visible Hand*, 4.
[38] Carlson, "Telephone as Political Instrument," 37–38.
[39] Merritt Roe Smith, *Harpers Ferry Armory and the New Technology: The Challenge of Change* (Ithaca, NY: Cornell University Press, 1977); Nathan Rosenberg, "Technological Change in the Machine Tool Industry, 1840–1910," *The Journal of Economic History* 23 (1963): 414–443;

For late nineteenth-century American mechanics and engineers, there were scarce alternatives to the American system of intrafirm de facto standardization. American statesmen of the era ensured that the federal government would not play an active or energetic role. For example, Congress established an Office of Weights and Measures in the 1830s, but for many years the office lacked the funds, personnel, or political backing to be effective. The formation of the National Bureau of Standards in 1901 expanded the federal government's role in industrial standardization, but the bureau never received the funds and personnel it would have needed to transform the American system of standardization into one that was dominated, or even heavily influenced, by the federal government.[40]

Norvin Green's campaign against government ownership of the telegraph was typical of the deep skepticism of nineteenth-century Americans toward a powerful federal government – a skepticism that was noted by Tocqueville and subsequent observers. Instead, Americans often expressed their collective energies by "constantly" forming associations in commercial, technical, and other realms. Some experiments with private associations occurred in sectors of the economy where market participants craved more information about the quality, price, and techniques used to produce commodities. Information exchanges that took place under the auspices of the Chicago Board of Trade and the New England Cotton Manufacturers' Association sustained long-term relationships among a wide variety of market participants.[41]

Voluntary organizations also coalesced around mechanical and scientific topics. One such group, the Franklin Institute for the Promotion of the Mechanic Arts, was founded in 1824. The Franklin Institute was at first dominated by Philadelphia-area mechanics – mostly social and commercial elites, rather than rank-and-file workers or artisans. The Franklin Institute pursued several strategies to raise its own profile – and the status of the mechanic arts – including popular lectures, exhibits, the establishment of a journal, and the provision of technical advice to the State of Pennsylvania, the U.S. Patent Office, and the U.S. Congress. As a result of these initiatives, Philadelphia's "philosopher

---

L. T. C. Rolt, *A Short History of Machine Tools* (Cambridge, MA: The MIT Press, 1965), 137–177; David A. Hounshell, *From the American System to Mass Production, 1800–1932* (Baltimore: The Johns Hopkins University Press, 1984); Donald R. Hoke, *Ingenious Yankees: The Rise of the American System of Manufactures in the Private Sector* (New York: Columbia University Press, 1990); John K. Brown, *The Baldwin Locomotive Works, 1831–1915: A Study in American Industrial Practice* (Baltimore: The Johns Hopkins University Press, 1995); Scranton, *Endless Novelty*.

40 John Perry, *The Story of Standards* (New York: Funk & Wagnalls Company, 1955), 56–72; Rexmond C. Cochrane, *Measures for Progress: A History of the National Bureau of Standards* (Washington, DC: Department of Commerce, 1966), 21–38; Thomas C. Lassman, "Government Science in Postwar America: Henry A. Wallace, Edward U. Condon, and the Transformation of the National Bureau of Standards, 1945–1951," *Isis* 96 (2005): 25–51.

41 Louis Galambos, *Cooperation and Competition: The Emergence of a National Trade Association* (Baltimore: The Johns Hopkins University Press, 1966), 20–30; Cronon, *Nature's Metropolis*, 104–142.

mechanics" made the Franklin Institute the leading technical society in the United States within two decades of its creation.[42]

Some of the Franklin Institute's activities concluded with specific recommendations for industrial practices and products. For example, it conducted an investigation between 1830 and 1837 that culminated in a *General Report on the Explosion of Steam Boilers*. Although the *General Report* provided clear safety recommendations, steamboat operators refused to adopt its stricter – and more costly – safety measures. These operators did not quite understand the harm that accidents imposed on their collective reputations, nor did they accept their collective responsibility to sacrifice profit to advance public safety and their private reputations. Moreover Congress, as historian Bruce Sinclair observed, "had yet to accept the principle that public safety demanded constraints on private industry." It took until 1852 for Congress to reverse course and pass legislation that backed the recommendations of the *General Report* with the force of law, thus requiring widespread adoption of safer steam boilers.[43]

This example highlights the major weakness of voluntary standards in nineteenth-century America: if private actors saw no advantage to proposed recommendations (such as new safety procedures), they could not be forced to adopt them. The leaders of the Franklin Institute seem to have learned this lesson, because their next major initiative to reform industrial practice charted a different strategic course. In 1864, William Sellers, Philadelphia's leading machinist and president of the Franklin Institute, presented a paper to the institute that proposed a new system for uniform American screw threads. Sellers's paper appealed to the practical shop values of American mechanics, who appreciated the simplicity of his system. Moreover, Sellers recognized that the Franklin Institute, dominated by a network of mechanics from Philadelphia's leading firms, could be an effective means for promoting the widespread adoption of his system. Sellers branded his proposal the Franklin Institute system and promoted its merits to powerful allies such as the secretary of the navy and leading firms in the railroad and machine tool industries. The eventual success of the Sellers screw, according to Sinclair, was a consequence of Sellers's adept use of institutions. Sellers used the Franklin Institute in particular as "an institutional framework for his system" – a framework that provided "a platform, a mechanism for its advancement, and an aura of objectivity."[44]

[42] Bruce Sinclair, *Philadelphia's Philosopher Mechanics: A History of the Franklin Institute, 1824–1865* (Baltimore: Johns Hopkins University Press, 1974).

[43] Sinclair, *Philadelphia's Philosopher Mechanics*, 170–191. See also Bruce Sinclair, *Early Research at the Franklin Institute: The Investigation into the Causes of Steam Boiler Explosions, 1830–1837* (Philadelphia: Franklin Institute, 1966).

[44] Bruce Sinclair, "At the Turn of a Screw: William Sellers, the Franklin Institute, and a Standard American Thread," *Technology & Culture* 10 (1969): 34.

This mix of technical, institutional, and cultural forces became the core features of a variety of professional engineering societies that emerged in the last decades of the nineteenth century. The earliest and most influential of these groups included

- the American Society for Civil Engineers (ASCE, founded in 1852, reorganized in 1867);
- the American Institute of Mining Engineers (AIME, founded in 1871);
- the American Society of Mechanical Engineers (ASME, founded in 1880);
- the American Institute of Electrical Engineers (AIEE, founded in 1884).

Taken together, the creation of these associations marks a fundamental transition in American engineering history. Their common conviction was that industrial society had become too chaotic, and that they, like other professionals, were obliged to lend their expertise to a "search for order" that could reform American society. In other words, their professionalization was part of a general response to the wasteful competition produced by the nineteenth-century American industrial economy.[45]

Standardization was an important function of these societies. As American engineering societies took shape, standards came to be seen as exemplars of an engineering ideal – an objective and impartial specification that had been created through joint effort, sound reasoning, and hard work. Standardization was thus an inherently social and power-laden activity, one that fostered rivalries and disputes as much as it advanced the values of cooperation and harmony.[46]

At a basic level, many engineers were drawn to standardization because it could support the ordering impulse inherent in the logic of engineering practice. Standards simply made sense: they had the power to rationalize and to harmonize basic engineering categories such as nomenclature and measurement. In an era during which engineers were increasingly aligned with American capitalist production, standardization also made good business sense. Some engineers, such as Charles Proteus Steinmetz at General Electric, advocated standardization because it facilitated greater coordination and systematization. Others, such as Chicago Edison boss Samuel Insull, saw standardization as a way to simplify factory operations and to reduce costs. Standardization thus combined an intellectual aspiration, an economic imperative, and, for some,

---

[45] Edwin T. Layton, *The Revolt of the Engineers: Social Responsibility and the American Engineering Profession* (Cleveland, OH: Press of Case Western Reserve University, 1971), 25–46; Terry S. Reynolds, "The Engineer in 19th-Century America," in Terry S. Reynolds, ed., *The Engineer in America: A Historical Anthology from Technology and Culture* (Chicago: University of Chicago Press, 1991); Wiebe, *Search for Order*.

[46] A. Michal McMahon, *The Making of a Profession: A Century of Electrical Engineering in America* (New York: Institute of Electrical and Electronics Engineers, 1984); Bruce Sinclair, *A Centennial History of the American Society of Mechanical Engineers, 1880–1980* (Toronto: University of Toronto Press, 1980), especially 46–60.

a personal obsession. In Sinclair's apt summary: "Standardization was to the engineer what administration was to the manager."[47]

Mechanical Engineers: Competing Cultures and Jurisdictions

Soon after they were formed, each of the professional engineering societies considered the establishment and publication of standards as a way to resolve technical problems and create a common basis for engineering practice. The first standards initiatives in the ASME, for example, explored systematic and rational definitions for the materials, methods, screws, nomenclature, and tools that were most important for the everyday labor of mechanical engineers.[48] Although most ASME members in the 1880s and early 1890s supported the rational appeal of standards, they were divided over the question of who should set them.

Some ASME members – representatives of the "shop culture" including James W. See, Oberlin Smith, and Coleman Sellers – called for a government bureau to record, publish, and therefore harmonize standards that had already emerged through shop practice. Other members – "school culture" engineers such as the ASME's "chief theoretician" and first president Robert Thurston – envisioned a central body that would not only record existing standards, but would also go one step further and use their expertise to determine which specifications would be superior. In doing so, mechanical engineers could embody a fusion of democratic and objective ideals and thus be uniquely prepared to enhance social welfare and elevate modern civilization.[49] Such centralization, for the defenders of shop culture, was dangerous because it threatened their dearest value: wisdom gained through practical experience and proved through open market competition. These engineers, as Sinclair has noted, were less interested in defining new scientific truths and far more concerned with "practical realities that emphasized acceptance as the crucial issue."[50]

The centralization and standardization of mechanical practice proved to be a fantasy. The reality was that industrial standardization in the late nineteenth century was a complicated and messy organizational field, where no central authority – neither ASME nor the industrial trade associations that were appearing throughout the American economy – was capable of coordinating national engineering practice. Instead, industrial standardization occurred under conditions of persistent instability, where competing communities of

---

[47] McMahon, *The Making of a Profession*, 88–98; Sinclair, *A Centennial History*, 50.

[48] Monte Calvert, *The Mechanical Engineer in America, 1830–1910* (Baltimore: The Johns Hopkins University Press, 1967), 169.

[49] Sinclair, *A Centennial History*, 59. On Thurston's role, see also Geoffrey W. Clark, *History of Stevens Institute of Technology: A Record of Broad-Based Curricula and Technogenesis* (Jersey City, NJ: Jensen/Daniels, 2000), 53–66.

[50] Sinclair, *A Centennial History*, 24–31, 57; Calvert, *The Mechanical Engineer*, 177–186; James W. See, "Standards," *Transactions of the American Society of Mechanical Engineers* 10 (1889): 542–575.

engineers struggled to extend their authority (or "jurisdiction," in Andrew Abbott's useful concept).[51]

The absence of order became manifest in an ever-expanding number of uncoordinated or loosely coordinated standards committees. The overlapping efforts led to a curious paradox for these imposers of rational order: the standard-setting process was itself irrational, wasteful, and disorderly. Work proceeded when representatives from various technical and industry groups met on an ad hoc basis to recruit allies, sort out lines of cooperation, and identify areas of potential conflict. For example, in the late 1880s, ASME created a Committee on Uniform Standards in Pipe and Pipe Threads composed of "men representative of pipe manufacturers and pipe users."[52] That committee, in turn, soon reached out to industry trade associations including the Pipe Manufacturers Association; the Manufacturers Association of Brass and Iron; Steam, Gas, and Water Works of the United States; and the Cast Iron Fittings Association. These trade associations were eager to set standards out of commercial self-interest rather than the lofty goals articulated by ASME and other professional engineering societies. Their motivation to cooperate was to facilitate trade in the domestic marketplace as well as exports to Canada and Latin America. In short, cooperation succeeded where engineers from different branches of American industry saw the potential to promote their own self-interest by cooperating with engineers and groups who, in other circumstances, they might see as rivals.[53]

The ASME style, typical of the other engineering societies, can be summarized as focused on practical needs, favorable toward methods proven in industry, tentative toward endorsing specific technologies as "standard," and active in its diplomacy with numerous other private organizations such as standard-setting committees and trade associations. Given the absence of any authority to compel or mandate the adoption of its standards, ASME developed an inclusive style of forming committees and making standards. This inclusive style, beyond its appeal to democratic ideals and progressive rhetoric, had two practical advantages. First, it encouraged engineers to voice their objections at an early stage so that they could be resolved before a technology became the industry standard – thus eliminating waste and preempting the destructive competition that deeply worried engineers and managers of the era. Second, an inclusive process could better encourage a priori commitments to adopting the

---

[51] Sinclair, *A Centennial History*, 41–46; Andrew Abbott, *The System of Professions: An Essay on the Division of Expert Labor* (Chicago: University of Chicago Press, 1988); Brent K. Jesiek, *Between Discipline and Profession: A History of Persistent Instability in the Field of Computer Engineering, circa 1951–2006*, PhD dissertation, Virginia Polytechnic Institute and State University, 2006; Paul J. DiMaggio and Walter W. Powell, "The Iron Cage Revisited: Institutional Isomorphism and Collective Rationality in Organizational Fields," *American Sociological Review* 48 (1983): 147–160.

[52] William Kent, quoted in Sinclair, *A Centennial History*, 54.

[53] Sinclair, *A Centennial History*, 55.

standard from a wide range of industrial partners. Both features of the ASME style increased the likelihood that the additional effort and resources required to define standards would not be wasted.

Electrical Standards: Elite Leaders and Evolving Rules

The early history of standardization activities in the AIEE reveals a similar pattern of development. The AIEE moved aggressively to set standards with practical importance for its members. Along the way, its leaders learned to negotiate and cooperate with similar efforts in domestic trade associations, such as the National Telephone Exchange Association and the National Electric Light Association, and foreign and international bodies such as the British Board of Trade, the British Association for the Advancement of Science, and the International Electrical Congresses.

After some early failures in creating committees to adopt a standard wire gauge in the mid-1880s, the AIEE formed a standing Committee on Units and Standards in 1889, with Edison Electric consulting engineer (and later Harvard and MIT professor) Arthur Kennelly as chairman. The committee approved and promoted standard nomenclature for electrical units, continued to study a standard wire gauge, and debated ratings and nomenclature for electrical apparatus.[54]

Within the course of these debates, members focused on technical subjects where they felt the AIEE could make a contribution. But they also discussed social aspects of the standards process, such as the appropriate membership and rules of the committee itself. In 1898, the debate focused on the question of whether membership should be open to representatives from manufacturers who had vital commercial interests at stake. Charles Steinmetz, the respected General Electric research engineer, tried to convince his colleagues – and perhaps himself – that they simply needed to trust the professionalism of the engineers involved: "The committee doing the work must be composed of men of such standing and reputation that, regardless of whether they are connected with manufacturing concerns or not, there can be no question that they will be impartial and not influenced by the fact that they are connected with this or that company."[55]

Elite participation in these committees lent prestige to the AIEE's efforts, a factor that should not be underestimated. Engineers and managers in the electrical industries were keen to avoid the perception that they were irresponsible, overly self-interested, or that electrical technologies would endanger the safety of the American public – thus also endangering the reputation of the electrical engineering profession. AIEE standards were more likely to be adopted because

---

[54] McMahon, *The Making of a Profession*, 79–92; Arthur E. Kennelly, "The Work of the Institute in Standardization," *Electrical Engineering* 53 (1934): 678.

[55] Steinmetz, quoted in McMahon, *The Making of a Profession*, 85; "Report of the Standardization Committee," *Transactions of the AIEE* (1899): 255–268.

TABLE 2.1 *Members of AIEE Standards Committee in 1898, 1902, and 1906*

- Francis Bacon Crocker, chairman, 1898, 1902, 1906
  • Consulting engineer and Columbia University
- Arthur E. Kennelly, 1898, 1902; secretary, 1906
  • Consulting engineer and Harvard University
- Arthur W. Berresford, 1906
  • Cutler-Hammer Manufacturing Co.
- Cary T. Hutchinson, 1898
  • Consulting engineer
- Dugald C. Jackson, 1906
  • Consulting engineer, University of Wisconsin, and Massachusetts Institute of Technology
- John W. Lieb, Jr., 1898, 1902
  • Edison Electric Illuminating Company and New York Edison
- C. O. Mailloux, 1902, 1906
  • Consulting engineer
- Robert B. Owens, 1906
  • McGill University
- Charles F. Scott, 1906
  • Westinghouse Electric
- Charles P. Steinmetz, 1898, 1902, 1906
  • General Electric
- Lewis B. Stillwell, 1898, 1902
  • Westinghouse Electric, Niagara Falls Power Company, and Rapid Transit Subway Company
- Henry G. Stott, 1906
  • Interborough Rapid Transit Company
- S. W. Stratton, 1906
  • National Bureau of Standards
- Elihu Thomson, 1898, 1902, 1906
  • General Electric

Source: "Standardization Rules of the AIEE," *Transactions of the AIEE* 35 (1916): 1551–1552.
Note: The committee had seven members in 1898 and 1902 and eleven in 1906. It expanded to sixteen members in 1911, with Comfort A. Adams appointed chairman and Kennelly staying on as secretary. Of the remaining fourteen members, all except Steinmetz were new.

many of the standard setters were academic elites, leading consulting engineers, or worked for the most powerful firms in the industry such as General Electric and Westinghouse (see Table 2.1 for a list of members from 1898, 1902, and 1906). The important function of consulting engineers – those who did not work for any single firm but instead were agents to spread knowledge and new practices to many different firms – was especially evident in the early AIEE standards committees. Their participation enhanced the AIEE's efforts while simultaneously ensuring that business had a voice in industrywide technical decisions. In spite of – or perhaps in addition to – Steinmetz's ideal of a committee of impartial professionals, AIEE standards became vehicles for dominant firms to maintain control and police the boundaries of the electrical industry.[56]

As the AIEE trusted this cohort to set standards, it tinkered with the structure of its committees and its process for evaluating technical reports. A stable structure and process for setting standards emerged between 1899 and 1916 as the AIEE revised the various standards and developed bylaws to govern their collaborations. The rules first appeared in a *Report of the Committee on Standardization* in the AIEE *Transactions* in 1899. Subsequent revisions culminated in 1916, when the AIEE published new bylaws for the Standards Committee, "the completion and clarification of the previous radical revision" with "a number of important additions." The justification of this work was twofold: "to crystallize the policy of the Standards Committee in its own activities, and in relation to similar committees of other engineering societies."[57]

Six of the nine AIEE Standards Committee bylaws published in 1916 defined aspects of the committee's "own activities," such as provisions for the circulation of minutes and amendments and additions to existing rules. The remaining three bylaws set the tone for subsequent engagements between the AIEE Standards Committee and similar committees elsewhere. Following a simple declaration, "Cooperation is desirable between the Standards Committee of the Institute and other standards committees," the AIEE's bylaws contained provisions to share meeting minutes with cooperating committees and to empower such committees to object (either in writing or in person) to actions of the AIEE Standards Committee. Many of the "other standards committees" that the bylaws mentioned were hosted by trade associations, thus reinforcing the AIEE's ties to collectives of industrial firms that banded together around their common commercial and technical interests.[58]

By this time, AIEE members had grown accustomed to working with colleagues in other standards bodies. Indeed, each of the six standing subcommittees established in 1913 – Rating, Telegraph and Telephone Standards, Railway Standards, Nomenclature and Symbols, Wires and Cables, and Rating and Testing of Control Apparatus – worked on technologies that had broad interest beyond the AIEE membership. In its 1916 "Standardization Rules," the AIEE Standards Committee acknowledged the "helpful cooperation" of no less

---

[56] C. E. Skinner, "The Present Status of Standards in the Electrical Industry," *Annals of the American Academy of Political and Social Science* 137 (1928): 151–156; Harold C. Passer, *The Electrical Manufacturers, 1875–1900: A Study in Competition, Technical Change, and Economic Growth* (Cambridge, MA: Harvard University Press, 1953); Steven W. Usselman, "From Novelty to Utility: George Westinghouse and the Business of Innovation during the Age of Edison," *Business History Review* 66 (1992): 251–304; Thomas P. Hughes, *Networks of Power: Electrification in Western Society, 1880–1930* (Baltimore, MD: Johns Hopkins University Press, 1983), 172–174; Layton, *Revolt of the Engineers*, 80; David F. Noble, *America by Design: Science, Technology, and the Rise of Corporate Capitalism* (New York: Alfred A. Knopf, Inc., 1977), 77–78.
[57] "Standardization Rules of the AIEE," *Transactions of the AIEE* 35 (1916): 1551–1557.
[58] "Standardization Rules of the AIEE," 1556–1557.

than eleven standards committees.[59] The bibliography of the "Rules" – which featured references to more than a dozen domestic publications and reports from Great Britain, Germany, and the IEC – also indicates the extent of the AIEE's engagement with outside standards committees, particularly committees organized by trade associations.[60]

To complement these formal channels for exchanging information about standards, engineers also developed informal social ties. New York City, the de facto capital of American engineering, was an important site where engineers could get to know one another better and endow American engineering with a more gentlemanly culture. In 1903, Andrew Carnegie sponsored the construction of the Engineering Societies' Building on West 39th Street, near Bryant Park in New York. For Carnegie and for the engineers who solicited his gift, the building stood as a testament to cooperation, a "great principle in which America led Europe."[61] When it opened in 1907, the building hosted the libraries and headquarters of the AIEE, ASME, and AIME (the ASCE joined in 1917) as well as an assembly hall that the *New York Times* called "one of the finest auditoriums of its kind in the city." The collegiality designed into this shared social space created new opportunities for engineers to exchange the lessons of "learning by doing" in standards committees – a factor that helps account for structural similarities between committees that grew within different professional and industrial settings.[62]

One such occasion for exchange was the 1916 visit of Charles le Maistre, the electrical secretary of the British Engineering Standards Committee and the general secretary of the IEC. Le Maistre spent several weeks in the United States advising groups such as the AIEE and National Bureau of Standards and developing relationships with his American colleagues. His diplomatic skills were on full display in his address to the AIEE's annual convention in June: "It

---

[59] They were: American Society for Testing Materials (Committee B-1), Association of Edison Illuminating Companies (Committee on Meters), Illuminating Engineering Society (Committee on Nomenclature and Standards), Electric Power Club (Committee on Engineering Recommendations; Standardization Committee), National Electrical Light Association (Committee on Meters; Committee on Apparatus), Association of Railway Electrical Engineers (Committee on Wires and Cables), American Electric Railway Engineering Association (Committees on Equipment and Distribution), Institute of Radio Engineers (Committee on Standardization), Society of Automobile Engineers (Standards Committee). "Standardization Rules of the AIEE," 1554–1555.

[60] *Standardization Rules of the American Institute of Electrical Engineers* (New York: American Institute of Electrical Engineers, 1914), 86.

[61] Charles F. Scott, "The Institute's First Half Century," *Electrical Engineering* 53 (1934): 660.

[62] Sinclair, *A Centennial History*, 26–27; Gano Dunn, "Early Headquarters of the Institute," *Electrical Engineering* 53 (1934): 685; Scott, "The Institute's First Half Century," 659–662; "Engineers Open Their New Home," *New York Times*, April 17, 1907: 18; Chi-nien Chung, "Networks and Governance in Trade Associations: AEIC and NELA in the Development of the American Electricity Industry, 1885–1910," *International Journal of Sociology and Social Policy* 17 (1997): 57–110; DiMaggio and Powell, "Iron Cage Revisited."

is by these meetings that we are gradually becoming more highly cosmopolitan in the truest sense of the word. We get our rough edges knocked off and begin to appreciate other people."[63] His American colleagues replied in kind. For example, Comfort A. Adams, who served as chairman and secretary of the AIEE Standards Committee from 1910 to 1919, noted that "the personal element is a most vitally important one.... [Le Maistre] entered into the spirit of our meetings and became one of us, giving generously of his time and experience; his personal assistance was of such an order that we shall hereafter feel any revision of our rules incomplete without his suggestions."[64]

AIEE members further displayed their gratitude toward their visitor by including text in the 1916 revision of their "Standardization Rules" that acknowledged le Maistre's contributions. As we will see, American standard setters in the 1910s and 1920s would continue to rely on le Maistre's advice and experience as they developed the capabilities to harmonize and centralize American industrial standardization. The British contributions to American standard setting went far beyond the "personal element" cited by the AIEE; they also included organizational practices that the Americans could use as precedents.[65]

One clear lesson that AIEE members learned from the first three decades of their experiences in setting standards was the value of maintaining an open and welcoming stance toward suggestions that came from fellow professionals. Comfort Adams, in his comments on le Maistre's 1916 paper, summarized these interactions as "a process of (sometimes stormy) mutual education."[66] There were, as Adams hinted, limits to the AIEE's openness: it neither welcomed nor expected suggestions from the general public, only from recognized communities of expert engineers. The AIEE made this stance explicit in a resolution approved in April 1916: "The Standards Committee will be pleased to receive from any of the engineering societies such standardization rules as they may care to have included.... Such rules will be included if they are found not to be incompatible with the Standardization Rules of the AIEE."[67]

Beneath this explicit acknowledgment of engineering societies, the AIEE Standards Committee's own history also indicates with great clarity the fundamental contributions to electrical standardization that came from trade associations, consulting engineers, and engineers who worked for General Electric, Westinghouse, and other leading manufacturers. Rather than attempting to

---

[63] Charles le Maistre, "Standardization," *Transactions of the AIEE* 35 (1916): 498.

[64] Comfort A. Adams, "Discussion on 'Standardization' (le Maistre)," *Transactions of the AIEE* 35 (1916): 500.

[65] "Standardization Rules of the AIEE," 1555; Charles le Maistre, "Standardization," 489–496; Charles le Maistre, "The British Engineering Standards Association," *Transactions of the American Society of Mechanical Engineers* 40 (1918): 863–868; Craig Murphy and JoAnne Yates, "Charles le Maistre: Entrepreneur in International Standardization," *Enterprises et Histoire* 51 (2008): 10–27.

[66] Comfort A. Adams, "Discussion on 'Standardization' (le Maistre)," 499.

[67] "Standardization Rules of the AIEE," 1557.

establish sole authority over electrical standardization, the AIEE tended to focus on "fundamental practices," and left the formal "establishment and maintenance of fundamental units" to groups such as the IEC and the National Bureau of Standards.[68] AIEE bylaws emphasized the desirability of cooperation with other standards committees and included provisions that opened the AIEE standards process to suggestions – and objections – from external expert communities. Through these (sometimes stormy) collaborations, American electrical engineers created new organizational capabilities that could sustain cooperation throughout the American and international electrical industries as well as throughout the various American engineering professions. No longer did electrical engineers have to rely on winner-takes-all battles between systems or waste time and resources waiting for a clear standard to emerge from market competition. The AIEE's committees and subcommittees presented engineers with alternatives: new coordination mechanisms that could establish industrywide standards and avoid the specter of monopoly.

Materials Testing: A Pluralist Solution to Mistrust

Electrical engineers, like mechanical engineers, first approached industrywide standardization as they began to define their professional identity and status. When controversies inevitably arose, electrical and mechanical engineers could rally around their aspirations to be, as Steinmetz put it, impartial professionals of unimpeachable standing and reputation. In other cases, such as the development of specifications for steel rails, engineers approached the standards process with higher levels of mistrust. Engineers in the steel and railroad industries – the realms of notoriously hardheaded capitalists – argued for years over the proper terminology, tests, and specifications to determine the hardness of steel rails. Here, as in the mechanical and electrical engineering societies, leading engineers experimented with new committees for settling disputes. And, once again, engineers sought to legitimize their technical work with a progressive rhetoric of cooperation, expertise, and social harmony.

Railroad engineers and managers had an intuitive grasp of the value of standardization for achieving regional and even transcontinental compatibility. Many ad hoc standards, including agreements for a standard width for railroad tracks and standard time, had proved to be vital for national expansion. Other standards, such as safety standards for brakes as well as standards that specified interfaces for axles, wheels, and rails, were developed through railroad trade associations such as the Master Car-Builders Association.[69]

[68] Skinner, "The Present Status of Standards."

[69] Douglas J. Puffert, *Tracks across Continents, Paths through History: The Economic Dynamics of Standardization in Railway Gauge* (Chicago: University of Chicago Press, 2009); Ian R. Bartky, *Selling the True Time: Nineteenth-Century Timekeeping in America* (Stanford, CA: Stanford University Press, 2000); Clark Blaise, *Time Lord: Sir Sanford Fleming and the Creation of Standard Time* (New York: Vintage, 2002); Steven W. Usselman, *Regulating Railroad Innovation: Business, Technology, and Politics in America, 1840–1920* (New York: Cambridge

The wear and failure of steel and iron rails, however, presented a different type of problem, one that railroad engineers could not resolve within their companies or through agreements with other railroads. Their central problem was that steel companies, not railroad companies, manufactured rails – and steel executives and managers were not known for their willingness to compromise. Moreover, the two industries lacked common terminology and criteria for evaluating rails: whereas engineers in the steel industry focused on the structural aspects of the shape and wear of rails, railroad engineers increasingly believed that chemical properties determined the strength and reliability of any given rail. Over time, it became clear to all parties that cooperation between the two communities would be necessary in order to improve rail quality and reliability. Between the 1870s and the 1910s, their cooperative efforts had two related features: institutions that could establish consistent methods for testing materials and rhetoric that could establish the ideological significance of the tests.[70]

At first, railroad engineers attempted to resolve the conflict through appeals to scientific analysis. Beginning in the late 1870s, a PhD chemist for the Pennsylvania Railroad named Charles Benjamin Dudley presented a series of papers to the AIME and the American Chemical Society on the results of his chemical analyses of steel rails. The main problem he addressed was the lack of consistent criteria for evaluating the technical characteristics of steel rails. Dudley also faced substantial resistance on all sides: steel manufacturers worried that specifications were an "unnecessary annoyance and interference with their works and processes," while railroad representatives were wary that the adoption of precise specifications would drive up the price of steel rails.[71] To make matters worse, engineers in railroad and steel firms were skeptical of Dudley's laboratory results because they contradicted hard-earned practical experience.[72]

University Press, 2002); Master Car-Builders' Association, *History and Early Reports of the Master Car-Builders' Association* (New York: Martin B. Brown, 1885).

[70] Usselman, *Regulating Railroad Innovation*, 215–261. On the market structure of the steel industry, see Naomi R. Lamoreaux, *The Great Merger Movement in American Business, 1895–1904* (New York: Cambridge University Press, 1985), 76–86.

[71] C. B. Dudley and F. N. Pease, "Chemistry Applied to Railroads. XXVI – How to Make Specifications," *The Railroad and Engineering Journal* 66 (1892): 160.

[72] Dudley's papers sparked unusually impassioned and lengthy debates among AIME members that filled hundreds of pages of the *AIME Transactions* between 1878 and 1883. See for example Charles B. Dudley, "The Chemical Composition and Physical Properties of Steel Rails," *AIME Transactions* 7 (May 1878–February 1879): 172–201; Charles B. Dudley, "Does the Wearing Power of Steel Rails Increase with the Hardness of the Steel?" *AIME Transactions* 7 (May 1878–February 1879): 202–205; and "Discussion of Dr. Charles B. Dudley's Papers on Steel Rails, Read at the Lake George Meeting, October 1878," *AIME Transactions* 7 (May 1878–February 1879): 357–413. See also Usselman, *Regulating Railroad Innovation*, 221–223.

In the 1880s and 1890s, Dudley and a small group of railroad engineers, including Robert Hunt and William R. Webster, engaged in a program of industrial diplomacy that ran parallel to Dudley's laboratory research. Their efforts spanned a number of institutions: Hunt worked within the ASCE and AIME in the 1890s and joined Webster in the late 1890s to become active in the American Railroad Engineering and Maintenance of Way Association (AREMWA). They all participated in the formative meetings of the American Society for Testing Materials (ASTM), beginning in 1898. Neither of these groups pretended to be impartial professional societies: AREMWA was dominated by railway engineers and ASTM by steel manufacturers. Yet Hunt and Webster found ways to persuade partisans from both industries to cooperate and, in the summary of historian Steven Usselman, "supplant the belligerent posturing of an interest group with the dispassionate aura of a community of engineering experts." The most striking symbol of their success in bringing these two communities together was the election of Dudley – a railroad man – as president of the steel-dominated ASTM in 1902. These strategic compromises set the stage for new specifications that, in turn, were written into contracts between steel and railroad companies.[73]

As ASTM president from 1902 to 1909, Dudley framed the ASTM's technical work as diligent and diplomatic professional labor. Dudley consistently returned to two rhetorical strategies to cultivate this image. First, he emphasized the moral and personal characteristics necessary for quality work. Testing chemists (as opposed to "routine chemists") needed to be skillful, careful, and approach their work with "a sincere disposition."[74] Such qualities would ensure that a specification would be a "very high order of work" that would take into account "the results of the latest and best studies of the properties of the material which it covers." Developed in such a way, Dudley concluded, a specification "combined within itself the harmonized antagonistic interests of both the producer and consumer."[75]

Dudley's language draws our attention to the broader social contexts in which technical experts were reframing their work as a pursuit of social

---

[73] Usselman, *Regulating Railroad Innovation*, 235–241.

[74] C. B. Dudley and F. N. Pease, "The Need of Standard Methods for the Analysis of Iron and Steel, with Some Proposed Standard Methods," *Journal of the American Chemical Society* 15 (1893): 506. See also Charles B. Dudley, "The Testing Engineer," *Proceedings of the Annual Meeting – American Society for Testing Materials* 5 (1905): 17–29.

[75] C. B. Dudley, "The Making of Specifications for Materials," *Proceedings of the Annual Meeting – American Society for Testing Materials* 3 (1903): 34. See also Usselman, *Regulating Railroad Innovation*, 253–260; and Ann Johnson, "Material Experiments: Environment and Engineering Institutions in the Early American Republic," *Osiris* 24 (2009): 53–74. On the values of precision in British metrology and electrical science, see Graeme J. N. Gooday, *The Morals of Measurement: Accuracy, Irony, and Trust in Late Victorian Electrical Practice* (New York: Cambridge University Press, 2004). For dozens of eulogies full of effusive praise for Dudley's character and accomplishments see *Memorial Volume Commemorative of the Life and Life-Work of Charles Benjamin Dudley, Ph.D.* (Philadelphia: American Society for Testing Materials, 1911).

harmony. At the turn of the twentieth century, the image of the American steel and railroad industries were more closely associated with cutthroat capitalism than they were with social harmony. By injecting progressive values of expertise, civic concern, public safety, and social responsibility into these industrial relations, Dudley articulated a critique of self-interested industrial capitalism. His critique is notable in part because it emerged not from a muckraking reformer but instead from a corporate insider who was deeply committed to his company's commercial success. In contrast to the antagonisms of the nineteenth century, Dudley recast representatives of steel and railroad corporations as trustworthy citizens and earnest collaborators who would defer to expert knowledge. Dudley's rhetorical campaign, like those occurring in other branches of American professional engineering, rested on his promotion of the practical, technical, and moral advantages of cooperation.[76]

The second pillar of Dudley's rhetorical strategy was his embrace of inclusive and pluralist procedures for creating and approving specifications. He stated the guiding principle – a principle that would echo throughout the twentieth century – clearly in his 1903 presidential address to the ASTM: "All parties whose interests are affected by a specification should have a voice in its preparation."[77] Perhaps the key factor in his progression from bilateral (producers and consumers) to multilateral ("all parties") negotiations was his openness to expert knowledge of materials that came from a number of different sources and experiences, such as "a chemical analysis or a microscopic examination, or a statement of the method of manufacture." Accordingly, Dudley concluded that "information from any source" could be incorporated into a specification, as long as it was found to be "useful or valuable in defining limitations, or in deciding upon the quality of material furnished."[78]

As a complement to this inclusive ideal for the creation of specifications, the ASTM amended its bylaws in 1908 to add a "procedure governing the adoption of standard specifications." The new requirement, which originated in the ASTM's Executive Committee and was adopted by the ASTM membership without comment or debate, laid out a three-step voting process for a specification to gain official standing. First, the specifications should be presented to the ASTM annual meeting, where it could be amended by a majority vote. The specification then needed the support of two-thirds of the meeting before being referred to a letter ballot. The final step required two-thirds of the letter votes to support the specification. The cumulative effect of Dudley's rhetoric and the revised ASTM bylaws was the formation of a more open model for creating

---

[76] For explorations of these issues within broader Progressive contexts, see Usselman, *Regulating Railroad Innovation*, 327–387; and Mark Aldrich, *Safety First: Technology, Labor, and Business in the Building of American Work Safety, 1870–1939* (Baltimore: The Johns Hopkins University Press, 1997).

[77] C. B. Dudley, "The Making of Specifications for Materials," 30.

[78] C. B. Dudley, "The Making of Specifications for Materials," 31.

industrial specifications, one that could accommodate more voices in the preparation of specifications and ground its legitimacy on formal procedures.[79]

Conclusions

The developments in the standards committees of the ASME, AIEE, and ASTM marked a fundamental shift in the American system of standardization. The new committees shared some basic characteristics. First, all participants joined standards committees as a response to the demands for efficiency, reliability, and profitability that emerged from their daily practice. The voluntary associations that Americans constantly formed indicate the private orientation of the American industrial economy of the nineteenth century, in which the federal government did not play as aggressive a role as it might have – or as other national governments did.

Second, committees grew to welcome all *interested* parties – that is, corporate engineers and managers who recognized that they would be directly and materially affected by the outcome. Engineers learned that they could accomplish more extensive and sophisticated levels of standardization through committees than they could through onetime market transactions or through the hierarchy of a single corporation. Experience taught them that the eventual success of a standard depended on the support of a strong network of manufacturers and buyers (or, in Dudley's language, producers and consumers). The commercial experience of consulting engineers, as well as engineers who participated in industry trade associations, was fundamental.[80]

Third, engineers who joined standards committees embraced increasingly formal rules and procedures to prevent any one interested party from dominating the process. In the early stages of any particular standards committee, engineers were eager to defer to practical experience and were mindful of jurisdictional conflicts and preexisting efforts in related technical areas. Standards committees constituted an expansive network of institutions between markets and hierarchies – a network where no one institution had complete control. Of course, not all nodes in this network were equally powerful: in electrical standardization, for example, the presence of General Electric and Westinghouse loomed large. But even in sectors where ownership was highly concentrated (such as the American telegraph industry after 1866), no single organization

---

[79] *Proceedings of the Annual Meeting – American Society for Testing Materials* 8 (1908): 10–11, 689.

[80] For accounts of standardization in the automobile industry that emphasize similar institutional factors, see George V. Thompson, "Intercompany Technical Standardization in the Early American Automobile Industry," *The Journal of Economic History* 14 (1954): 1–20; Jeffrey Robert Yost, *Components of the Past and Vehicles of Change: Parts Manufacturers and Supplier Relations in the U.S. Automobile Industry* (PhD dissertation, Case Western Reserve University, 1998), 236–296; and James J. Flink, *The Automobile Age* (Cambridge, MA: The MIT Press, 1990).

monopolized standardization. Despite Western Union's dominant position, the variety of standards developed around telegraphy indicate that control over communication networks was a diffuse and complicated affair – far beyond the capabilities of any single entity, even powerful institutions such as Western Union or the U.S. Congress.

Fourth, standards committees grew as important functions of their professional societies, thus creating permanent links between standardization and the values of expertise and professionalism. Elite engineers led standard-setting committees in the mechanical, electrical, and material testing fields, which they understood as an essential part of their professional self-fashioning. The moral aspects of standard setting were especially significant, and several articulate engineers cleverly linked standards to progressive values of fair competition and public safety. Leaders such as Thurston, Steinmetz, and Dudley developed an elaborate rhetoric of cooperation that emphasized their professionalism and trustworthiness; they also laid the foundations of a new political philosophy that would govern their interactions in standards committees. The first generations of rules and bylaws in ASME, AIEE, and ASTM reflected a broader progressive shift away from the chaos and mistrust of the market and toward (sometimes stormy) widespread voluntary cooperation and deference to the orderly designs of expert minds. Their work led to a new generation of problems and, by the 1920s, a new generation of institutional and ideological innovations that anchored the vast expansion of standardization throughout the American and global economy.

The institutional innovations developed by standards engineers in the late nineteenth century were implicit critiques of the existing order, where the only alternatives to monopoly capitalism were either the chaos of "wasteful competition" or the possibility of government ownership of the means of production – a possibility that was never popular among Americans or their elected representatives. When engineers were successful in creating standards committees in the late nineteenth century, it was because they were able to cast their committees as new alternatives to existing modes of industrial control. By the 1920s, as we see in Chapter 3, standards engineers put forth a bolder claim and explicit critique of the status quo: the "consensus" committee method, according to its advocates such as Paul Gough Agnew, was in fact a superior form of rule making when compared to traditional democratic institutions such as legislative bodies and regulatory commissions. Democracy in America – as well as technological and social progress – was not to be found in its government or its corporations; it was to be found in the voluntary associations of private citizens. In Agnew's vision, we see a clear example of institutional innovations that take shape as ideological critique: instead of complaining about the existing world, Agnew and his colleagues simply worked to build a new one.

3

Ideological Origins of Open Standards II: American Standards, 1910s–1930s

> We do not leave to Congress, or to the vote of 110,000,000 people, the decision whether a bridge shall be built in the city of Oshkosh. We leave it to the people of Oshkosh, who will walk over it and ride over it, and who will have to pay for it. Why should not the very limited groups directly interested in each of the innumerable industrial problems with which they are faced, themselves solve these problems through coöperative effort?
>
> – Paul Gough Agnew
> "A Step Toward Industrial Self-Government," 1926[1]

In the late afternoon of June 15, 1922, Herbert Hoover dropped in to a meeting of forty-four engineers who had assembled in a conference room at his Department of Commerce in Washington, D.C. Hoover, who earned international fame as the "Great Humanitarian" by guiding relief efforts during World War I, turned his talents and energies toward the American economy when he was named secretary of commerce by Warren Harding in 1921. Hoover was, in the pre-Depression era, a living and breathing icon of the triumph of rationality, organization, and the progressive spirit of the engineering profession – the "engineering method personified" in the words of one admiring colleague.[2]

Several of Hoover's contemporaries – such as the social theorist Thorstein Veblen and the radical engineers Howard Scott and Morris Cooke – believed that technical expertise could reform society and engineer a new age of

---

[1] P. G. Agnew, "A Step toward Industrial Self-Government," *The New Republic* (March 17, 1926), 95.

[2] "Minutes, Joint Meeting of the AESC and the Executive Committee," June 15, 1922, Hoover Commerce Papers, Herbert Hoover Presidential Library, West Branch, Iowa, Box 23, AESC Folder [hereafter, Hoover Commerce Papers]. "Engineering method personified" is an accolade from Morris L. Cooke, quoted in Edwin T. Layton, *The Revolt of the Engineers: Social Responsibility and the American Engineering Profession* (Cleveland, OH: Press of Case Western Reserve University, 1971), 179.

efficiency and abundance. But the meeting at the Department of Commerce was not some sort of revolutionary "Soviet of Technicians." Instead, it was a meeting of the leaders of the American Engineering Standards Committee (AESC), a private group of engineers who had a more modest and pragmatic goal: to order the inconsistent patchwork of codes, tests, and standards used in American industrial practice. Although the AESC sought (and found) support from Hoover and other government officials, their method of work – consistent with the American mentality that Tocqueville observed in the mid-nineteenth century – depended on private cooperation to create standards that would, in turn, be adopted on a voluntary basis throughout American industry. Unlike their peers abroad, they did not depend on kings (Tocqueville's "men of rank") or presidents ("men of government"). Instead, their success would hinge on their ability to convince industrial managers and executives that industry standards would make business less wasteful and more profitable.

Hoover's presence at the AESC meeting was a sign that their cause – the voluntary coordination of industrial standards – had come of age. Indeed, such a meeting would have been unimaginable only forty years earlier. Between the 1880s and the 1920s, all of the vital ingredients of this meeting came into being: the category of a "professional" engineer, the progressive ideology of engineering, and institutions and rules that could govern collaborations among industrial engineers on a national scale. In the process, Americans grappled with a fundamental question of their industrializing society: Who should have the authority and responsibility to settle the mounting problems of industrial America? Progressives invited public control over these issues through democratic reforms such as referenda, direct elections, regulatory commissions, and state and municipal administration. Efficiency, equity, health, safety, and expertise were the animating spirits of these reforms; their underlying message was that traditional forms of power were failing to meet the challenges of the industrial age. Reformers often disagreed on the best path forward, but there existed an overarching consensus that the vitality of the American experiment would require new institutions, new values, and a renewed commitment to democracy.[3]

Where did technical standards fit into this picture? Chapter 2 describes how engineers embraced the Progressive spirit of experimentation, cooperation, and social responsibility as they created new institutions and fashioned new professional identities. The standards committees in groups such as the American Society of Mechanical Engineers (ASME), American Institute of Electrical Engineers (AIEE), and American Society for Testing Materials (ASTM) became

---

[3] Louis Galambos, "Technology, Political Economy, and Professionalization: Central Themes of the Organizational Synthesis," *Business History Review* 57 (1983): 471–493; Steven J. Diner, *A Very Different Age: Americans of the Progressive Era* (New York: Hill and Wang, 1998); Daniel T. Rodgers, *Atlantic Crossings: Social Politics in a Progressive Age* (Boston: Harvard University Press, 2000).

a fusion between two extreme visions of what engineers should be. They were not simply venues where engineers could carry out the wishes of their capitalist masters, nor were they incipient "Soviets of Engineers" that could be some sort of a staging ground for a revolution that would see engineers become the technocratic masters of modern society. Instead, these committees were conservative responses devised from within industry itself, not imposed by outsiders. Engineers believed these committees were, in the words of the standards engineer Paul Agnew, a "step toward industrial self-government." Agnew left little doubt about the political character of the steps he recommended: "The movement must continue to develop along conservative lines, since each problem has to be solved by men responsible for continuity of the industrial processes involved. In other words, the movement is evolutionary and not revolutionary."[4]

The leaders of the new standards committees did not argue that capitalists should be left alone in their selfish pursuit of profit; rather, they believed that some problems of industrial society could be resolved more efficiently by cooperation among experts. Starting in the 1880s, their cooperative methods began to bear fruit in the shape of standards for nomenclature, materials, and testing methods. Some difficulties in standards committees were technical or scientific in nature, but standards engineers also devoted their time to creative solutions for institutional and procedural problems. Ironically, the propagation of standards committees had begun to undermine their underlying rational and cooperative impulse. Confusion and incompatibility persisted in technologies where four or five different committees issued different standards – for example in electrical machinery, screw threads, and pipe threads – without any systematic or formal channels of communication or coordination. The consequences of incompatibility could be catastrophic: a tragic example was the 1904 fire that destroyed downtown Baltimore, which might have been prevented if fire departments from neighboring cities had hoses that fit Baltimore's fire hydrants. The Baltimore fire highlighted the moral and legal responsibilities of professional engineers, who felt a duty to ensure the health and safety of Americans in the industrial age.[5]

American engineers confronted these problems through continued organizational innovation. Inspired by their counterparts in Great Britain, fifteen engineers gathered in 1918 to form the AESC, a group that intended to coordinate the more than 100 organizations that were publishing standards and introduce

4 Agnew, "A Step toward Industrial Self-Government," 95. See also Layton, *Revolt of the Engineers*; David F. Noble, *America by Design: Science, Technology, and the Rise of Corporate Capitalism* (New York: Oxford University Press, 1977).
5 Comfort A. Adams, "Industrial Standardization," *Annals of the American Academy of Political and Social Science* 82, Industries in Readjustment (1919): 292–296; Henry A. Rowland, "Screw," *Encyclopedia Britannica Ninth Edition*, Volume XXI (1900), 506–511. On the 1904 Baltimore fire, see Rexmond C. Cochrane, *Measures for Progress: A History of the National Bureau of Standards* (Washington, DC: Department of Commerce, 1966), 84–86.

"systematic methods of cooperation." The AESC had an ambitious mission that included the unification of methods for setting standards, the promotion of American standards abroad, domestic and international cooperation among standards organizations, and the collection and distribution of information and data about standardization. Most important, the AESC did not intend to formulate standards itself; it only aimed to "secure cooperation between various interested organizations in order to prevent duplication of work and promulgation of conflicting standards." In other words, the AESC's boldest ambition was jurisdictional. Its founders endeavored to create a federalist structure and process that could resolve jurisdictional conflicts at different levels – at the boundaries between competing firms, between overlapping standards committees, between industry and American government, and between Americans and the rest of the world.[6]

The leaders of the AESC bolstered their program of organizational cooperation with rhetoric that cast their new organization as an exemplar of collaboration among experts in all walks of industrial life. Because the AESC was a private institution – not a government body – its defining challenge was to convince engineers and corporate managers that it had both the authority and the legitimacy to establish American standards. Paul Agnew, who was the AESC secretary from 1919 to 1947, led this campaign by blending political and economic language that portrayed the group as progressive, responsible, and democratic. With his use of catchphrases such as the "consensus principle" and "industrial legislatures," Agnew tied the unglamorous work of standardization to predominant trends in 1920s American political culture. In some ways, Agnew's emphasis on cooperation and industry self-regulation complemented mainstream Progressivism as well as the "associationalist" approach of Commerce Secretary Herbert Hoover. In other ways, however, Agnew advanced a subtle and novel critique of Progressivism by portraying traditional forms of democracy – including courts, Congress, and majority voting ("the vote of 110,000,000 people") – as outdated and inferior to the cooperative methods developed by the AESC.

The AESC played an important role in the coordination of standards for the American industrial economy, but its methods were no panacea for the manifold problems of American industrial society. Critics were right to chastise the AESC for being too slow and, at times, too beholden to commercial interests. Over the long term, however, the AESC's cooperative method matured and spread throughout all sectors of American industry. By the late twentieth century, its institutional and rhetorical innovations had been reformulated to constitute the core practices and values of open standards and open systems. The AESC's history is important because it demonstrates that the defining principles at work in the infrastructure of the digital age – openness, transparency,

---

[6] *Annual Report of the American Engineering Standards Committee, 1920* (New York: American Engineering Standards Committee, 1920), 1.

cooperation, balance, and inclusiveness – cannot be seen as values that are somehow determined by or inherent in the decentralized architecture of digital networks. These values, and the institutions to sustain them, emerged instead from a Progressive response to engineer new solutions to incompatibility, inefficiency, and waste in early twentieth-century industrial America.

Creation of the American Engineering Standards Committee, 1910–1922

Beginning around 1910, elite American engineers recognized that an alliance in standards work between the national engineering societies would be a "most desirable thing."[7] There was less agreement, however, on who would control the alliance and how much authority it would have. The catalyst for these discussions was a 1910 visit to London by Henry Hess, a German-born machine tool expert who owned a business in Philadelphia and was active in the American mechanical and automotive standards communities. While in London, Hess met with leaders of the British Engineering Standards Committee (BESC). The BESC, founded in 1901, was a body designed to unify standardization efforts in British government and private industry. Convinced that Americans would benefit by following the British example, Hess initiated discussions with American engineers that dragged on for eight years before bearing fruit. Throughout this period, American engineers looked to the BESC – and especially to its secretary Charles le Maistre – for advice on how Americans could establish a national standardization committee as the British had done.[8]

In 1911, Hess became chairman of the Joint Engineering Standards Committee of the ASME and promptly reached out to his counterpart at the AIEE, the Harvard professor and consulting engineer Comfort A. Adams. Hess told Adams of his vision for a Joint Engineering Standards Committee that would function as a hub for communication among domestic and international standard setters.[9] Adams responded enthusiastically to Hess's proposal, but he also felt moved to relay to Hess some lessons he had learned during the creation of the International Electrotechnical Commission (IEC) in the previous decade. It would, Adams warned Hess, be "a very tedious and difficult matter to get such an international organization into successful operation."[10]

---

[7] Henry Hess to Calvin W. Rice, quoted in Clifford B. LePage, "Twenty-Five Years – The American Standards Association (1. Origins)," *Industrial Standardization* (1943): 318.

[8] On the early history of the BESC (renamed British Engineering Standards Association in 1918), see Charles le Maistre, "Summary of the Work of the British Engineering Standards Association," *Annals of the American Academy of Political and Social Science* 82 (1919): 247–252; JoAnne Yates and Craig N. Murphy, "Charles Le Maistre: Entrepreneur in International Standardization," *Enterprises et Histoire* 51 (2008): 10–57.

[9] Henry Hess to F. L. Hutchinson, quoted in LePage, "Twenty-Five Years – The American Standards Association (1. Origins)," 318.

[10] Comfort A. Adams to Henry Hess, quoted in LePage, "Twenty-Five Years – The American Standards Association (1. Origins)," 319.

Perhaps daunted by the obstacles ahead, this initial burst of enthusiasm faded and the initiative stalled until 1916, when Arthur Kennelly, a professor at Harvard and MIT and Adams's successor as chairman of the AIEE Standards Committee, reached out to Hess. Kennelly proposed that the four major national engineering societies – AIEE, ASME, the American Society for Civil Engineers (ASCE), and the American Institute of Mining Engineers (AIME) – send representatives from their respective standards committees to create a "standing federated Engineering Standards Committee," which would share equally in the meeting costs and rotate the chairmanship on an annual basis. Kennelly hoped that this institutional design would avoid rivalries and tensions between the four professional societies. "The main idea," Kennelly concluded, "is that no one society shall seek to dominate the situation, but that each should endeavor to assist all the others."[11]

By the end of the year, the idea finally began to come to life as a new institution. The first meeting of the Joint Conference Committee on American Engineering Standards took place on December 29, 1916, with Comfort Adams presiding as chairman. Following the meeting, Adams and his colleagues asked leaders of the ASTM to join them in forming a permanent American Engineering Standards Committee.[12] The Joint Committee also authorized Adams to gauge the interest of officials in the federal government, including the director of the Bureau of Standards and the secretaries of war and navy. Adams had already established collegial relations with two employees of the Bureau of Standards, Paul Agnew and Edward Rosa, and found his correspondents at the Departments of War and Navy – Colonel Warren R. Roberts and Franklin Delano Roosevelt, respectively – willing to follow the bureau's lead. Their negotiations set the pattern of government participation as a partner, not a leader or adversary, in American industrial standardization. Indeed, government agencies were already recognized as significant consumers of industrial products, and their procurement power was already understood to be a decisive factor in the adoption of dominant designs and product specifications.

With five engineering societies and three government departments committed to collaboration, the most vexing questions facing the newly formed AESC were not scientific or technical in nature. Instead, the group struggled to define its organizational character and jurisdiction: what would be the constitution and bylaws of the new group? How would it classify and

[11] Arthur Kennelly to Henry Hess, quoted in LePage, "Twenty-Five Years – The American Standards Association (1. Origins)," 321. Charles le Maistre, "Standardization," *Transactions of the AIEE* 35 (1916): 489–500; Comfort A. Adams, "National Standards Movement – Its Evolution and Future," in Dickson Reck, ed. *National Standards in a Modern Economy* (New York: Harper, 1956).

[12] The AESC changed its name to the American Standards Association (ASA) in 1928, then to the United States of America Standards Institute (USASI) in 1966, and then to its present name, the American National Standards Institute (ANSI) in 1969.

approve standards? Most important, what would be the relationship between these new procedures and the various existing procedures in place at the founder societies?

These questions took almost two years to settle. Looking back, Agnew commented wryly that the five societies engaged in "endless discussions" and wrote "innumerable drafts of constitutions and methods of procedure" before they finally approved the AESC Constitution and Rules of Procedure in 1918. Adams, the chairman of the effort, later recalled, "What happened during those two years would be a long, but very interesting story, if it were related in full.... Fear and jealousy, as well as ignorance, were the chief obstacles which had to be overcome during two years of the hardest kind of work for the relatively small group that carried the load."[13] The heart of the problem, articulated most sharply and persistently by ASTM representative Edgar Marburg, was a fear that the new collective organization would usurp the hard-earned prestige and authority of the member societies. When the AESC issued its founding Constitution and Rules of Procedure in 1918, this jurisdictional tension was not completely resolved but rather deferred and therefore destined to resurface time and again as a fundamental problem.

One thing was clear from the start: private institutions, not the federal government, would maintain leading roles in American industrial standardization. The American emphasis on private-sector leadership – built on the foundation of private-sector collaboration that emerged in the late nineteenth century – thus continued to distinguish the American system of standardization from those developed by their European counterparts.[14] At the same time, the potential benefits of cooperation with government authorities became clear to American industrialists and engineers during the nation's unprecedented mobilization for World War I. Several prominent wartime initiatives demonstrated the benefits of collaborative and progressive engineering, including Howard Coffin's leadership of the Naval Consulting Board at the outset of the war, Bernard Baruch's management of the War Industries Board during the war, and Herbert Hoover's oversight of the provision of food for eleven million Belgian refugees during and after the war. In each of these examples, the purchasing power of the federal government and its urgent wartime demands highlighted the need for private companies to streamline and coordinate the goods they produced. Each of these efforts shared the same impulse in wartime that American standard setters hoped to nurture in peacetime: America and the world could benefit from

[13] P. G. Agnew, "Twenty Years of Standardization," *Industrial Standardization* (1938): 229; Comfort A. Adams, "How the AESC Was Organized," *Industrial Standardization* (1938): 237–238.

[14] Jay Tate, "National Varieties of Standardization," in Peter A. Hall and David Soskice, eds., *Varieties of Capitalism: The Institutional Foundations of Comparative Advantage* (New York: Oxford University Press, 2001), 442–473.

the systematic links between scientific knowledge, industrial experience, and bureaucratic skill.[15]

The experience of wartime cooperation was encouraging, but it would be a mistake to see the growth of the AESC simply as a natural progression from it. Indeed, the longtime AESC secretary Paul Gough Agnew later argued that the "British influence" – specifically le Maistre's advice – was far more influential as an inspiration for the AESC than anything that "came out of Washington or even from civilian engineers' stays in Washington." Moreover, the template for cooperation among American standards engineers had already been drawn before the war through the ASME and AIEE standards committees as well as through discussions among Hess, Kennelly, Adams, and le Maistre that started back in 1911. Their plans for the AESC happened to be timed very well, during the period after World War I when Americans reconstructed their political economy and consolidated the ascendancy of corporate liberalism.[16]

The first meeting of the AESC took place on October 18, 1918, at the Engineering Societies' Building on West 39th Street in New York City. Comfort Adams was elected as chairman, and the ASME volunteered the services of Clifford B. LePage as acting secretary. LePage was replaced in 1919 by Agnew, who moved from the Bureau of Standards to become the AESC's full-time secretary. Agnew would continue as secretary until 1947 – an unusually long tenure that makes him the single most important figure in the history of American industrial standardization. To his peers, who recognized his decades of service with the first American standards Association Standards Medal in 1951, Agnew became "Mr. Standards." In a 1949 biographical sketch, one of Agnew's

---

[15] Robert D. Cuff, *The War Industries Board: Business-Government Relations during World War I* (Baltimore: The Johns Hopkins University Press, 1973), 15–30. See also Ronald C. Tobey, *The American Ideology of National Science, 1919–1930* (Pittsburgh: University of Pittsburgh Press, 1971), xii, 3–96; and Noble, *America By Design*, 79–81.

[16] Paul G. Agnew, "Historical Memoranda to H (for Mrs. Moffett), 9/3/48, in P. G. Agnew, *Historical and Policy Papers* (New York: American Standards Association, 1920–1952), 335. Agnew's recollection is consistent with Henry May's insight that fundamental changes in American society were well under way before the beginning of the world war. See Henry May, *The End of American Innocence: A Study of the First Years of Our Own Time, 1912–1917* (New York: Knopf, 1959). In contrast to the AESC, the National Research Council (NRC) was formed to respond to military needs during World War I. The NRC, focused more on research (as opposed to standardization), grew rapidly in the 1920s only to lose most of its funding in the 1930s. See Robert H. Kargon, "Introduction," in Robert H. Kargon, ed., *The Maturing of American Science* (Washington, DC: American Association for the Advancement of Science, 1974), 1–29; and Theda Skocpol et al., "Patriotic Partnerships: Why Great Wars Nourished American Civic Voluntarism," in Ira Katznelson and Martin Shefter, eds., *Shaped by War and Trade: International Influences on American Political Development* (Princeton, NJ: Princeton University Press, 2002), 134–180. See more generally Stephen Skowronek, *Building a New American State: The Expansion of National Administrative Capacities, 1877–1920* (New York: Cambridge University Press, 1982); and Martin J. Sklar, *The Corporate Reconstruction of American Capitalism, 1890–1916: The Market, the Law, and Politics* (New York: Cambridge University Press, 1988).

FIGURE 3.1   Paul Gough Agnew.
Source: American Standards Association, *Industrial Standardization* (December 1943), 324.
Courtesy of the American National Standards Institute.

colleagues stated simply, "When speaking of standards one cannot help but think of P. G. Agnew."[17]

Agnew (1881–1954) earned a master's degree from the University of Michigan in 1902 and, after teaching high school physics for three years, joined the staff of the Bureau of Standards in 1906 (Figure 3.1). At the bureau he performed research and published several important papers on electrical instrumentation and measurement methods. He also enrolled as a graduate student in physics at The Johns Hopkins University, which awarded him a PhD in 1911.[18]

Agnew's fascination with industrial standardization blossomed during World War I, when he was the technical assistant to the Bureau of Standards chief

[17] S. P. Kaidanovsky, "Personalia: Dr. P. G. Agnew," *Standards World* 1 (1949): 113–114.
[18] See for example P. G. Agnew, "A Study of the Current Transformer with Particular Reference to Iron Loss," *Bulletin of the Bureau of Standards* 7 (1911): 423–474; P. G. Agnew, "A Device for Measuring the Torque of Electrical Instruments," *Bulletin of the Bureau of Standards* 7 (1911): 45–48; P. G. Agnew, W. H. Stannard, and J. L. Fearing, "A System of Remote Control for an Electric Testing Laboratory," *Scientific Papers of the Bureau of Standards*, No. 291 (Washington, DC: Government Printing Office, 1916); and P. G. Agnew, "A New Form of Vibration Galvanometer," *Scientific Papers of the Bureau of Standards*, No. 370 (Washington, DC: Government Printing Office, 1920).

physicist Edward B. Rosa as well as a liaison between the Bureau of Standards and the War Industries Board.[19] Agnew later provided important assistance for the federal government during the Second World War as an emissary between the government, military, and industry in the development of American War Standards used by the armed services. Agnew first became involved with the AESC in 1919, when he represented the Bureau of Standards in discussions about the inclusion of government departments and trade organizations within the AESC. Agnew left the bureau to join the AESC later that year, tempted in part by a salary increase of at least $1,000.[20]

Agnew maintained a personal interest and involvement in almost every aspect of the AESC's operations. He was deeply involved with a vast range of topics related to standardization, from the technical details of electrical, photography, and building standards to the philosophical, financial, and legal dimensions of industry, government, and international standardization. Beginning in the late 1920s, he was a leader in the effort to create ratings, certifications, and quality standards for consumer products such as clothing, bedding, and food.[21] Moreover, he represented Americans in efforts to create international alliances for standardization, including the International Standards Association and a series of meetings with Latin American countries, both in the late 1920s, and was a "dominant figure" in the negotiations that led to the establishment of the International Organization for Standardization (ISO) in 1946.[22] Even when Agnew was not a protagonist in standards work, he was always involved: with few exceptions, he was present at every meeting of the AESC Main Committee and, through his role as minute taker, generated its documentary history and served as the scribe of its institutional memory.

The key to Agnew's success and longevity with standards was the spirit in which he carried out his work. Where many of his peers in the early twentieth century chose to become entrepreneurs, executives, or industrial researchers,

[19] P. G. Agnew, "The Work of the Bureau of Standards," *Annals of the American Academy of Political and Social Science* 82 (1919): 278–288.

[20] According to Rexmond Cochrane, "industry paid close to twice the Bureau salary at every level of training and experience." Agnew would have earned between $2,240 and $4,000 a year after over ten years at the bureau; his 1921 salary at the AESC was $5,000. Cochrane, *Measures for Progress*, 223; *Work of the American Engineering Standards Committee (Year Book)* (New York: American Engineering Standards Committee, 1921), 13.

[21] The American Home Economics Association joined the AESC in 1929. See "Home Engineering," *New York Times* (March 7, 1929), 17; "Standardized Bedding," *New York Times* (April 1, 1929), 19; and P. G. Agnew, "The Movement for Standards for Consumer Goods," *Annals of the American Academy of Political and Social Science* 137 (1934): 60–69. For tributes to Agnew's formative role in standards for photographic equipment and consumer goods, see Paul Arnold, "American Standards in Complementary Industries," and Irwin D. Wolf, "Consumer Goods Standards – the Retailer's Viewpoint," both in Reck, ed., *National Standards in a Modern Economy*, 125–136 and 269–274.

[22] Kaidanovsky, "Personalia: Dr. P. G. Agnew," 113–114.

Agnew chose not to confine his loyalties to any individual firm or technology. Instead, he devoted his career to the more altruistic goal of sustaining the growth of American industrial standardization. One eulogy noted: "Dr. Agnew loved the standardization movement and the organization for which he worked. He also loved a good fight on its behalf." Agnew frequently reminded his colleagues that, in the standards-setting process, "the human difficulties are usually much more serious than the technical ones."[23] If the work needed to ensure the creation of industry and national standards had more in common with diplomacy than it did with laboratory science or factory engineering, then Paul Gough Agnew was its most skilled, most articulate, and most experienced diplomat. To understand the character and significance of his contributions, it is necessary to explore the growth of the organization to which he devoted the bulk of his life's work.

The Growth of the AESC: Funding, Participation, and Consensus

The founders of the AESC distributed power across four basic units of the organization – the Member Bodies, Main Committee, Sponsors, and Sectional Committees – illustrated in Figure 3.2.

The original AESC Member Bodies were the five founding societies and, after the first meeting, the Departments of War, Navy, and Commerce. Between 1918 and 1927, sixteen additional organizations joined these original eight as AESC Member Bodies. As Table 3.1 illustrates, many of the new Member Bodies that joined after 1918 were trade associations or groups focused narrowly on a particular sector of the economy. Representatives of industry thus took their place alongside the elite engineers from the founding professional societies and the government representatives who hoped to rationalize their procurement of industrial goods.

Member Bodies chose up to three individuals to serve as representatives on the Main Committee, which identified itself as "solely an administrative and policy-forming body, [which] does not concern itself with technical details." A Member Body paid $500 in dues for each representative it named to the AESC Main Committee.[24] Although expansion meant a diffusion of power and responsibility, the leaders of the AESC came from its most powerful members: of the four AESC chairmen elected in its first ten years, two were representatives from the AIEE (Comfort Adams in 1918–1919 and Charles Skinner in 1925–1927), one was from the ASTM (A. A. Stevenson in

---

[23] Eulogy for Paul Gough Agnew prepared by the American Standards Association, 1954. MB6, American Institute of Physics, Niels Bohr Library, College Park, Maryland; P. G. Agnew, "Standardization," *Encyclopedia Britannica*, 14th ed. (1940), reprinted as "Standards in Our Social Order," *Industrial Standardization and Commercial Standards Monthly* 11 (1940): 7.

[24] *Annual Report of the American Engineering Standards Committee, 1920*, 5.

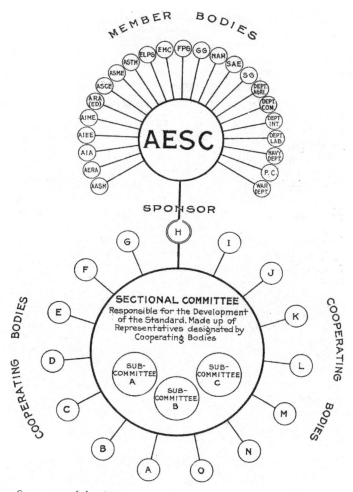

FIGURE 3.2 Structure of the AESC.
Source: American Engineering Standards Committee, *American Engineering Standards Committee Year Book* (New York: American Engineering Standards Committee, 1923). Courtesy of the American National Standards Institute.

1920–1922), and the other was from the National Safety Council (Albert Whitney in 1922–1924).

The detailed work of setting standards through the AESC occurred in a less centralized manner. Any individual project to create an American standard began with a request from an organization or coalition of organizations (known as Sponsors). Such a request would prompt the Main Committee to ensure that all interested parties ("cooperating bodies") were aware of and

TABLE 3.1 *Member Bodies of the AESC, 1927*

American Electric Railway Association
American Institute of Architects
American Institute of Electrical Engineers
American Institute of Mining and Metallurgical Engineers
American Mining Congress
American Railway Association – Engineering Division
American Society for Testing Materials
American Society of Civil Engineers
American Society of Mechanical Engineers
Association of American Steel Manufacturers
Electric Light and Power Group (consisting of the Association of Edison
    Illuminating Companies and the National Electric Light Association)
Fire Protection Group (consisting of the Associated Factory Mutual Fire Insurance
    Companies, the National Board of Fire Underwriters, the National Fire
    Protection Association, and Underwriters' Laboratories)
Gas Group (consisting of the American Gas Association, the Compressed Gas
    Manufacturers' Association, and the International Acetylene Association)
National Electrical Manufacturers' Association
Panama Canal
Safety Group (consisting of the National Bureau of Casualty and Surety
    Underwriters and the National Safety Council)
Society of Automotive Engineers
Telephone Group (consisting of the Bell Telephone System and the United States
    Independent Telephone Association)
U.S. Department of Agriculture
U.S. Department of Commerce
U.S. Department of Labor
U.S. Department of the Interior
U.S. Navy Department
U.S. War Department

Source: AESC Year Book, 1927.

invited to participate in the initiative.[25] Sponsorship entailed administrative duties rather than the primarily financial obligations that, in more recent times, we have come to associate with the term. The Sponsor led the development of the standard within one of the AESC Sectional Committees, which were organized along industry lines. In 1921, the Sectional Committees were as follows: (A) Civil Engineering and Building Trade; (B) Mechanical; (C) Electrical; (D) Automotive (Aircraft and Automobile); (E) Transport; (F) Ships and Their

[25] See "Rules of Procedure (Revised October, 1920)," *Annual Report of the American Engineering Standards Committee, 1920*, 7–8.

Machinery; (G) Ferrous Metals; (H) Non-Ferrous Metals; (K) Chemical, Including Chemical Engineering; (L) Textiles; (M) Mining; (N) Agriculture; (X) General.[26]

To ensure that any given Sectional Committee would be "authoritative and adequately representative," the Main Committee required Sectional Committees to maintain a balance of "producers, consumers, and general interests."[27] No single group could form a majority on a Sectional Committee without the consent of the other two groups. These provisions represent a crystallization of traditions developed by ASME engineers in the late 1880s and by Charles Dudley and his colleagues in the ASTM in the 1900s. Much more than a trivial bureaucratic detail, the mandate for balance between producers, consumers, and general interests was the foundational principle and essence of the enterprise. It was an acknowledgment that successful standards emerge from the needs both of makers and users. Over the long term, this mandate for balance became the hallmark of the AESC process and the foundation of its claims to have a fair, representative, and democratic means for establishing standards.

The framers of the AESC Constitution also took special care to be precise with terminology and procedure. As the Sectional Committees worked, they would refer to a proposed standard as a "Recommended Practice" or "Tentative Standard." When a Sectional Committee finished its work, it would submit its recommendation to the AESC Main Committee. The Main Committee's role was judicial, not technical. It did not examine the technical content of the proposed standard. Rather, it only checked to see if the Sectional Committee followed a fair and representative process that addressed the concerns of all interested parties – in short, to verify that no legitimate opposition had been ignored or silenced. The AESC summarized this functional separation between the Main Committee and the Sectional Committee in a common description of its "Method of Work" section of the 1920 Annual Report: "Each industry, or branch of industry, is wholly autonomous in its standardization work, the function of the Main Committee being merely to assure that each body or group concerned in a standard shall have opportunity to participate in its formulation."[28]

Once a Sectional Committee reached consensus regarding a Tentative Standard, it submitted it to the Main Committee for a vote. If 90 percent of the

---

[26] *Work of the American Engineering Standards Committee (Year Book)*, 1921, 23–27. The AESC also formed General Correlating Committees in areas of industrial practice where substantial effort to create standards already existed. In the early 1920s these committees included the National Safety Code Committee and the General Correlating Committee for Mining Standardization. Agnew characteristically maintained personal involvement by serving as secretary for both committees. See for example *Proceedings of the Third National Standardization Conference of the American Mining Congress* (Washington, DC: American Mining Congress, 1923).

[27] *Annual Report of the American Engineering Standards Committee, 1920*, 8.

[28] *Annual Report of the American Engineering Standards Committee, 1920*, 5–6.

votes were in favor, the specification would be published by the AESC as an "American Standard."[29] Between 1919 and 1923, there was a rapid increase in the number of projects operating under this procedure: 9 at the end of 1919, 49 at the end of 1920, 77 at the end of 1921, 121 at the end of 1922, and 132 at the end of 1923. These numbers indicate the enthusiasm that industrial engineers and managers had for such a cooperative endeavor on a national scale. A majority of the American Standards approved during this period were related either to safety codes or methods for testing materials, but the diversity of AESC projects is suggested by a random sample of ongoing work described in the 1924 *AESC Year Book*: A-13 Identification of Piping Systems; C-11 Electrical Properties of Aluminum; C-14 Specifications for Wood Poles; D-3 Colors for Traffic Signals; E-1 Specifications for Railroad Ties.[30]

Two persistent problems faced the leaders of the AESC in its first decade: financing the group's activities, and encouraging American industrial firms to accept the authority of its American Standards. The AESC's solution to these two problems was to bring them together within an overarching model of "voluntary consensus" standardization, in which the authority of American Standards became a function of widespread participation and support. Comfort Adams advocated this approach at the AESC's first meeting in 1918 by arguing that a radical increase in membership – an expansion from the five founding organizations to, eventually, up to 1,000 organizations – would be necessary if the AESC was to achieve a truly national and comprehensive scope. The challenge, in other words, was to create a bandwagon effect that would convince industrial engineers to lend their time, expertise, and capital to the AESC.[31]

The first and most pressing problem for the AESC was to finance its operations. Forging industrial standards was labor-intensive and costly work, for which thousands of dollars needed to be spent on travel, meetings, correspondence, and the costs of maintaining an office in New York. At first, Agnew and other AESC officials were optimistic that federal funding could be secured through an official yet "quasi-governmental" relationship with the Department of Commerce, to be administered through the Bureau of Standards. Agnew met personally with Commerce Secretary Hoover in March 1921 to discuss this possibility

---

[29] In 1919, Comfort Adams summarized the entire process with admirable brevity: "Standard assigned by main committee to sponsor body. Sponsor body appoints a thoroughly representative sectional committee, subject to approval of main committee. Sectional committee prepares standard and submits to sponsor body which then submits the standard with its approval to the main committee. The standard is then published by the sponsor body and labeled 'American Standard.'" Adams, "Industrial Standardization," 298. In its "Rules of Procedure," the AESC "requested that cooperating bodies do not use the term 'American Standard' in their publications except in connection with a standard that has received approval of the Main Committee as such." *Work of the American Engineering Standards Committee (Year Book)*, 1921, 17.

[30] *American Engineering Standards Committee Year Book* (New York: American Engineering Standards Committee, 1924).

[31] AESC Minutes, 18 Jan. 1919, page 3.

and left so enthused about his prospects that he drafted a congressional joint resolution for Hoover's consideration. Agnew's resolution was a bold act of policy entrepreneurship: he hoped that Congress would name government representatives to AESC committees, authorize appropriations from the Treasury Department, and mandate AESC specifications as "the basis for Government purchases of materials, apparatus, and supplies," so long as such actions would not create disadvantages for the government.[32] It was an ambitious fundraising pitch on Agnew's part, but did not succeed as he had hoped. Hoover responded by acknowledging the importance of the AESC's work, but, characteristically, preferred voluntary cooperation with his Department of Commerce to the pursuit of such broad Congressional mandates and expenditures.[33]

The AESC found more success with campaigns to finance its operations through its primary constituency, American industrial firms in the private sector. One fund-raising tactic was introduced as an institutional innovation in 1922, when the AESC created a class for individual firms called Sustaining Members. Sustaining Members of the AESC agreed to pay a fee that was determined in proportion to the gross annual revenues of their company. In exchange, the member received a subscription to the extensive AESC "information service," disseminated in periodic bulletins, that included a full list of standards – more than 2,000 in number – approved by standardization bodies both in the United States and abroad. For Agnew and other AESC leaders, better and more complete information was not only "necessary as a basis of sound work" but also a vital "basis for that cooperation which is essential to real standardization." The Sustaining Member category allowed firms that lacked the personnel or funds to become full AESC Member Bodies nevertheless to benefit from – and contribute to – AESC activities. The initiative met with immediate success: 228 companies joined as Sustaining Members in 1923, and, by the end of 1924, the AESC had received more than $27,000 from sustaining memberships, making it the greatest single source of income for the organization.[34]

The AESC's second major problem in its early years was to encourage industrial firms – as well as existing standards-setting organizations – to participate in its committees and to accept the legitimacy of its American standards. Here again, AESC leaders such as Arthur Kennelly and Comfort Adams built on their experience with electrical standardization to conclude that the success of

---

[32] P. G. Agnew to Herbert Hoover, March 31, 1921, Hoover Commerce Papers, AESC 1921 Folder, Box 23.

[33] Herbert Hoover to P. G. Agnew, April 12, 1921, Hoover Commerce Papers, AESC 1921 Folder, Box 23. See also William R. Tanner, "Secretary of Commerce Hoover's War on Waste, 1921–1928," in Carl E. Krog and William R. Tanner, eds., *Herbert Hoover and the Republican Era: A Reconsideration* (New York: University Press of America, 1984), 1–35.

[34] AESC Minutes, 14 May 1918, page 1; *American Engineering Standards Committee Year Book* (1924), 17–18, 56–57; *American Engineering Standards Committee Year Book* (New York: American Engineering Standards Committee, 1925), 62. In 1924, member dues generated $20,000.

the standardization process depended on broad participation. To accomplish their goals, they could not rely on familiar coordination mechanisms such as market exchanges or corporate hierarchies – they needed to devise new modes of economic and technological coordination that would be both effective and legitimate. The links between participation and legitimacy became evident in the difficulties that American engineers found as they undertook their first major project, the harmonization of industrial safety codes.

Safety codes presented a typical example of the coordination problems of the era: rather than a lack of any standards, there was instead an abundance of codes and guidelines that lacked any form of overarching coordination or authority to make them universal. During the 1910s, a variety of organizations – including state commissions, safety groups such as Underwriters' Laboratories, the National Safety Council, the American Mining Congress, and federal authorities such as the Bureau of Standards – issued a variety of conflicting and overlapping codes. Industry executives, confused by the situation, were further irritated by safety rules published by the Bureau of Standards in 1914 and 1915 that they perceived as an unwarranted encroachment of federal regulators. Congress might have clarified matters with strong leadership, but it refused to allocate funds for an activity that many congressmen believed could be accomplished by industry itself.[35]

Stranded between unfriendly executives and unsupportive politicians, officials in the Bureau of Standards realized they needed a different approach to break the stalemate. The bureau sponsored two conferences on safety codes in 1919, in which more than 100 representatives from industry, insurance companies, government, and engineering societies discussed their common problems. After much deliberation, the delegates concluded that the best way to develop national safety codes was to follow the Rules of Procedure developed by the AESC. The engineers agreed that the AESC, and not the Bureau of Standards, was the only organization capable of fostering cooperation between representatives from industry and government to develop a set of consensual (rather than adversarial) safety codes.

When the AESC began its work, it was immediately besieged by protests from industrial executives who complained that the organization was "entirely too undemocratic and narrow in its set-up to be entrusted with so important a program."[36] The AESC's response to this political critique – the charge that

---

[35] Broader considerations of safety codes in this era include Mark Aldrich, *Safety First: Technology, Labor, and Business in the Building of American Work Safety, 1870–1939* (Baltimore: The Johns Hopkins University Press, 1997); and Scott Gabriel Knowles, *The Disaster Experts: Mastering Risk in Modern America* (Philadelphia: University of Pennsylvania Press, 2011).

[36] Cochrane, *Measures for Progress*, 121–122; M. G. Lloyd, "The Safety Code Work of the Bureau of Standards," *Industrial Standardization* (1933): 203–206; David van Schaack, "Development of Standard Safety Codes," *Safety Engineering* 56 (September 1928): 83–86; Aldrich, *Safety First*, 103; P. G. Agnew, "Twenty-Five Years – The American Standards Association (2. Development of the ASA)," *Industrial Standardization* (1943): 322.

it was too undemocratic to be trusted – was to invite trade associations, safety groups, and insurance organizations to participate in its committees. The AESC thus deepened its founding, pluralist commitment to achieve a balance of interests by ensuring that all interests had a voice. By the end of 1920, twenty-two AESC projects were under way to publish safety codes for such diverse areas as aviation, ladders, lightning protection, machine tools, mechanical refrigeration, and industrial sanitation. The eventual adoption of dozens of AESC safety codes by state governments, insurance companies, and manufacturers indicates the success of the AESC's effort to open itself to the groups that would, in turn, be expected to adhere to American Standards.[37]

The AESC's entry into safety standards – an already crowded and complex organizational field – illustrates a central theme of the American system of standardization that developed in the wake of World War I. Private actors stepped into a void left by federal authority and created a new, supple organization that distributed power among a variety of industrial constituencies. The federalist character of the AESC indicates its core principle: standards should arise from those who are materially involved in their production and use. Standards should not be subject to remote or disinterested oversight. Agnew's 1926 publication on the AESC's safety code work summarized its governing concept and jurisdictional claim: "Thus, the work of drawing up national safety codes is in the hands, jointly, of those who are responsible for the administrative and legal aspects of the problems involved, of those who have to face the technical, industrial and financial sides of the problems, and of those who have to face the hazards to life and limb."[38]

Agnew's description indicates the importance of the outreach efforts he performed as part of his duties as AESC secretary. As the organization struggled with its practical and procedural problems in the mid-1920s, Agnew led a public relations campaign that cast his organization as a model of cooperative governance: he wrote articles in business publications, gave lectures at academic conferences, and testified to congressional committees. One goal of his campaign was simply to introduce the AESC process to a broader audience, but Agnew also used the opportunity to mount a political and ideological defense of the AESC's cooperative method. Rather than dwelling on the technical challenges that faced American engineers, Agnew emphasized the open and participatory character of the AESC's political process. For Agnew, the experiments in the AESC could shed new light on the core political questions of American

---

[37] For a list of the organizations participating in AESC projects, see *Annual Report of the American Engineering Standards Committee, 1920*, 11. On industry adoption, see P. G. Agnew, "The National Safety Code Program," *Annals of the American Academy of Political and Social Science* 123 (1926): 51–54; P. G. Agnew, "Twenty Years of Standardization," 232–233; "States Curb Loss of Life by Using American Standard Safety Codes," *Industrial Standardization* 6 (1935): 266–270; and Leslie Peat, "The Place of Safety Codes in the Industries," *Safety Engineering* 70 (1935): 173–174.

[38] Agnew, "The National Safety Code Program," 52.

society: who has the right to vote, what rules govern their participation, and how do they settle disputes?[39]

Agnew often compared the AESC's Sectional Committees to a variety of institutions that were the traditional forums for democratic rule making. For example, in an article published in *The New Republic* in 1926, he argued that the AESC's method of cooperative standardization had "all the directness and vitality of elementary local self-government." The comparison, he argued, was valid because each AESC Sectional Committee was working to solve specific problems that directly affected their daily existence.[40]

As he publicized the AESC's work, he began to articulate a concept – the consensus principle – that would soon become the guiding philosophy of the AESC and American industrial standards more generally. His 1926 *New Republic* article, "A Step Toward Industrial Self-Government," began with the claim that laws were but one type of a "real consensus, which means a common understanding, common purpose, and common will." Yet the common-law and statute-law methods, Agnew continued, were proving to be too cumbersome, slow, and inexact to confront "the complexity of modern life and industry." He asked his readers to consider an alternative: "In most industrial problems, what is needed is not a consensus of the whole population (which the legislative method is theoretically supposed to bring about) but only a consensus of those directly affected and competent to handle the problem." Where legislative bodies were capable of "settling" disputes quickly by majority vote, the voluntary cooperative method of standardization by small groups of experts was more capable of creating a real consensus. The advantages, for Agnew, were clear: "Normally this means that a reasonably satisfactory solution, and not a mere compromise, has been reached on fundamentals."[41]

By 1928, Agnew was able to convince his peers that the AESC should utilize the language of consensus as they began to undertake a major revision of the AESC's structure and process. Until this point, the AESC tended to refer to a "balance of interests" in the makeup of Sectional Committees, and did not use the term "consensus" in its annual reports, bylaws, or constitution. The revised Procedure approved on March 8, 1928, began with a new declaration

---

[39] Publications from AESC officers and staff included Albert Wurts Whitney, *The Place of Standardization in Modern Life* (Washington, DC: Government Printing Office, 1924); P. G. Agnew, "Results of Standardization of Supplies," *Annals of the American Academy of Political and Social Science* 133 (1924): 269–271; Stuart Chase and F. J. Schlink, "A Few Billions for Consumers," *The New Republic* (December 20, 1925): 153–155; Stuart Chase and F. J. Schlink, "A Few Billions for Consumers," *The New Republic* (January 6, 1926): 180–182; P. G. Agnew, "The National Safety Code Program," 51–54; P. G. Agnew, "A Step toward Industrial Self-Government," 92–95; and P. G. Agnew, "Can Industry Make Its Own Law?" *Mining Congress Journal* 13 (1927): 257–259.

[40] Agnew, "A Step toward Industrial Self-Government," 95.

[41] Agnew, "A Step toward Industrial Self-Government," 92–93, 95. Agnew frequently revisited the distinction between "compromise" and "solution" in subsequent publications.

and statement of purpose: "A national standard implies a consensus of those substantially concerned with its scope and provisions."[42] After 1928, Agnew regularly identified this procedural trait as one of the core values of American industrial standardization. In 1947, he stated the point with a clarity that did not exist in the organization's initial phase in the early 1920s: "The most fundamental principle underlying ASA [American Standards Association] work is the consensus principle. This has been so from the start."[43] A model of effective rhetoric, Agnew's brevity obscured the complex process by which he and other AESC leaders struggled to formalize their vague notions of a "balance of interests" into a set of principles that they could present as uncontroversial and ever present.

In the mid-1920s, Agnew began to deploy another term – "industrial legislatures" – to point out what he saw as the distinctive and advantageous aspects of the AESC's style of developing standards through its Sectional Committees. Agnew chose the term carefully to illustrate how American standards fit in with the prevailing political and philosophical preferences for industry self-regulation. "Each of these sectional committees," he explained, "is essentially a miniature industrial legislature organized upon a subject basis instead of upon a geographical basis."[44] By casting standards committees as "legislatures," Agnew was able to emphasize the requirement for balance that governed AESC Sectional Committees and made them appear as inclusive and legitimate rule-making forums. Yet, because committees operated on the basis of consensus solutions – as opposed to adversarial compromises – Agnew believed that the AESC's industrial legislatures were superior to government bodies such as courts and regulatory commissions.

Indeed, Agnew's enthusiasm for the AESC's cooperative method was at times matched by his critical appraisal of existing modes of democratic governance: "Experience in diverse fields has amply shown that the [cooperative] method combines many of the advantages of the common-law and the statutory-law methods ... while it avoids many of the limitations and abuses that have grown up about the legislative process."[45] An illustration that accompanied Agnew's

[42] *American Engineering Standards Committee Year Book* (New York: American Engineering Standards Committee, 1928), 68.

[43] P. G. Agnew to C. L. Warwick, "Letter to ASTM, 9/10/47 – ASA Policy in the Initiation of Projects (BD 244)," in Paul Gough Agnew, *Historical and Policy Papers*, 566.

[44] P. G. Agnew, "Work of the American Engineering Standards Committee," *Annals of the American Academy of Political and Social Science* 137 (1928): 13–16. This description also appeared in numerous editions of the AESC *Year Books*, which indicates that Agnew was a lead author (if not the sole author) of these annual reports.

[45] Agnew, "A Step toward Industrial Self-Government," 95. Agnew and his colleagues persisted in this promotion of voluntary standards as a viable alternative to control by legislature and common-law rule. See Agnew, "Standards in Our Social Order"; Howard Coonley and P. G. Agnew, *The Role of Standards in the System of Free Enterprise: A Study of Voluntary Standards as Alternative to Legislative and Commission Control. Prepared by Request of the Temporary National Economic Committee* (New York: American Standards Association, 1941); and Whitney, "The Place of Standardization in Modern Life."

1925 article "How Business Is Policing Itself" captured the essence of his critique (see Figure 3.3). At one side of the illustration (by the artist Emmett Watson), a lawyer peers through thick glasses as he reads from one of a mountain of legal books. His oratory is failing to hold the attention of the judge and a clerk, both of whom are nodding off or already fast asleep. In the middle of the illustration, a young, well-dressed businessman flees the courtroom, briefcase in hand, and enters a boardroom. He is joined there by two colleagues, one who is offering a match to light the young businessman's cigar, the other who is intently reading a document that the three men appear to share. The caption tells us that "The business man is turning from the courts with their numberless laws and intricate procedure to boards and reforms of his own choosing."[46]

Agnew's advocacy was not so intense that he could not admit the limitations of the AESC's "coöperative method." These limitations, in his summary, included the absence of important groups from the consensus process, the "frequent short-sighted jockeying for immediate commercial advantage," and the "endless jealousies and bickerings" between competing interests.[47] Critics of the AESC, most notably Lyman Briggs (director of the National Bureau of Standards from 1933 to 1945), lamented "the output in quantity [of the AESC and ASA] has not been all that some of its friends had hoped" – although he also concluded grudgingly that the AESC's procedure, "while painful and sometimes slow, is nevertheless fundamentally sound."[48]

Agnew and his colleagues were less willing to discuss the systematic exclusion of American industrial workers from the AESC process. For the most part, the AESC's stance toward labor was limited to a paternalistic concern with worker safety. The AESC departed only marginally from the prevailing anti-labor sentiment of industrial leaders in the early 1920s. It is not surprising that W. H. Barr, the chairman of AESC's Committee on the Foundry Code and president of the Lumen Bearing Company in Buffalo, New York, did not want John P. Frey, vice president of the American Federation of Labor, to participate in the committee's work. It is perhaps more surprising, and indicative of the AESC's deep commitment to inclusivity, that the AESC allowed Frey to join the committee over Barr's objections.

Barr's position was representative of a wider range of industrial executives who opposed the presence of organized labor in the AESC. In 1921, AESC chairman A. A. Stevenson proposed a compromise: union officials could serve on safety code committees, not as representatives of a particular union but rather under the flag of the U.S. Department of Labor. According to Agnew, this compromise was tenable enough to allow the committee to proceed with its work

---

[46] P. G. Agnew, "How Business Is Policing Itself," *The Nation's Business* (December 1925), 41–43.

[47] Agnew, "A Step toward Industrial Self-Government," 95.

[48] Lyman J. Briggs, "The American Standards Association and The National Bureau of Standards," *Industrial Standardization* (1938): 239.

FIGURE 3.3 "From the courts ... to reforms of his own choosing." "The business man is turning from the courts with their numberless laws and intricate procedure to boards and reforms of his own choosing."

Source: P. G. Agnew, "How Business Is Policing Itself," *The Nation's Business* (December 1925): 41–43. Courtesy of the U.S. Chamber of Commerce.

while avoiding the greater underlying tensions between capital and labor. With Stevenson's diplomatic solution, Agnew recalled in 1949, "The union had the substance of representation but not the form." The Department of Labor joined the AESC as a Member Body the next year (in 1922), but industrial unions continued to weaken under the strain of the postwar depression as managers adopted the discourse of the "open shop" to justify the exclusion of unionized labor. Labor's informal, indirect, and sporadic inclusion continued until the late 1930s, when, in the wake of the Wagner Act, the ASA changed its rules to allow direct representation of labor interests in all Sectional Committees.[49]

The potential for standardization to have negative social consequences in America – either for workers on the factory floor or for middle-class residents of "Main Street" and "Middletown" – seems to have been far beyond the concern of the elite engineers and executives in the AESC.[50] Agnew and his allies did, at times, show sensitivity to derogatory criticisms that implied standardization would produce a dull and mediocre world.[51] They recognized that the term "standardization" could have a numbing and confining effect and thus developed an elaborate philosophical response that set the tedious work of standardization into a broader and more progressive context. Agnew often seemed to be sparring against unnamed critics who cast standardization as the antithesis of innovation and progress. Surely it was he who, in 1925, began the tradition of opening AESC publications with an anonymous epigraph: "Standardization is dynamic, not static; it means not to stand still, but to move forward together."[52] In this regard, Agnew was inspired by Albert Whitney, a mathematician who was AESC chairman from 1922 to 1924. Whitney pursued a similar evolutionary interpretation in his 1924 essay, "The Place of Standardization in Modern Life." He wrote,

Variation is creative, it pioneers the advance; standardization is conservational, it seizes the advance and establishes it as an actual concrete fact.... Standardization is thus the liberator that relegates the problems that have already been solved to their proper place, namely to the field of routine, and leaves the creative faculties free for the problems that

[49] P. G. Agnew, "Labor Representation and Membership in ASA," XX 766, April 20, 1949, *Historical and Policy Papers*, 483–485; David Montgomery, *Workers' Control in America: Studies in the History of Work, Technology, and Labor Struggles* (New York: Cambridge University Press, 1979), 113–134. See also Amy E. Slaton, *Reinforced Concrete and the Modernization of American Building, 1900–1930* (Baltimore: The Johns Hopkins University Press, 2001); and Gregory J. Downey, *Telegraph Messenger Boys: Labor, Technology, and Geography, 1850–1950* (New York: Routledge, 2002).

[50] Marina Moskowitz, *Standard of Living: The Measure of the Middle Class in Modern America* (Baltimore: The Johns Hopkins University Press, 2004).

[51] One critic – an economist at the War Trade Board writing in 1919 – objected to standardization on aesthetic as well as economic grounds. See Homer Hoyt, "Standardization and Its Relation to Industrial Concentration," *Annals of the American Academy of Political and Social Science* 82, Industries in Readjustment (1919): 271–277.

[52] *American Engineering Standards Committee Year Book* (1925), 1.

are still unsolved. Standardization from this point of view is thus an indispensable ally of the creative genius.[53]

In Whitney's worldview, the liberatory essence of standardization could be generalized far beyond the progressive spirit of engineering. "In a very real sense," he continued,

all the conservational forces of civilization are within the field of standardization, institutions, customs, laws, literature, and other forms of art, science – they all involve the fixation of advances which have been made into a better understanding of the world, and such advances are in turn points from which to make fresh advances.[54]

With these sorts of rhetorical flourishes, Whitney, Agnew, and their colleagues justified their style of standardization as an inclusive, democratic, economical, and progressive practice. They hoped to move past the chaos and disorder of industrial capitalism and instead create a world where civic-minded engineers could foster cooperation and efficiency. In short, they aimed for nothing less than a transformation in American political culture. It was no coincidence that they found support from the Great Engineer, Herbert Hoover, who was using his position in the federal government to mount his own "crusade for standards."[55]

American Standards and Herbert Hoover's Associative State

The successful growth of the AESC depended not only on the refinements to its structure and process described previously but also on its relationship with the American federal government. Herbert Hoover's historical reputation has been tarred by his inadequate response as president at the start of the Great Depression, but historians who focus primarily on his tenure as commerce secretary from 1921 to 1928 have found more to admire. As commerce secretary, Hoover pursued his vision of "progressive democracy" that featured cooperation among various segments of the American industrial economy – the "associative state" in historian Ellis Hawley's influential interpretation. Missing from the histories of Hoover in the 1920s, however, are accounts

---

[53] Whitney, *The Place of Standardization in Modern Life*, 5.

[54] Whitney, *The Place of Standardization in Modern Life*, 5. Henry Ford, among others, concurred with Whitney's reasoning. In 1930, he wrote, "No product ever remains standard. It has to be kept standard.... It is the fault of the managers in thinking that one design or one method can for long continue.... Standardization, instead of making for sameness, has introduced unheard-of variety into our life. It is surprising that this has not been generally perceived." Henry Ford, with Samuel Crowther, *Moving Forward* (Garden City, NJ: Doubleday, Doran & Co., 1930), 28–29.

[55] Herbert Hoover, "The Crusade for Standards," in Dickson Reck, ed., *National Standards in a Modern Economy* (New York: Harper and Brothers, 1956).

of the mutually reinforcing links between his associative ideology and the cooperative approach of the American Engineering Standards Committee.[56]

Hoover hoped to harness "American individualism" to increase industrial efficiency, social harmony, and American power in the international arena. The most prominent articulation of this new social philosophy came in 1921 with the publication of *Waste in Industry*, a report written by the Committee on Elimination of Waste in Industry that was appointed by Hoover in his capacity as president of the Federated American Engineering Societies. The preface and Hoover's own foreword to the report boasted that it represented "the combined effort of about eighty engineers and their associates" – an exemplar of coordinated and applied expertise, "carefully planned and rapidly executed" in only five months. After identifying the sources and causes of waste, the report outlined recommendations for every branch of civil society: management, labor, owners, the public, trade associations, and government. Standardization in various guises – in factory equipment and production, through trade associations, and with the assistance of government – featured prominently in the recommendations. The report also spoke of "the duty of the engineer" to use his skills and social standing to overcome economic conflict and eliminate waste in industry. *Waste in Industry* used a rhetorical strategy – which became increasingly common in the 1920s – that utilized estimates of millions and in some cases billions of dollars that could be saved through simplification, standardization, and the elimination of waste.[57]

Cooperative institutions in general, and trade associations in particular, played key roles in Hoover's associative vision. Two studies published in the early 1920s justified cooperation within trade associations by distancing these organizations from earlier incarnations of business associations – namely the trusts, pools, cartels, monopolies, and collusive behavior that had attracted the contempt of the public and the unwelcome scrutiny of government lawyers emboldened by the Sherman Act.[58] Advocates of trade associations emphasized

[56] See Ellis W. Hawley, "Herbert Hoover, the Commerce Secretariat, and the Vision of an 'Associative State,' 1921–1928," *Journal of American History* 61 (1974): 116–140; Ellis W. Hawley, "Three Facets of Hooverian Associationalism: Lumber, Aviation, and Movies, 1921–1930," in Thomas K. McCraw, ed., *Regulation in Perspective: Historical Essays* (Boston: Harvard University Press, 1981), 95–123; David M. Hart, *Forged Consensus: Science, Technology, and Economic Policy in the United States, 1921–1953* (Princeton, NJ: Princeton University Press, 1998); and David M. Hart, "Herbert Hoover's Last Laugh: The Enduring Significance of the 'Associative State,'" *Journal of Policy History* 10 (1998): 419–444.

[57] Federated American Engineering Societies, *Waste in Industry* (New York: McGraw-Hill, 1921), vi, ix, 33; Layton, *Revolt of the Engineers*, 189–205; Chase and Schlink, "A Few Billions for Consumers"; Ray M. Hudson, "Organized Effort in Simplification," *Annals of the American Academy of Political and Social Science* 137 (1928): 1–8.

[58] Franklin D. Jones, *Trade Association Activities and the Law: A Discussion of the Legal and Economic Aspects of Collective Action through Trade Organizations* (New York: McGraw-Hill, 1922); National Industrial Conference Board, *Trade Associations, Their Economic Significance and Legal Status* (New York: National Industrial Conference Board, 1925).

that these groups did not restrict production or fix prices but, instead, merely exchanged information about cost and accounting methods, trade statistics, cooperative advertising, cooperative industrial research, and standards for products, nomenclature, and practice. In 1922, an exchange of letters between Secretary of Commerce Hoover and Attorney General Harry M. Daugherty clarified the legality of such activities, and encouraged trade associations to "further and extend standardization activities."[59] In its 1924 Annual Report, the AESC noted the correspondence between Secretary Hoover and Attorney General Daugherty and offered a cheerful – if exaggerated – interpretation: "Fortunately it is everywhere recognized that standardization is a legitimate and constructive association activity."[60]

The U.S. Supreme Court validated Hoover's view of trade associations in two cases decided in June 1925.[61] In drafting his opinions for the 6–3 majorities in both cases, Justice Harlan Fiske Stone denied that the exchange of information by itself constituted a price-fixing arrangement: "Competition does not become less free merely because the conduct of commercial operations becomes more intelligent through the free distribution of knowledge of all the essential factors entering into the commercial transaction."[62] One political scientist put the decisions in more blunt and dramatic terms: "Intelligence, the Supreme Court declared, is not necessarily a crime." Standardization, like other exchanges of statistical and economic data, thus found new stability in American law.[63]

Hoover's policy entrepreneurship and the court's decisions in June 1925 had profound legal and ideological consequences: they provided the trade associations with an ironclad defense from allegations of anticompetitive conduct, and, at the same time, helped them frame their actions in the cooperative ideology of industrial simplification, economic efficiency, and social stability. The only remaining legal questions surrounding trade associations concerned price fixing, but because technical standardization in the AESC never broached the dangerous topic of prices, the process of setting standards remained safe from

---

[59] *American Engineering Standards Committee Year Book* (New York: American Engineering Standards Committee, 1923), 6.

[60] *American Engineering Standards Committee Year Book* (1924), 18.

[61] *Maple Flooring Mfrs. Assn. v. United States*, 268 U.S. 563 (1925); *Cement Mfrs. Protective Assn. v. United States*, 268 U.S. 588 (1925).

[62] *Maple Flooring*, 268 U.S. 563, 583.

[63] Gilbert H. Montague, "New Opportunities and Responsibilities of Trade Associations as a Result of Recent United States Supreme Court Decisions," *Proceedings of the Academy of Political Science in the City of New York* 11 (1926): 27. See more generally Louis Galambos, *Cooperation and Competition: The Emergence of a National Trade Association* (Baltimore: The Johns Hopkins University Press, 1966), 80–101; William J. Barber, *From New Era to New Deal: Herbert Hoover, the Economists, and American Economic Policy, 1921–1933* (New York: Cambridge University Press, 1985), 7–13; and M. Browning Carrott, "The Supreme Court and Trade Associations, 1921–1925," *Business History Review* 44 (1970): 320–338.

antitrust prosecution.[64] In 1951, Agnew commented on this happy aspect of his organization's history: "In the 32 years in which ASA has been in existence, no question or suspicion of violation of anti-trust laws has been raised in connection with ASA operations, conferences, or committees."[65]

Hoover also took a personal interest in the AESC and maintained constant contact through his staff. Evidence of this interest survives through Hoover's correspondence with AESC chairmen and its secretary, P. G. Agnew, as well as in the minutes of AESC meetings that Hoover attended in person or by proxy. Records from AESC meetings in 1922 document increasing levels of cooperation, starting with a small, high-powered meeting in Washington in early 1922 between Chairman Stevenson, Chairman-elect Albert Whitney, Secretary Hoover, Bureau of Standards director Samuel Stratton, and Chief of the Division of Simplified Practice A. W. Durgin. At the March 9, 1922, meeting of the Main Committee, the AESC approved measures to appoint formal liaisons between the Division of Simplified Practice and the AESC in order to coordinate their efforts "as specific cases arise." Stevenson was appointed as the AESC's representative for work with the Department of Commerce. In turn, Durgin would work "as Mr. Hoover's representative" at the New York AESC offices as the liaison officer for the Bureau of Standards and Federal Specifications Board, an organization that was responsible for setting interagency specifications within the government.[66]

Secretary Hoover personally addressed his kindred spirits and admiring colleagues at the AESC Main Committee meeting of June 15, 1922. He noted that the government "can lend a certain prestige" to help overcome the hesitancy of some manufacturers and trade associations in joining the work of collaborative standardization. He also urged the AESC to continue to move the engineering profession to reduce industrial waste through standardization and simplification, which were, in his mind, aspects essential in protecting "the standards of living and wages in this country" from sliding back to prewar levels. After his brief speech, Hoover concluded by asking the engineers a simple question: "Are there any special points on which I could be of any use?" Rather than lecture the audience of engineers, Hoover preferred to hear the results of a survey that, at his request, had been sent to hundreds of organizations and companies to solicit ideas and targets for simplification and standardization.

[64] Trade associations became the object of controversy once again during the National Recovery Administration, but collaborative standardization remained uncontroversial and has only rarely – and indirectly – been the focus of antitrust scrutiny. See Simon N. Whitney, *Trade Associations and Industrial Control: A Critique of the NRA* (New York: Central Book Company, 1934), 32–60; and *American Society of Mechanical Engineers, Inc. v. Hydrolevel Corp.*, 102 S. Ct. 1935 (1982).

[65] P. G. Agnew, "Policy Questions Concerning the Proposed Congressional Charter for the American Standards Association," January 31, 1951 (Xxb1108), *Historical and Policy Papers*, 195.

[66] "Minutes," March 9, 1922, Hoover Commerce Papers, Box 23, AESC Folder.

Hoover's performance was somewhat rushed, but it provided a model of his preference for a collaborative and associative style of work.[67]

Ultimately, Hoover's Department of Commerce lent "a certain prestige" to the AESC, but not much more. Instead, the AESC grew despite a certain tension with groups such as the Bureau of Simplified Practice and the Bureau of Standards. On the surface, the respective missions of the AESC and Bureau of Standards appeared to be complementary: whereas the Bureau of Standards was concerned primarily with scientific research for fundamental standards and units for weights and measurement, the AESC worked to establish agreement regarding standards for applications in industrial production. The coexistence of the two groups was further facilitated by the formal liaisons and informal meetings and correspondence described earlier. To a great extent, both organizations benefited from cooperation. The AESC provided an avenue for the bureau to extend its influence beyond its modest size, and in return the AESC benefited from the technical expertise and political legitimacy carried under the banner of the bureau and the Department of Commerce.[68]

These complementary missions were sustained by a rhetoric of cooperation, such as in Agnew's publications and in the AESC *Year Books*, that emphasized productive and positive aspects of the relationship. AESC engineers, led by Agnew, appropriated and mobilized Hoover's cooperative rhetoric to justify their organization's work. For example, each AESC annual report (called *Year Books*) contained a section on "Government Cooperation" that detailed projects and committees sponsored or cosponsored by government departments and celebrated examples of AESC standards in use by various government departments.[69] By 1926, the AESC reported that "one or more arms of the government are cooperating in nearly all projects." Significantly, however, Congress refused to authorize government departments to pay the customary $500 membership fee – an indication of Congress's continued reluctance to dedicate federal funds to support industrial standardization.[70]

The public rhetoric of cooperation, however, cleverly obscured the persistent tensions and jurisdictional conflicts at the boundaries of the AESC and the Bureau of Standards in the 1920s and beyond. "The two organizations

[67] "Minutes, Joint Meeting of the AESC and the Executive Committee," June 15, 1922, Hoover Commerce Papers, Box 23, AESC Folder.

[68] For a broader account of the Bureau of Standards in the 1920s, see Cochrane, *Measures for Progress*, 220–298.

[69] See for example *Work of the AESC* (1921), 7–8, 32; and *American Engineering Standards Committee Year Book* (1923), 2.

[70] In 1924–1925, representatives from the Bureau of Standards were involved in 67 of the AESC's 212 projects in progress; only the ASTM was involved in a comparable proportion of the AESC's work. *American Engineering Standards Committee Year Book* (1925), 54; *American Engineering Standards Committee Year Book* (New York: American Engineering Standards Committee, 1926), 5–6. In its *Year Books*, the AESC consistently noted Congress's refusal to release membership fees for government participation. See for example *Annual Report of the AESC* (1920), 6; and *American Engineering Standards Committee Year Book* (1924), 17.

engaged in a sometimes awkward but almost always outwardly civil dance of ostensible mutual support," summarized sociologist Marc Olshan. "But it was a dance in which the AESC increasingly took the lead."[71] In this and other senses, then, the AESC's rise was a sign of the times in the business-friendly 1920s. The AESC provided industry executives with the opportunity to dilute regulations that might otherwise have been difficult or costly to implement, and, ironically, workers who had to face "hazards to life and limb" (Agnew's phrase) were for the most part absent from AESC discussions. The collective judgment of interested experts – rather than the decisions of elected officials or the vote of 110,000,000 Americans – assumed jurisdiction over American industrial standards.

The AESC Becomes the American Standards Association

In 1928, AESC leaders decided to undertake the most radical reforms in their organization's brief history and reconstitute the group under a new name: the American Standards Association. The reformation of the AESC represented a response to three distinct sets of pressures from inside and outside the group: the rising influence of trade associations, the increased interest in standardization from industrial executives, and the desire of AESC leaders to sanction a wider variety of methods for creating American standards.[72]

Thanks to the actions of Secretary Hoover and the Supreme Court in the mid-1920s, trade associations became more powerful within American industry in general as well as in the detailed work of industrial standardization. Of the 350 organizations participating in AESC activities in 1928, almost 300 were trade associations.[73] Two symbolic changes confirmed the enhanced role of trade associations in the reconstituted ASA. First, William J. Serrill, a representative of the Gas Group (a coalition of trade associations), was elected AESC chairman in 1928. After the change from AESC to ASA, Serrill became ASA president and Standards Council chairman until 1930.

The second – and more visible – symbolic change was the group's new name, adopted in 1928: the American Standards Association. With the conspicuous removal of the word "engineering" from the title of the organization, the ASA indicated a major step in its development, moving away from a committee to harmonize engineering standards toward a broader and more inclusive

---

[71] Marc A. Olshan, "Standards-Making Organizations and the Rationalization of American Life," *The Sociological Quarterly* 34 (1993): 319–335.

[72] In July 1928, the AESC Main Committee voted unanimously to support the proposed reforms. The proposal was then submitted to the thirty-seven Member Bodies, whose unanimous approval was announced in November 1928. "Scientific Events: The American Standards Association," *Science*, New Series, Vol. 68, No. 1751 (July 20, 1928), 53; "The American Standards Association," *Science*, New Series, Vol. 68, No. 1768 (November 16, 1928), 473–474.

[73] "Standards Group to Broaden Scope," *New York Times* (July 8, 1928), 40.

association of all industrial standardization in the United States. The 1929 *American Standards Year Book* noted that a good example of this change was the initiation of committees to set standards in areas such as household goods, although it hastened to add, somewhat awkwardly, that the ASA "limits itself strictly to those fields in which engineering methods apply, not concerning itself in any way with questions of style or personal taste."[74]

One measure of the success of the "crusade for standards" in the 1920s was increasing interest and participation of industrial executives in AESC activities. In 1925, the AESC formed an Advisory Committee of Industrial Executives in response to a request from the Conference of Industrial Executives. It was a powerful group of men that included a vice president of New York Edison and the presidents of U.S. Steel, Consolidated Gas Company of New York, the Delaware & Hudson Company, and General Electric.[75]

Executives assumed an even greater role when the AESC became the ASA in 1928. As part of the reorganization of the group, the AESC Main Committee was dissolved and its functions split between two new groups: the board of directors and the Standards Council. The board of directors consisted of twelve executives who were selected by AESC Member Bodies to assume the administrative and financial responsibilities that previously had fallen to the AESC Main Committee. The creation of the board of directors, according to the 1929 *American Standards Year Book*, reflected a rising awareness for managers and executives of the importance of standardization, including "purchasing, production, accounting, inspection, service, sales, and in every other department of manufacturing organizations." When the *New York Times* described the change, it noted that control within the reconstituted organization had moved from engineers and scientists to "the executives of railroad, public utility companies and industrial concerns."[76]

This outreach effort, combined with a fund-raising campaign from the board's finance committee, ensured the financial stability of the ASA in the early years of the Great Depression. The Underwriters' Fund, led by AT&T vice president Bancroft Gherardi and Bethlehem Steel vice president Quincy Bent, successfully raised money through direct contributions from industry: in 1929 alone, this fund added $74,000 to the ASA's annual income of $54,000; by 1930 they had obtained the means to increase the ASA budget by $500,000 over the next three years.[77] Not only had executives been persuaded of the value of adopting American standards in their

[74] *American Standards Year Book* (New York: American Standards Association, 1929), 7.

[75] *American Engineering Standards Committee Year Book* (1926), 4.

[76] *American Standards Year Book*, 7; "Executives to Direct Standards Body," *New York Times* (July 8, 1929), 36.

[77] "Unity of Standards Sought by Industry," *New York Times* (July 10, 1929), 50; "Plan to Enlarge Standards Work," *New York Times* (January 5, 1930), N21; William J. Serrill, "President's Report," *American Standards Year Book* (New York: American Standards Association, 1930), 9–10.

products and on their shop floors, executives such as Gherardi – AT&T's chief engineer whom we will meet again in Chapter 4 – also decided that they should invest their company's money and their own time to sustain the ASA's momentum.

Just as the ASA reached out to executives to ensure its financial viability, so too did it reach out to a diverse constituency of engineers to ensure its technical relevance and legitimacy. With the board of directors taking responsibility for administrative and financial duties, the Standards Council focused on making refinements to the ASA's process for determining if a proposed standard enjoyed sufficient support and "consensus" to qualify as an American standard.[78] The Standards Council's jurisdiction was defined clearly in the new ASA Constitution of 1929:

The functions of the Council shall be to formulate rules for the development of standards and for the constitution of committees; to approve, on behalf of the Association, such standards as it may find to be supported by a consensus, affirmatively expressed, of those substantially concerned with the standard; but not to formulate standards.[79]

The ASA Procedure made explicit the judicial – as opposed to technical – role of the Standards Council by beginning the revised version with these words.

101. A national standard implies a consensus of those substantially concerned with its scope and provisions. A chief function of the American Standards Association is the judicial one of determining whether a national consensus has been reached.[80]

A qualitative political judgment – the "consensus of those substantially concerned" – thus became enshrined as the leading principle of the ASA. To become an American Standard, a technical specification would first have to pass through this political process and judicial test. By creating separate spheres for the "technical" and the "judicial," the ASA tenuously charted a course that preserved the autonomy of its members while simultaneously setting conditions that would ensure the expert consensus would not be dominated by one powerful actor. The formalization of the split between technical and judicial tasks was an important step in addressing the reservations of Member Bodies – especially the ASTM – that were eager to preserve their own authority and status. At the same time, the reframing the ASA as a mere court of appeal indicates its desire to move away from costly lawsuits that arose at the uneven – and at times dangerous – edges of American industrial growth. In Chapter 4 we will see officials within the monopoly Bell System, especially its chief engineer

---

[78] The Standards Council, following the custom of the AESC Main Committee, consisted of up to three representative from each Member Body. "Standards Group to Broaden Scope," *New York Times* (July 8, 1928), 40; "Scientific Events: The American Standards Association," *Science*, New Series, 68 (July 20, 1928): 53–54.

[79] "Constitution," *American Standards Year Book* (1929), 82.

[80] "Procedure," *American Standards Year Book* (1929), 84.

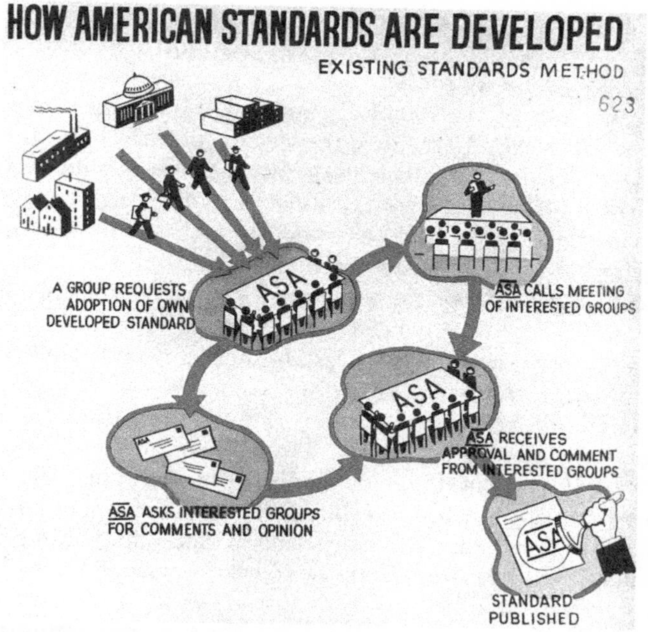

HOW AMERICAN STANDARDS ARE DEVELOPED

EXISTING STANDARDS METHOD

623

A GROUP REQUESTS ADOPTION OF OWN DEVELOPED STANDARD

ASA CALLS MEETING OF INTERESTED GROUPS

ASA RECEIVES APPROVAL AND COMMENT FROM INTERESTED GROUPS

ASA ASKS INTERESTED GROUPS FOR COMMENTS AND OPINION

STANDARD PUBLISHED

FIGURE 3.4 ASA Existing Standards Method after 1929.
Source: American Standards Association, *Voluntary Standards* (New York: American Standards Association, 1946).
Courtesy of the American National Standards Institute.

Gherardi, also experimenting with organizational solutions that could replace or preempt an influx of lawyers and lawsuits.

A further reflection of the ASA's eagerness to defer to the authority of its member organizations can be seen in the development of new methods for creating American Standards. In 1928, the reconstituted ASA outlined four distinct methods that could be used to create American standards: the Sectional Committee Method, Existing Standards Method, Proprietary Standards Method, and General Acceptance Method. Of these four, only the Sectional Committee Method was in use before 1928 in the AESC.

The new Existing Standards Method (see Figure 3.4) provided the clearest evidence of the ASA's determination to cater to the concerns of its Members. The ASA would approve standards through the Existing Standards Method if the submitting body could demonstrate it had already obtained an industrywide consensus on its own. Although not advertised as such, the adoption of the Existing Standards Method was an acknowledgment of the need to recognize the autonomy of powerful Member Bodies – particularly the ASTM, which had consistently opposed the construction of the AESC as a strong central body. (Agnew later referred to this tension, which had been present since before the first meeting of the AESC, as the "states' righters versus federalists" controversy.)[81] The ASTM was the most productive constituent in the AESC federation by the late 1920s: it sponsored somewhere between one-third to one-half of the standards under development in the AESC. The Existing Standards Method permitted the blessing of ASTM (or other standard-setting bodies deemed to be sufficiently representative) standards as American Standards. Although the real pressure for reform came from the ASTM, ASA leaders justified the new method by appealing to the core values of industrial standardization: after all, the new method could avoid the costs and redundancies of convening a Sectional Committee because, ostensibly, the work of consensus building had already been completed elsewhere.

The remaining two new methods established greater flexibility and speed in the procedure for approving American Standards. The Proprietary Method required a standard's sponsor to demonstrate the support of "a consensus of those substantially concerned with its development and use." Such a method invited the participation of dominant or monopoly firms into the ASA by providing a streamlined path for translating their de facto standards into consensus American Standards. The new General Acceptance Method provided a way to make standards from "simple projects" developed in the industry or academic conferences that became commonplace in the 1920s. It also provided a path to standardization for other uncontroversial, nonexclusionary, or ad hoc forums that did not merit the effort or cost required to organize a Sectional Committee. For all four of the methods outlined in the new ASA Procedure, the Standards Council was responsible for gauging the consensus obtained by a standard developed through each of the new methods of standardization. Any controversies over judicial decisions taken by the Standards Council could be appealed on a case-by-case basis, where each appeal would be presented by the ASA secretary, Paul Gough Agnew.[82]

When the reorganized American Standards Association was announced in 1928, *Science* reported accurately that the changes were approved by unanimous votes, first of the AESC Main Committee and then of all thirty-seven Member Bodies.[83] Although this spectacle of unanimity was vital for advancing the cooperative goals of the group, it also reveals the extent to which the leaders of industrial standardization kept internal disputes out of the public view. Records from AESC and ASA meetings and some frank recollections from Agnew and other ASA officials indicate that the ASA's harmonious public self-representation

---

[81] See Agnew, "Twenty Years of Standardization," 230–232; and Adams, "How the AESC Was Organized," 237–238.

[82] *American Standards Year Book* (1929), 8, 83–87.

[83] "Scientific Events: The American Standards Association," *Science*, New Series, 68 (July 20, 1928): 53; "The American Standards Association," *Science*, New Series, 68 (November 16, 1928): 473–474.

TABLE 3.2 *Membership, Participation, and Products in the AESC and ASA*

Year	Member Bodies	Participating Organizations	American Standards
1923	23	235	48
1928	36	350	111
1938	67	~600	667
1945	96	650	1,507

Sources: American Engineering Standards Committee *Yearbook*, 1924; American Standards Association, "Voluntary Standards," 1946.

intentionally obscured heated internal disputes. In 1938, Agnew recalled how J. A. Capp, longtime ASTM representative and head of the AESC Rules Committee from 1921 to 1925, declared at one point that the name "American Standards Association" would be "adopted only over his dead body." Capp later relented and in fact seconded the motion for adoption of the reorganization plan – which passed with the unanimous vote that was duly noted in *Science*, the *New York Times*, and the 1929 *American Standards Year Book*.[84]

The success of these reforms may be seen in the ASA's continued growth during the Depression, even as circumstances took their toll on ASA members and the money they were willing to dedicate to creating American Standards.[85] When Agnew resigned as ASA secretary in 1947, his staff which began with only two people in 1918 had boomed to eighty-five. The ASA grew steadily throughout the 1930s and, by the early 1940s, emerged as an important source of cooperation and expertise for the coordinated mobilization of the military and American industry for World War II. In the years immediately following the war, the ASA – and the consensus principle that it embodied and promoted across an increasingly wide range of American industries – continued to flourish. Throughout this period, consistently strong support from the private sector – as measured in financial contributions, participating organizations, and the number of American Standards published – were the keys to the survival and success of the ASA (see Table 3.2).[86]

When Herbert Hoover accepted the ASA's prestigious Howard Coonley Medal in 1951 – an award named for the ASA's three-time president who had served on its board of directors for more than twenty years – he used the

---

[84] Agnew, "Twenty Years of Standardization," 232. As Edwin Layton noted in his history of the American engineering profession, "It is considered bad form to publicize the inner workings of engineering societies." Layton, *Revolt of the Engineers*, 15.

[85] "Group Hears of Gain in Standards Work," *New York Times* (December 1, 1932), 38.

[86] For assessments of the vitality of the ASA and affiliated industry standardization groups in the 1940s and 1950s, see Reck, ed., *National Standards in a Modern Economy*.

occasion to reflect on the ASA's decades of success and unique contribution to the American political economy:

This work of your Association has brought another invaluable accomplishment. Only a few of the literally tens of thousands of standardizations or simplifications have been imposed by law. The vast numbers of them have been the result of spontaneous, voluntary, yet organized cooperation within highly individualized industry. To secure general acceptance of any one of them has been tedious at times. But their adoption has been playing a real part in the creation of the cooperation vital among a free people.[87]

Conclusions

The most significant accomplishments of the first generations of standards engineers were not technical or scientific in nature; they were organizational, ideological, and political. Historians have found these accomplishments easy to overlook, and usually choose to focus on the astonishing advances in scientific and technological research and development that occurred between the 1880s and 1920s. Consequently, historical accounts of this era tend to focus on professional managers, scientists, and engineers who worked within corporate hierarchies to exploit technical knowledge for profit.[88] Yet, as this chapter has shown, increasing numbers of engineers confronted inconsistencies – such as the standard sizes of wire gauge, the quality of steel rails, and conflicting safety codes – that they could not resolve within their corporate hierarchies or laboratories. Standards committees in professional engineering societies, the AESC, and the ASA grew around these sorts of problems that could not be settled by any single authority in the government or private sector. Herbert Hoover recognized the ASA's vital institutional contribution in 1951: "The public mostly assumes that this progress has come from scientific discovery of natural laws, new materials, inventions, and increasing skills. But you and I know that the increase in our living standards and comfort has received an enormous contribution from these related ideas of standards, of simplifications and specifications."[89]

The steady growth of standard-setting bodies in the decade after World War I inspired dozens of scholarly assessments of their performance. One such analysis, a 1929 report by the National Industrial Conference Board titled *Industrial Standardization*, evaluated the ASA in light of existing alternatives, including national standardizing bodies in Europe. The study concluded that

---

[87] Hoover, "Crusade for Standards," 4.
[88] Canonical texts include Alfred D. Chandler, Jr. *The Visible Hand: The Managerial Revolution in American Business* (Cambridge, MA: The Belknap Press of Harvard University Press, 1977); Thomas P. Hughes, *American Genesis: A Century of Invention and Technological Enthusiasm, 1870–1970* (New York: Penguin, 1990); and Noble, *America by Design.*
[89] Hoover, "Crusade for Standards," 3–4.

where German and other European bodies used state authority to produce and promulgate standards, "in America the object of the central organization is to *provide a procedure* by which national standards may be established." Although the study's author, Robert Brady, misstated the character of the British and German standardization processes (which were not as state-directed as he suggested), his conclusion indicates the extent to which Americans emphasized that their system of standardization created a rigorous procedure that did not need state authority to sanction its effectiveness and legitimacy.[90]

The ASA's distinctive procedure – in which participants developed standards by adhering to the consensus principle within miniature industrial legislatures – represented more than a clever bureaucratic innovation. It was an ideological and organizational critique of existing modes of industrial organization and government regulation. These traditional forms of control were incapable of coordinating the development of standards at the boundaries and interfaces of American industrial practice. The engineers who created and led the AESC responded by devising more sophisticated and orderly means to share information, resolve technical conflicts, and therefore reduce waste and inefficiency. The AESC experiment was successful to the extent that it showed that standards did not need to come from one powerful institution; rather, they could be negotiated by anyone with an interest in the outcome, so long as they adhered to a procedure with provisions for fairness, balance, and consensus.

The institutional features of the AESC, in turn, were the culmination of experience from a wide range of American institutions including universities, manufacturing firms, private technical societies, industry associations, and government agencies. Whereas most standards in the nineteenth century emerged through proprietary (or de facto) usage, increasing numbers of industrial standards in the early twentieth century were developed through a more open process, in which representatives from different sectors of the American economy were invited to participate, contribute, and, when necessary, object.

Beyond these institutional innovations, AESC engineers – led by Agnew – insisted that their procedure was superior to existing forms of rule-making institutions. Agnew advanced the clearest and most sophisticated critique of majority-rule democracy in the industrial age. Unlike radical reformers such as Morris Cooke, Charles Steinmetz, and Howard Scott, Agnew did not look to empower engineers solely on the basis of *expertise*; he recognized that participation in the standardization process also was a function of *interest*. By participating in standards committees, engineers could exercise their inherent right to represent their own interests. The AESC's cooperative method, in Agnew's view, was therefore more democratic, inclusive,

---

[90] Robert A. Brady, *Industrial Standardization* (New York: National Industrial Conference Board, 1929), 101 (emphasis added). See also Norman F. Harriman, *Standards and Standardization* (New York: McGraw-Hill, 1928); and more than three dozen articles in the May 1928 issue of the *Annals of the American Academy of Political and Social Science*.

and consistent with the American principle of self-government than were the existing alternatives.[91]

Agnew's rhetoric helps us locate the history of the AESC within a broader and longer historical context. In many ways the AESC's pluralist orientation can be seen as a mirror of the liberal international order that emerged in the late nineteenth century and matured in the early twentieth century. Like their counterparts in the AESC, American diplomats experimented with new institutions and policies – such as the Open Door and the League of Nations – that they hoped would avoid costly conflicts and settle disputes through negotiations between interested experts.[92]

Beyond its resonance with aspects of American political history of the early twentieth century, the long-term significance of the experiments in the AESC was its establishment of an institutional framework that could direct and sustain the process of technological change. Most important for the purposes of this book was the AESC's creation of the key principles and formative practices – due process, consensus, and a balance of interests – of the "open systems" that emerged near the end of the twentieth century. Engineers in the telephone and telegraph industries did not, at first, embrace cooperative standardization. After all, Western Union (and, as we will see in Chapter 4, AT&T) developed sophisticated and elaborate mechanisms to create and enforce standards in their monopoly communication networks. Chapter 4 describes AT&T's unprecedented development of organizational capabilities to oversee standardization in the monopoly Bell System, as well as some critical problems that traditional forms of hierarchical management were powerless to master. Chapter 5 then describes assaults on AT&T's intricate system of centralized control and the emergence of alternative mechanisms, designed in the middle decades of the twentieth century, to coordinate new standards for new networks.

[91] Layton, *Revolt of the Engineers*; Peter Meiksins, "The 'Revolt of the Engineers' Reconsidered," *Technology and Culture* 29 (1988): 219–246; William E. Akin, *Technocracy and the American Dream: The Technocrat Movement, 1900–1941* (Berkeley: University of California Press, 1977).

[92] Frank Ninkovich, "Ideology, the Open Door, and Foreign Policy," *Diplomatic History* 6 (1982): 185–208.

4

Standardization and the Monopoly Bell System, 1880s–1930s

> I think that if we don't consider a distinction in our use of the word standard and standardization we are very likely to get into trouble ... outsiders will have reason to assert that what we are attempting to do is to muzzle all possible development. What we are actually trying to do by our standardization work is to develop the telephone art in the best way that we know how to develop it.[1]
>
> – Frank Jewett, Assistant Chief Engineer, Western Electric, 1915

We have seen how organizations such as the American Standards Association (ASA) arose as creative responses to the status quo – that is, to the inadequacy of existing markets and hierarchies to provide effective mechanisms to coordinate technical standardization in American industry. The emergence of the "voluntary consensus" model of committee standardization, significant as it was, begs a related question: How did monopoly firms develop standards, and how did their standardization efforts fit within the collaborative, consensus-based model developed in the late nineteenth and early twentieth centuries within trade associations and engineering societies?

Conventional answers to these questions point to the power of the managerial hierarchies that exist within monopoly firms. In the conventional view, monopolies create standards through a hierarchical, closed, and proprietary process as part of a broader strategy to erect barriers to entry and maintain centralized control over a given market or markets. They capture the network effects that standardization generates. Monopolies, in this view, act in a monolithic and almost petulant manner: their goal is to reduce variety, stifle outside innovations, lock in users, and preserve their control.[2]

---

[1] Frank Jewett, "Discussion of Mr. McQuarrie's Paper," *Western Electric Company, Manufacturing and Engineering Conference*, Chicago, Illinois, May 24–28, 1915.

[2] Carl Shapiro and Hal R. Varian, *Information Rules: A Strategic Guide to the Network Economy* (Boston: Harvard Business School Press, 1999), 103–172; Jonathan E. Nuechterlein

From the late nineteenth century to the present day, critics of the monopoly Bell Telephone System have found ample evidence to support this conventional view of monopoly standardization. For most of the twentieth century, executives in AT&T presided over an association of research, manufacturing, and operating companies – known as the Bell System, or, more colloquially, Ma Bell – that achieved unparalleled success in dominating the American telephone industry. By the mid-twentieth century, Bell officials had successfully cultivated a certain "network mystique" that flourished within the monopoly Bell System. Engineers in the Bell System valued most highly uniformity, predictability, and reliability. Outside advances "not invented here" usually represented unwelcome distractions or intrusions.[3]

Although it contains seeds of truth – telephone executives and engineers did indeed develop standards to bring greater efficiency and market power through centralized control – the caricature of monopoly standardization as sluggish, arrogant, and solipsistic paints a distorted picture of the Bell System's struggle to achieve standardization. In particular, the caricature neglects two important themes.

First, it glosses over the difficulties and contingencies that AT&T executives and engineers faced in bringing their centralizing, universalizing vision to fruition. Standardization in the Bell System, as the many examples in this chapter demonstrate, was not simply a matter of issuing commands to subordinates. Standards did not flow easily down the lines of its vast organizational charts. As the historian Richard John has suggested, the concept of a system does not adequately capture the fluid set of relationships among the different entities involved in providing telephone service in this era: relational concepts such as network or association much more accurately convey the persistent instability in which negotiations over technical standards took place.[4]

As with other large technical systems, standards were both causes and consequences of systematization. By the early twentieth century, AT&T executives found increasing levels of success in consolidating their control over telephone

and Philip J. Weiser, *Digital Crossroads: American Telecommunications Policy in the Internet Age* (Cambridge, MA: The MIT Press, 2005), 1–30.

3 Louis Galambos, "Looking for the Boundaries of Technological Determinism: A Brief History of the Telephone System," in Renate Mayntz and Thomas P. Hughes, eds., *The Development of Large Technical Systems* (Boulder, CO: Westview Press, 1988), 143; Alvin von Auw, *Heritage and Destiny: Reflections on the Bell System in Transition* (New York: Praeger, 1983), 6; Tim Wu, *The Master Switch: The Rise and Fall of Information Empires* (New York: Knopf, 2010); Lillian Hoddeson, "The Emergence of Basic Research in the Bell Telephone System, 1875–1915," *Technology and Culture* 22 (1981): 512–544.

4 Richard R. John, *Network Nation: Inventing American Telecommunications* (Boston: Harvard University Press, 2010), 8–10; Thomas P. Hughes, "From Firm to Networked System," *Business History Review* 79 (2005): 587–593. For a discussion of earlier episodes of innovation and standardization in the Pennsylvania Railroad that parallels and anticipates many aspects of the Bell System story, see Steven W. Usselman, *Regulating Railroad Innovation: Business, Technology, and Politics in America, 1840–1920* (New York: Cambridge University Press, 2002).

networks across the country. AT&T president Theodore Vail coined a phrase to justify the extension of centralized control: "One System, One Policy, Universal Service."[5] This slogan bundled together Vail's critique of the cutthroat capitalism of the past, a business plan for the present, and a vision for a harmonious future. Advocates of the Bell System touted it as a progressive monopoly: a civic-minded corporation that would make deep investments in research and development, maintain reliable and high-quality service to the American public for a reasonable price, and pay regular – but not excessive – dividends to shareholders. Their success on all fronts had the unfortunate side effect of obscuring the day-to-day difficulties they faced in reconciling the diversity of local conditions with their designs on centralized control. In order to recover a sense of these difficulties, this chapter investigates the work of Bancroft Gherardi, AT&T vice president and chief engineer from 1920 to 1938. Gherardi, who is absent from existing histories of the early Bell System, was a chief architect of AT&T's methodical and deliberate style of "normal design" that regulators eventually came to view with suspicion.[6]

The prevailing caricature of standardization in the Bell monopoly as closed and monolithic also suffers from a second weakness: it fails to account for activities that occurred across the boundaries of the Bell System. Boundary activities, as we have seen, are crucial sites where managers and engineers decide through managerial hierarchies what they can make or decide inside their firm, and what they need to do with respect to markets and organizations that exist outside the firm. In some cases, these decisions can be understood in terms of economic efficiency, using the economist Ronald Coase's concept of transaction costs.[7] The problem with the concept is that it tends to reduce – or ignore altogether – strategic, political, and cultural factors that are in many cases decisive. We deceive ourselves if we pretend that decisions to build from within or purchase from without are made solely on the grounds of economic efficiency.

In the late nineteenth and early twentieth centuries, Bell engineers were deeply suspicious of the "outsiders" described by Jewett in 1915 as those who were looking for "reason to assert that what we are attempting to do is to muzzle all possible development." Bell managers and engineers had established a tradition of one-way information flows across the boundaries of their firm: they regularly purchased patents and hired outside experts, but they rarely disclosed their technical work to the outside world. In the 1920s and 1930s, however, Bell System engineers reached out and joined industry standards

[5] *Annual Report of the Directors of American Telephone and Telegraph Company to the Stockholders for the Year Ending December 31, 1909* (Boston: Geo. H. Ellis Co., Printers, 1910), 18.

[6] Walter G. Vincenti, *What Engineers Know and How They Know It: Analytical Studies from Aeronautical History* (Baltimore: The Johns Hopkins University Press, 1990), 7–9.

[7] Ronald H. Coase, "The Nature of the Firm," *Economica* 4 (1937): 386–405.

committees, including those organized by the American Institute of Electrical Engineers (AIEE) and the American Standards Association. In some cases their participation was meant simply to promote their own professional status and civic credentials, but they also found utility in industry standards committees as well. Gherardi's career is again suggestive of the growth of AT&T's commitment to industrywide standardization: while he was vice president and chief engineer of AT&T, he also served terms as president of the AIEE (1927–1928) and the ASA (1931–1932).

A historical tour of the Bell System from the vantage point of standardization, then, can help correct the prevailing myths and caricatures of monopoly standardization. By moving from the general matters of standard-setting bodies to the specific decisions of a single company, we can see how standards functioned as a crucial mode of governance both within the firm and across its boundaries with the outside world. This chapter examines standards for a few mundane artifacts (such as wood poles and metal washers) as well as the processes that Gherardi and his colleagues developed to create Bell System standards. It also follows Gherardi as he traveled outside the Bell monopoly and into industry standards committees in the mid-1920s and early 1930s. By the time Gherardi's career reached its peak in the early 1930s, standardization had become an obsession for Bell System executives and engineers. It is instructive that Gherardi's work blurred the boundaries of technical and diplomatic work and those between the internal operation of the Bell System and the outside world.

A close consideration of standardization in the monopoly Bell System also serves as a transition in the narrative arc of *Open Standards and the Digital Age*. In the Bell System, we can see the tentative intermingling of two stories that are usually seen as distinct: the history of the collaborative form of "consensus" standardization, and the history of communication networks. In the process, we can see with greater clarity the impressions that the engineering and business practices of the nineteenth century made on the monopoly Bell System in the twentieth; we will also be better prepared to understand how these developments shaped the twenty-first-century digital age.

"Universal Service": Standardization and the Creation of the Bell System

In the 1880s, there were no "telephone standards" in any formal sense of the term. Telephone technology was in an experimental phase, and telephone systems, like telegraph systems in the 1840s and 1850s, were administered by a handful of scattered entrepreneurs who licensed the Bell patents and were on their own to finance, build, and sell telephone service. The telephone patents themselves imposed a degree of stability and basic level of technical standardization, but they did not in any way dictate a comprehensive approach to the construction and operation of telephone networks. The telephone business may not have been as chaotic as the "methodless enthusiasm" of telegraphy in the 1850s, but there certainly was no overarching national system. This situation began to change in the 1880s, when managers in the American Bell Telephone Company[8] began to articulate and impose a vision of centralized control and interconnection.

Administrative standardization preceded and facilitated technical standardization. As Bell managers sought to establish their company, they found that standardization could help them define and redefine the scope of their control and to integrate functions of manufacturing, engineering, and administration on a local, regional, and eventually national scale. Although Bell managers eventually presided over a nationwide monopoly, it is a mistake to see standardization as a consequence of monopoly. At the turn of the twentieth century, the causal relationship worked the other way: standards emerged first as part of a strategy to dominate the telephone industry and to create, justify, and sustain hierarchical control that could generate advantages in a competitive market.

The first American telephone networks were built by entrepreneurs who licensed Bell's patents. These systems were not interconnected. Bell licenses were geographically exclusive, and it was impossible to place a call from one city to another. One consequence of geographical decentralization was considerable diversity in early telephone systems, particularly in complex components such as switchboards. Between 1877 and 1881, up to five different manufacturers supplied telephone equipment to Bell licensees with no overarching strategy to impose standardized equipment or procedures. The state of affairs changed in 1881, when Bell managers, led by Theodore N. Vail and Gardiner Hubbard, began to consolidate manufacturing within one firm, Western Electric – a crucial first step in the creation of a "Bell System" (although it is anachronistic to use the term in this context, since a national system was not envisioned at the time). Momentum toward consolidation and standardization was slow. Even where standards did exist, they were inconsistent across the diffuse operating units of Western Electric and the Bell licensees. One difficulty was Western Electric's reliance on a contract system – where specific jobs would be

---

[8] The Bell Telephone Company was incorporated in 1877, and kept that name until 1879. From 1879 to 1880 it was the National Bell Telephone Company, which was reincorporated as the American Bell Telephone Company (1880–1899). The American Telephone and Telegraph Company was incorporated as a long-distance firm in 1885, and became the parent company for the Bell interests on December 31, 1899. On the early history of American telephony and the Bell System, see George David Smith, *The Anatomy of a Business Strategy: Bell, Western Electric, and the Origins of the American Telephone Industry* (Baltimore: The Johns Hopkins University Press, 1985); Robert W. Garnet, *The Telephone Enterprise: The Evolution of the Bell System's Horizontal Structure, 1876–1909* (Baltimore: The Johns Hopkins University Press, 1985); Neil H. Wasserman, *From Invention to Innovation: Long-Distance Telephone Transmission at the Turn of the Century* (Baltimore: The Johns Hopkins University Press, 1985); and M. D. Fagen, ed., *A History of Engineering and Science in the Bell System: The Early Years* (New York: The Laboratories, 1975).

completed in different ways by different contracting foremen – that inhibited the production of uniform components.[9]

The two executives most responsible for promoting a culture of standardization and centralization were Edward J. Hall and Theodore Vail, the founding general manager and president, respectively, of AT&T.[10] Hall, as president of Southern Bell (1894–1909) and AT&T vice president (1887–1914), was a champion of the managerial and technological benefits of standardization. He believed it could simplify management while at the same time push local operating companies to adopt newer and more powerful transmission technologies such as the loading coil. Vail, on the other hand, was one of several telephone managers who had gained experience working within the railroads and the Post Office Department – large bureaucracies that provided organizational templates for many aspects of the telephone business. Hubbard hired Vail to be the general manager of American Bell in 1878; Vail went on to serve as president of AT&T from its founding in 1885 until his first attempt at retirement in 1899. Vail returned in 1907 before retiring for good in 1919.[11]

Managers in many Bell operating companies resisted the Vail-Hall push for centralization and standardization, sometimes out of a desire to remain autonomous, other times because of concerns about the cost of standardization. There was, to put it mildly, a vigorous debate among telephone executives and managers about the degree to which centralization and standardization should be imposed.[12] Anxious to avoid adversarial confrontations with the local companies – and lacking the financial power or administrative machinery to enforce compliance in a hierarchical manner – Hall pursued a more collaborative approach to creating and implementing technical standards throughout the operating companies that licensed the Bell patents.

One significant consequence of their efforts was a cooperative process to set standards through the committees of the National Telephone Exchange Association (NTEA) during the 1880s and 1890s. Meeting under the auspices of NTEA committees, telephone engineers developed technical specifications based on the operational experiences culled from their everyday work. In these

[9] Smith, *The Anatomy of a Business Strategy*, 5–7, 12, 47–48, 57–58, 69–71, 87, 130–132, 162; Garnet, *The Telephone Enterprise*, 15–17, 22–23.

[10] Kenneth Lipartito, *The Bell System and Regional Business: The Telephone in the South, 1877–1920* (Baltimore: The Johns Hopkins University Press, 1989), 116–168; Garnet, *The Telephone Enterprise*, 75–99; Smith, *The Anatomy of a Business Strategy*, 127–133; Richard R. John, "Vail, Theodore Newton," *American National Biography Online* (February 2000), http://www.anb.org/articles/10/10-01671.html (accessed August 4, 2007).

[11] Lipartito, *The Bell System and Regional Business*, 141–148; Garnet, *The Telephone Enterprise*, 37, 70–73, 87–89; Smith, *The Anatomy of a Business Strategy*, 57–60; Robert MacDougall, "Long Lines: AT&T's Long-Distance Network as an Organizational and Political Strategy," *Business History Review* 80 (2006): 305–307; Wasserman, *From Invention to Innovation*, 4–6; Hoddeson, "The Emergence of Basic Research in the Bell Telephone System."

[12] See Garnet, *The Telephone Enterprise*, 83–99; and Lipartito, *The Bell System and Regional Business*, 116–124.

conferences, engineers from Bell and other telephone companies established baseline specifications that could then be adopted on a voluntary basis and adapted by companies who, if circumstances warranted, could adjust the specifications to meet local conditions.[13]

Hall and the brilliant young telephone engineer John J. Carty pursued a similar cooperative strategy in the 1890s and early 1900s as they promoted functional specialization and standardization in other realms of the telephone business. Their goal was to make it easier for employees with experience in particular tasks – such as switchboard equipment, network operations, accounting, and auditing – to undertake the slow and steady work of identifying and communicating standard practices across the Bell operating companies. One important example of this process was the introduction and gradual adoption of a new accounting system that generated more uniform financial information, which made analysis easier both for managers and investors. Like the technical standards created through the NTEA, the introduction of accounting standards occurred not by dictate but rather through consultation and negotiation with staff in the Bell operating companies (often called Associated Companies). After all, the operating companies would be forced to bear the costs of introducing new methods and retraining or recruiting people to manage it.[14]

It was in this methodical, iterative, and cooperative manner that the culture and practices of standardization came to assume a central place in the creation of an increasingly integrated Bell system at the turn of the twentieth century. Standardization provided solutions for the problems of scale and scope, and had the potential to contribute to the crucial matters of reliability and performance. As engineers in AT&T and the Bell operating companies developed and applied the principles of standardization in a variety of realms – equipment manufacturing, long-distance transmission, and financial accounting – they learned that standards were essential for facilitating a style of flexible yet

[13] Garnet, *The Telephone Enterprise*, 83–99; Lipartito, *The Bell System and Regional Business*, 66–81 and 116–124; MacDougall, "Long Lines," 303–309; John, *Network Nation*, 220–221. On the NTEA's lengthy consideration of standard wire gauges, see *Sixth Annual Meeting of the National Telephone Exchange Association* (Brooklyn, NY: Eagle Book and Job Printing Department, 1884), 141–143; and *Seventh Annual Meeting of the National Telephone Exchange Association* (Brooklyn, NY: Eagle Book and Job Printing Department, 1884), 20–45. The NTEA's structure and goals closely resemble the Association of Transport Officers of the Pennsylvania Railroad, which was formed in the late 1870s. See Usselman, *Regulating Railroad Innovation*, 189–191 and 335–371.

[14] Garnet, *The Telephone Enterprise*, 96–126; Frank B. Jewett, "John Joseph Carty, 1861–1932," *National Academy of Science Biographical Memoirs* (New York: Columbia University Press, 1936), 2–27. See also Paul J. Miranti, Jr., "Probability Theory and the Challenge of Sustaining Innovation: Traffic Management at the Bell System, 1900–1929," in Sally H. Clarke, Naomi R. Lamoreaux, and Steven W. Usselman, eds., *The Challenge of Remaining Innovative: Insights from Twentieth-Century Business History* (Stanford, CA: Stanford University Press, 2009), 114–131.

centralized control in a way that suited their overarching strategy of building a high-quality, unified national telephone system.

A national, so-called universal telephone system emerged in the early twentieth century as a compromise strategy between the totalizing goals of centralized control and the inescapable conditions of local variation and competition. With the expiration of Bell's initial telephone patents in 1894, AT&T's financial and technological position in the industry weakened considerably under competitive pressure from independent telephone companies. Under intense pressure from AT&T's powerful team of Boston investors to reinvigorate the company, Theodore Vail returned to AT&T in 1907 for a second term as president. One pillar of Vail's response was to centralize and streamline the technical and engineering operations of the Bell-affiliated companies, which were at that point spread among departments in New York, Boston, and Chicago. He named John J. Carty, previously chief engineer of New York Telephone (the largest operating company in the country), to be AT&T's chief engineer. Carty, in turn, closed down labs in Chicago and Boston, consolidated the Engineering Departments of Western Electric and AT&T in New York, and centralized the Bell System technical operations under his control in AT&T's Engineering Department in New York. The new, combined Engineering Department effectively became the intermediary – an obligatory point of passage – between the operating companies and the Manufacturing Department of Western Electric, which remained in Chicago. Carty designated Western Electric, in turn, as the sole site for the design and standardization of equipment for use in Bell operating companies. Carty's reforms, in summary, introduced hierarchical and centralized control over standardization in the Bell System.[15]

As Carty struggled to consolidate Bell's unwieldy processes of design and manufacturing, Vail provided rhetorical and conceptual support with his introduction of a slogan that defined the modern Bell System: "One System, One Policy, Universal Service."[16] Although "Universal Service" eventually became shorthand for government policies to extend telephone service to all Americans, the phrase originally drew the rhetorical distinction between the reality of "dual service" (or noninterconnection) in the competitive era (1894–1913) and Vail's ambition for Bell to assume complete control over American telephony. Scholars have interpreted this slogan in a variety of ways, referring to it as the largest public relations campaign of its kind, a manifestation of Vail's civic values and administrative experience, a strategy to outmaneuver competitors, a response to the creation of state and federal regulatory commissions, an

---

[15] Louis Galambos, "Theodore N. Vail and the Role of Innovation in the Modern Bell System," *Business History Review* 66 (1992): 99–103; Oliver E. Buckley, "Frank Baldwin Jewett," *National Academy of Science Biographical Memoirs* (New York: Columbia University Press, 1952), 239–264.

[16] *Annual Report of the Directors of American Telephone and Telegraph Company to the Stockholders for the Year Ending December 31, 1909*, 18.

extension of Hall's southern strategy to the entire country, and, for critics, an element of the Bell strategy of public indoctrination.[17]

The vantage point of standardization – and the context of AT&T's struggle to centralize control over telephony – suggests a different interpretation: "universal service" was a standardization strategy, one that should be considered among the most comprehensive and successful in the history of American business. Vail's slogan was a concise articulation of what we might think of as the Bell ideology of standardization that was cultivated in the 1880s and 1890s. By the first decades of the twentieth century, this ideology anchored the creation of the modern Bell System. In the Bell ideology of standardization, the administrative, political, and technological aspects of the telephone business were inseparable – and the best way to oversee all three aspects was for AT&T to assume total control, free from the inefficiencies of competition and adversarial regulation. The public would benefit from consistent and high-quality service; investors would benefit from the payment of regular dividends; and the managers of the Bell System, protected from competition, would be able to oversee a national telephone network in an orderly, rational, and centralized manner – in short, as a benevolent and progressive monopoly. Vail's strategy was not only a clear response to the immediate problems facing AT&T, it was also an articulation of the organizational learning that had been occurring throughout American industry during the past several decades. The Bell ideology of standardization, distilled into the "universal service" slogan, shaped and limited the realm of the possible in a way that recalls a famous phrase written in 1911 by one of Vail's contemporaries, Frederick Winslow Taylor: "The system must be first."[18]

Through the ideology and rhetoric of universal service, Vail and his colleagues persuaded regulators to appreciate the value of an interconnected monopoly telephone system – one that operated more like a public utility than one of several firms in a competitive market. Vail's slogan was, fundamentally, a critique of wasteful competition, adversarial regulation, and the existing hodgepodge of proprietary networks that precluded interconnection among all telephone users. Vail's public relations campaign finally succeeded on a national scale in 1913, when Nathan Kingsbury, AT&T's vice president

[17] Roland Marchand, *Creating the Corporate Soul: The Rise of Public Relations and Corporate Imagery in American Big Business* (Berkeley: University of California Press, 1998), 48–87; Richard R. John, "Theodore N. Vail and the Civic Origins of Universal Service," *Business and Economic History* 28 (1999): 71–81; Milton Mueller, *Universal Service: Competition, Interconnection, and Monopoly in the Making of the American Telephone System* (Cambridge, MA: The MIT Press and American Enterprise Institute, 1997), 92–135; John, *Network Nation*, 340–345; Lipartito, *The Bell System and Regional Business*, 139; Federal Communications Commission, *Investigation of the Telephone Industry of the United States* (Washington, DC: Government Printing Office, 1939), 475–485.

[18] Frederick W. Taylor, *The Principles of Scientific Management* (New York: Harper & Brothers Publishers, 1911), 7.

of advertising and publicity, came to an agreement with James C. McReynolds, attorney general for President Woodrow Wilson, that endorsed Bell control over a single interconnected telephone network. The Willis-Graham Act passed by Congress in 1921 further confirmed AT&T's control by setting generous conditions to exempt the company from antitrust review. A corporatist settlement in the shape of federal regulation thus emerged as the preferred alternative to "cutthroat" competition, antitrust prosecution, or government ownership. At the same time, the appealing (and vague) rhetoric of universality became a central feature of the discourse and practice of American communication technology.[19]

Bancroft Gherardi and the Engineering of the Present

Standardization continued to be the sustaining ideology of the Bell System in the late 1910s and the 1920s. AT&T's leading advocate of standardization during this period was its vice president and chief engineer, Bancroft Gherardi (1873–1941). Gherardi's role in the development of the Bell System is obscured in most histories by the accomplishments of his peers in Bell Labs, especially Frank Jewett and John Carty, but he takes center stage when we examine telephone history from the vantage point of standardization. Such a shift in focus is necessary when one follows historian Kenneth Lipartito's insight that Bell Labs "owed its success not to pure research or an academic culture but to the careful integration of research, development, manufacturing, and operations."[20] Because standardization was essential for the successful integration of these aspects of the telephone business, there is a pressing need to know more about Gherardi and others like him whose careers were devoted to system architecture and integration.

Gherardi was a paradigmatic "organization man," a thoughtful professional who embodied the fusion of engineering and corporate management that occurred in the early twentieth century. In a memoir published by the National Academy of Science, the electrical engineer and Bell Labs president Oliver Buckley wrote that "Bancroft Gherardi was a great engineer … one of the most eminent engineers of his time."[21]

---

[19] Mueller, *Universal Service*, 129–145; Lipartito, *The Bell System and Regional Business*, 185–207; John, *Network Nation*, 340–369.

[20] Kenneth Lipartito, "Rethinking the Invention Factory: Bell Laboratories in Perspective," in Clarke, Lamoreaux, and Usselman, eds., *The Challenge of Remaining Innovative*, 133. In many ways my discussion of the Bell System strategies for standardization parallels Lipartito's account of Bell System strategies for research.

[21] Oliver E. Buckley, "Bancroft Gherardi, 1873–1941," *National Academy of Science Biographical Memoirs* (New York: Columbia University Press, 1957), 157. Buckley continued, "His major interest throughout his long professional career was in the design, construction, and operation of what has become probably the greatest, most complex and most completely integrated machine of our time." See also Lipartito, "Rethinking the Invention Factory."

Gherardi began his career as a telephone engineer in 1895 after graduating from Brooklyn Polytechnic and Cornell University. He quickly earned a reputation for his skill and diligence as a traffic engineer and for his thorough technical investigations of cables and standards of transmission. As a testing engineer, he learned to embody the same moralistic attitudes toward his work that we saw exemplified in Charles Benjamin Dudley in Chapter 2. In 1901, Gherardi became chief engineer of the New York and New Jersey Telephone Company, where he presided over a phase of rapid technological and organizational change: the company grew from 30,000 telephones in 1901 to more than 110,000 in 1906. Gherardi was particularly adept at working within the constraints of a large, complex, and relentlessly expanding technological system. He thrived in the midst of Bell's systematic and scientific approach to engineering and demonstrated his abilities in a number of projects including extensive studies of telephone traffic, the creation of new floor plans for central offices, experiments to test cable durability and safety, and the creation of technical standards for switchboard components and signal transmission.[22]

John J. Carty, Gherardi's supervisor in New York, saw potential in the young engineer. Gherardi demonstrated his worth to Carty on a number of occasions, including the collaboration between the two men to supervise the first commercial application of Michael Pupin's loading coils in New York in 1902.[23] As Carty rose into the upper echelons of the Bell System, so did Gherardi. When Carty was named chief engineer of AT&T in 1907, he brought Gherardi with him as his right-hand man. In 1909, Gherardi was promoted to AT&T engineer of plant development and standardization, responsible for these areas throughout the entire Bell System. In this capacity, Gherardi worked closely with Carty and Frank Jewett, a PhD physicist whom Carty had moved to New York as a research manager for Western Electric.[24]

The triad of Carty, Jewett, and Gherardi developed an effective partnership as they guided the technological trajectory of the Bell System over the next thirty years. As system architects, their major technical challenge was to balance two conflicting priorities: the development of new technologies to make the system more efficient and profitable, and the maintenance of quality and consistency throughout the entire system. This technical challenge was inseparable from their chief managerial challenge: to reconcile the diversity of local

---

[22] Gherardi earned a BS in 1891 from the Brooklyn Polytechnic Institute, and an ME (1893) and MME (1894) from Cornell University. Buckley, "Bancroft Gherardi," 157–177; "List of Some Items of Scientific and Engineering Work Done by Mr. Bancroft Gherardi," July 20, 1932, Box 1133, AT&T Archives and History Center, Warren, New Jersey [hereafter AT&T Archives, Warren]; "Biographical Statement of Bancroft Gherardi [circa 1932]," Box 1133, AT&T Archives, Warren; and "Bancroft Gherardi Dies; Phone Pioneer, 68," *New York Times* (August 16, 1941).
[23] Bancroft Gherardi, "The Commercial Loading of Telephone Circuits in the Bell System," *Transactions of the AIEE* 30 (1911): 1743–1773.
[24] Buckley, "Frank Baldwin Jewett"; Buckley, "Bancroft Gherardi."

conditions and equipment with their desire to centralize hierarchical control within an expanding system.

Experience taught the triad that systemwide harmonization along these technical and managerial lines often called for rational discussion with engineers in Western Electric and the Bell operating companies. In 1912, Gherardi noted a paradox in the need for flexibility in the establishment of systemwide standards: "In applying transmission standards, it must be recognized that they cannot be considered as hard and fast rules which must be followed in all cases ... in other words, to attain the standards might necessitate expenditures not warranted."[25] Western Electric engineer J. L. McQuarrie echoed this sentiment at a 1915 technical summit of Bell engineers (the first such meeting in the history of the Bell System) when he noted that "it is not the policy of the A. T. & T. Company to use force in compelling the associate companies to follow their standards." Instead, persuasion and consensus were the preferred tactics: "I think their policy is to set up their cases in such a manner that the associate companies will see for themselves that that is the thing they ought to do."[26] Subsequent technical meetings, such as a 1916 conference on telephone transmission chaired by Carty and Gherardi, provided venues where Bell managers and engineers could debate, exchange information, and establish consistent, systemwide practices and procedures.[27]

Changes in AT&T's corporate organization between 1918 and 1920 propelled Gherardi into a commanding role over standardization throughout the Bell System. While Carty and Jewett directed their attention to the American mission in World War I, Gherardi remained in New York and assumed a less publicly visible role as acting chief engineer of AT&T. President Vail retired in June 1919 and was replaced by his protégé and close friend, Western Electric president Harry Thayer. Thayer immediately promoted Gherardi to chief engineer of AT&T and promoted him again in 1920 to vice president of AT&T. Gherardi held both positions until his retirement in 1938.[28]

Gherardi assumed the role of chief engineer at the beginning of a new technological and organizational era for AT&T. At the beginning of Gherardi's tenure, he and Thayer split the Engineering Department into two: the Operations

[25] Bancroft Gherardi, "Discussion of Transmission – Cooperation of Departments," *Telephony* 62 (1912): 468–470.

[26] J. L. McQuarrie, "Discussion of Mr. McQuarrie's Paper," *Manufacturing and Engineering Conference*. This tension between centralized control and regional autonomy is the central theme of Lipartito, *The Bell System and Regional Business*. See especially chapter 4, "Regional Change and Technological Conflict"; chapter 7, "A Merging of Interests"; and chapter 9, "Regional and Corporate Cultures."

[27] American Telephone and Telegraph Company, *Telephone Transmission: Meeting of the Technical Representatives of the Bell System* (New York City: December 11–12, 1916).

[28] Buckley, "Bancroft Gherardi," 165–166; "List of Some Items of Scientific and Engineering Work Done by Mr. Bancroft Gherardi," July 20, 1932, Box 1133, AT&T Archives, Warren; Leonard S. Reich, *The Making of Industrial Research: Science and Business at GE and Bell, 1876–1926* (New York: Cambridge University Press, 1985), 180–181.

and Engineering Department and the Development and Research Department. The purpose of this reorganization, according to Thayer, was to allow Bell engineers to "differentiate in our work between the *engineering of the present* and the *engineering of the future*."[29] Under this new regime, Carty and Jewett turned to the engineering of the future, first in the Development and Research Department and, in 1925, in the newly created and much-celebrated Bell Labs.

Gherardi took over the less glamorous duties of the "engineering of the present" within the Department of Operations and Engineering. His jurisdiction was vast: he directed a staff of hundreds of engineers who collectively were responsible for developing engineering methods, operating plans, and methods for analyzing and comparing different types of service. Further, Gherardi's staff was responsible for advising the technical staffs of the other units of the Bell System, including Western Electric, the regional operating companies, and the Long Lines division, as well as for the overall technical coordination of the system.

Standardization was Gherardi's primary tool to manage complexity. He had learned from his own experience and in working with Jewett and Carty that standards were not simply technical prescriptions. They documented years of experimentation and provided a record of the technical know-how and organizational capabilities of thousands of individuals who worked for the Bell System. As such, standards were effective organizational tools that could help resolve – or at least minimize – conflicts between managers in different branches of the Bell System. As Jewett summarized in 1915, "A standard piece of apparatus or a standard practice at any time so far as the Bell System is concerned represents the best total of all the experience of all concerned."[30]

In order to collect, standardize, and share their combined experience, Bell System engineers experimented with methods of documentation and correspondence. In 1905, AT&T chief engineer Hammond Hayes began to use the term "General Engineering Circular" (GEC) at the top of technical correspondence with around two dozen recipients, including engineers at Western Electric and the chief engineers of regional Bell operating companies. Some GECs were requests for comments on issues such as underground construction in central offices; others described field experiments conducted in one part of the country; still others were draft proposals for standard designs for manhole frames and covers, private branch exchanges in apartment buildings, insulation and other protection from electric shock, and mountings for telephone jacks.

---

[29] See Harry B. Thayer, "The Development of Development and Research," *Bell Telephone Quarterly* 4 (1925): 6 (emphasis added); "Centralizing Bell System Researches," *Science*, New Series, 79 (1934): 366–367; and "Notes on Recent Occurrences: Bancroft Gherardi Retires as AT&T Vice President and Chief Engineer after a Distinguished Career of 43 Years of Bell System Service," *Bell Telephone Quarterly* 17 (1938): 139–144.

[30] Jewett, "Discussion of Mr. McQuarrie's Paper," *Manufacturing and Engineering Conference.*

Between 1905 and 1914, more than 400 GECs had been exchanged among Bell System engineers.[31]

By the time Gherardi became AT&T's chief engineer in 1919, many GECs referred to Engineering Specifications that AT&T began to distribute throughout the Bell System in 1918. Many GECs, in fact, were simply cover letters to specifications that explained their origins and clarified ways in which the specifications should be used. Bell System engineers published and circulated thousands of Engineering Specifications for the widest imaginable variety of items, including porcelain knobs, shellac, shellac thinner, machine bolts, medical supplies such as a rhubarb-soda mixture, witch hazel, ammonia inhalants, and so on. After 1920, each specification was drafted and published by AT&T's Department of Development and Research – the group that Thayer said was responsible for the "engineering of the future" and was thus best positioned to summarize the Bell System's most complete and sophisticated knowledge of any aspect of the telephone art.[32]

The D&R Bulletins, as these specifications came to be known, were the product of a dialectical process between the executives in the engineering departments in New York and the engineers who worked in Western Electric and the Bell operating companies throughout the country. The orders came from New York, but the ideas often came from the Bell System's rank and file and were transmitted up the organization chart. A typical example of Gherardi's approach can be seen in a series of letters among Gherardi and the presidents of the Bell operating companies. In May 1922, Gherardi wrote to the presidents to "make a special review" of the "Bell System standards and methods of construction, maintenance and operation which will be as economical as practicable [*sic*] consistent with good and continuous service and adequate durability." Gherardi, writing less than a year after the Federated American Engineering Societies published its landmark study, *Waste in Industry*, acknowledged that his main interest was "the possibility of reducing costs without losing too much in other directions. "Nothing could be more helpful to us here," he continued, "than frank criticism from those using the standards. We want to leave nothing undone which if done might produce better and more economical results."[33]

Throughout the summer of 1922, Gherardi was flooded with responses from operating companies from all over the country on a predictably wide range of topics, such as plant construction, pole line specifications, insulating sleeves, hand lanterns for manhole and repair use, different varieties of wood

[31] A copy of the GECs are held in Collection 6 AT&T Corp., RG 4 Corporate Functions, Boxes 1–4, AT&T Archives and History Center, San Antonio, Texas [hereafter AT&T Archives, San Antonio].

[32] A copy of Engineering Specifications are held in Collection 6 AT&T Corp., RG 4 Corporate Functions, Box 10, AT&T Archives, San Antonio.

[33] Bancroft Gherardi to George McFarland, May 10, 1922, Collection 3 Pacific Telesis Group, RG 5 Predecessors, Box 17, AT&T Archives, San Antonio.

for pole construction, and billing methods. In August, Gherardi wrote again to thank his colleagues for their replies so far. He also requested that any subsequent replies be sent by the middle of September, "in order that we may study them and have an opportunity to discuss" any "general matters" at the weeklong AT&T Conference to Discuss Economy and Efficiency in Operation that was planned for the middle of October. The conference was the fourth such national conference of Bell System technical staff that had been held since 1915, all with the intention of facilitating better communication and nurturing a stronger sense of "team work" – in effect, to imagine and produce a more cohesive community – among telephone employees working under the Bell umbrella.[34]

These surviving documentary traces – GECs, Bell System Specifications, memoranda between Gherardi and presidents and chief engineers at the various Bell operating companies, and published proceedings of Bell technical conferences – do not support a top-down interpretation of standardization in the monopoly Bell System. Instead, they sketch a more human picture of where Gherardi, an earnest and intelligent engineer-executive, struggled to maintain clear lines of communication under conditions of almost unimaginable complexity. The Bell System Standards, circulated in the D&R Bulletins, continued to grow in number and detail as the Bell System itself grew. In the 1920s, the document series that carried Bell System Standards was renamed the Bell System Practices. Like the GECs and other forms of systemwide communication that preceded it, the Bell System Practices distributed the latest technical information to engineering departments at all of the Bell operating companies. Standards appear here (and elsewhere) as hierarchical when we see them in hindsight, but when we look closely at standards-in-the-making, we are able to discern more collaboration, more negotiation, and less hierarchy.

It was under Gherardi's watchful eye as chief engineer, according to a history of the Bell System Practices published in *Bell Telephone Magazine* in 1952, that standardization became firmly and finally embedded in the culture of the Bell System: "In the decade following World War I, there were more developments in the Practice situation than in any other similar period in Bell System history." By 1929, AT&T had created standards for an astonishing variety of functions, including telephone plant design; underground cables; raw materials; manufacture, distribution, installation, inspection, and maintenance of new equipment; business and accounting methods; nontechnical supplies (such as office furniture, appliance, janitors' supplies, cutlery, and china); and provisions for

---

[34] Bancroft Gherardi to George McFarland, August 22, 1922, Collection 3 Pacific Telesis Group, RG 5 Predecessors, Box 17, AT&T Archives, San Antonio; J. C. Nowell to Bancroft Gherardi, September 1, 1922, Collection 3 Pacific Telesis Group, RG 5 Predecessors, Box 17, AT&T Archives, San Antonio; American Telephone and Telegraph Company, *Conference to Discuss Economy and Efficiency in Operation*, Shawnee, Pennsylvania, October 18–25, 1922; Bancroft Gherardi, "Conference of Personnel Group," *Bell Telephone Quarterly* 1 (July 1922): 39–43.

safety, health, and even sleet storms. By the 1980s, the index alone of the Bell System Practices filled 969 pages; the volumes filled more than 80 cubic feet.[35]

Standards helped Gherardi and all Bell System engineers cope with the increasing scale, scope, and complexity of the system. During Gherardi's tenure as chief engineer between 1920 and 1938, the number of telephones in the Bell System grew from 7.7 million to more than 19 million. With the rapid expansion of telephone service, the explosive growth of radio transmission, and the introduction of television technology, the 1920s and 1930s were exciting times to be on the front lines of electrical communication. As in the initial phase of his career in New York, Gherardi flourished when faced with the unusual problems of engineering a massive and rapidly evolving system. Gherardi was a key member of the AT&T technical community during this period – for example, he played a leading role in AT&T's opening of transatlantic radio telephone service in January 1927, served on the board of directors of Bell Labs, served on the editorial board of the *Bell System Technical Journal*, and coauthored essays on "Telephone Progress" for the annual editions of the *Encyclopedia Americana* between 1923 and 1936. Nevertheless, he did not always confront technological change with a sense of inevitable progress. In 1920, on reviewing reports of systemwide growth that exceeded estimates by 40 percent, Gherardi worried that the "effect of these very large gains will be to create extraordinary demands for all classes of labor and material," demands that would exceed the Bell System's capacity to act. "A continued growth such as we are now having," Gherardi growled, "will necessarily bring about shortly a most serious state of affairs." Uncontrolled growth, in other words, was dangerous – particularly when it undermined careful and orderly planning.[36]

A brief consideration of Gherardi's role in the Bell System's transition from manual to mechanical switching – an event in telephone history that attracted the close scrutiny of regulators, critics, and historians – illustrates his cautious approach to innovation and standardization. The independent inventor Almon Strowger invented the first mechanical switch in 1891, but Bell experts did not

[35] Bancroft Gherardi and Frank B. Jewett, "Telephone Communication System of the United States," *The Bell System Technical Journal* 9 (1930), 1–100; Harold S. Osborne, "The Fundamental Role of Standardization in the Operations of the Bell System," *American Standards Association Bulletin* (September, 1931), 3; O. C. Lyon, "Standardization of Non-Technical Telephone Supplies," American Telephone and Telegraph Company, *Plant and Engineering Conference of the Bell System*, New York City, December 6–10, 1920, Section IV, 97–103; Fagen, *Engineering and Science in the Bell System*, 638; Buckley, "Bancroft Gherardi," 166–167; A. B. Covey, "The Bell System's Best Sellers," *Bell Telephone Magazine* (Summer 1952), 90; and Harold S. Osborne, "Abstract of Discussion of Osborne Paper on Standardization in the Bell System," *American Standards Association Bulletin* (October 1931), 27–28.

[36] Bancroft Gherardi, "Voices Across the Sea," *North American Review* 224 (1927): 654–661; Reich, *The Making of American Industrial Research*, 170–238. For Gherardi's concerns about rapid growth and expansion of telephone service, see Bancroft Gherardi to George McFarland, February 27, 1920, Collection 3 Pacific Telesis Group, RG 5 Predecessors, Box 17, AT&T Archives, San Antonio.

see a way to integrate the device into their operations and declined to buy or even license Strowger's patent. In the subsequent decades, Bell officials kept a close eye on improvements to Strowger's device and developed their own "panel type" mechanical switch but did not introduce mechanical switching into large exchanges before 1920. In his study of AT&T's transition from manual to machine switching, historian Kenneth Lipartito explained this reluctance by pointing to the difficulties and uncertainties of introducing a new innovation in a system context. Many factors, including conflicts over the efficacy and economy of machine switching technology as well as AT&T's sunk cost in its existing "techno-labor system," convinced AT&T managers to pursue incremental innovations within the existing paradigm of manual switching. Only when they faced successive crises of system growth and labor shortages due to the American mobilization for World War I did Bell executives and managers decide to embrace machine switching and invest decisively in the new technology.[37]

Gherardi was the prime mover behind this deliberate transition to machine switching. He first presented to Carty a study on technological and human aspects of machine switching in 1917. On his promotion to chief engineer in 1919, Gherardi made a formal recommendation on the subject to AT&T executives that called for the gradual replacement of manual switching. He was careful to note that significant obstacles needed to be overcome, including negotiating with the Automatic Electric Company to use the patents for the Strowger switch, engineering the production of panel switches at Western Electric, educating the engineers at regional operating companies, and convincing the general public to change their dialing habits.[38]

On his promotion to chief engineer, Gherardi became personally responsible for implementing his own plan. He pointedly preferred the term "machine switching" to "automatic switching" because of what he saw as the misleading connotations of the latter term: the new mechanical equipment developed for telephone switches was anything but automatic. The work he directed was both time consuming and labor intensive. As early as 1920, Western Electric added five buildings to its Hawthorne Plant and drew up plans to add three more. In a published overview of the project, Gherardi and AT&T engineer Harry Charlesworth claimed that AT&T and Western Electric had combined to make "three thousand new piece parts involving some thirty-six thousand

---

[37] Kenneth Lipartito, "When Women Were Switches: Technology, Work, and Gender in the Telephone Industry, 1890–1920," *American Historical Review* 99 (1994): 1074–1111; Venus Green, "Goodbye Central: Automation and the Decline of 'Personal Service' in the Bell System, 1878–1921," *Technology and Culture* 36 (1995): 912–949; John, *Network Nation*, 232–233.

[38] Bancroft Gherardi and Harry P. Charlesworth, "Machine Switching for the Bell System," *Telephone Review* Vol. 2, Supplement (April 1920): 1–12; "List of Some Items of Scientific and Engineering Work Done by Mr. Bancroft Gherardi," July 20, 1932, Box 1133, AT&T Archives, Warren; Green, "Goodbye Central," 939–941.

manufacturing and inspecting operations," including new working drawings, manufacturing methods, new tools and machinery, testing gauges, and the shipment of factory-assembled switch frames that would facilitate their installation throughout the Bell System. Thanks to sufficient planning and coordination, Gherardi and Charlesworth promised that the introduction of the new machine switching system would "be carried on in the usual Bell way, that is, in an economical and orderly manner, without inconvenience to the subscriber, and without derangement or interruption of service."[39]

The first mechanical exchange opened successfully in Omaha in December 1921, with Gherardi personally on hand to "cut it over." Gherardi, reporting triumphantly to Bell executives around the country, declared that the Omaha exchange was "operating in a most satisfactory manner." Subscribers were "pleased with the service" and, better yet, "all newspaper accounts are favorable." Gherardi concluded by noting the organizational factor behind their success: "The numerous opportunities for trouble in any cut over, especially of new types of apparatus, were avoided by the skillful manner in which the work was supervised and done by the Associated Company people and the Western Electric Company."[40]

Once the first mechanical exchanges were up and running in Omaha and elsewhere, Gherardi presided over cost studies to estimate when the cutover to machine systems would be economical, the situations where it would not, and the amount of time it would take for AT&T to recover its investment in the new switching system. Based on the data collected in these studies, Gherardi proceeded gradually and, characteristically, did not force change too quickly: by the time he retired in 1938, human operators still were using manual switches to route nearly half of all American telephone calls.[41]

---

[39] Carty endorsed Gherardi's suggestion to use the term "machine switching" instead of "automatic switching" at a 1916 conference. Carty, "Universal Service," in AT&T, *Telephone Transmission*, 14. Gherardi and Charlesworth, "Machine Switching for the Bell System," 11–12; Gherardi and Jewett, "Telephone Communication System of the United States," 14, 26. Bancroft Gherardi, "Remarks on Machine Switching Cutovers," R. F. Estabrook, "Machine Switching Service Observing," and W. E. Farnham, "The Preparation for a Machine Switching Cutover," all in AT&T, *Plant and Engineering Conference of the Bell System*, Section IV. See also Bancroft Gherardi, "Machine Switching Substations – Installing Subscriber's Sets with Dials," GEC 1013, January 17, 1921, Collection 6 AT&T Corp., RG 4 Corporate Functions, Box 3, AT&T Archives, San Antonio; and Bancroft Gherardi, "Outline of Special Work in Connection with the Cutover of a Machine Switching Office," GEC 1017, January 25, 1921, Collection 6 AT&T Corp., RG 4 Corporate Functions, Box 3, AT&T Archives, San Antonio.

[40] "Say It with Fingers to Get 'At' With Numbers," *Omaha World-Herald* (December 11, 1921); A. E. Van Hagen, "The Dial Office 'Cutover,'" *Bell Telephone Quarterly* 8 (1929): 96; Bancroft Gherardi to George McFarland, December 12, 1921, Collection 3 Pacific Telesis Group, RG 5 Predecessors, Box 17, AT&T Archives, San Antonio.

[41] Bancroft Gherardi, "Engineering Considerations," in American Telephone and Telegraph Company, *Conference to Discuss Economy and Efficiency in Operation*, Shawnee, Pennsylvania, October 18–25, 1922; Buckley, "Bancroft Gherardi," 169–170.

The cutover to mechanical switching illustrates Gherardi's managerial style, which emphasized the advantages of having a "centralized staff" in the AT&T General Department. Such a department, as he described in a 1925 speech at the Bell System Educational Conference, "does those things which can best be done at one central point rather than for each Associate Company to do such work itself." Centralized control was not absolute, but it was the best way to approach "the study of all questions which are common to all or a number of the companies in the system," problems that ranged from "development and research, general engineering studies, the study of operating methods, a comparison of operating results from different companies, [and] a study of auditing and bookkeeping methods."[42]

In sum, Gherardi presided over the defining phase of the Bell System's centralized style of standardization in its technical and organizational work. Standardization, as many scholars have noted, was a key element in AT&T's systematic and deliberate style of innovation. However, Gherardi's approach to innovation was not an unqualified pursuit of technological progress. Because his position as chief engineer required him to consider the systemwide implications of any new technology or procedure, Gherardi often used standardization as a means to *slow* and *tame* innovation. In the toolkit of a skilled engineer and manager such as Gherardi, standardization was a strategy not just to integrate state-of-the-art technologies into a large system but also to *limit* and *suppress* the rate of change so that he and his colleagues could maintain control over the systemwide alliances they so expertly cultivated. This rational (if somewhat languid) approach to innovation and standardization anchored the Bell System's successful growth in the 1920s and 1930s. But, as Jewett warned presciently in 1915, it was open to a powerful reinterpretation by critics of the Bell System who might conclude that centralized control over standardization could suppress invention, force customers to buy outdated equipment, and discourage AT&T engineers from taking advantage of outside technical improvements.[43]

Gherardi and his subordinates would have preferred to treat telephone transmission as a closed system that was insulated from outside forces and therefore more easily predicted and controlled. But the reality was more complex. Gherardi recognized that the Bell System was open to outside forces that he could not control by means of the vast managerial hierarchy under his command. Here, as in other areas, Gherardi learned to succeed: he explored the outside factors that introduced problems and inefficiencies into his system and, over time, he learned to experiment with organizational strategies to attack critical problems and extend order in ways that he could not otherwise achieve,

[42] Bancroft Gherardi, "The Bell System," *Bell Telephone Quarterly* 4 (1925): 255–265.
[43] Galambos, "Innovation in the Modern Bell System"; United States Congress, *Report of the Federal Communications Commission on the Investigation of the Telephone Industry in the United States* (Washington, DC: Government Printing Office, 1939).

despite his position at the commanding heights of the monopoly Bell System hierarchy.

Gherardi, the Bell System, and Consensus Standardization

Throughout its early history, AT&T did not encourage its engineers to collaborate openly in industry associations. For example, even though AT&T had tremendous expertise in electrical engineering, only four telephone engineers – Bell, Carty, Jewett, and Gherardi – served as president of the American Institute of Electrical Engineers between 1884 and 1930. Furthermore, no employees of AT&T were members of the AIEE Standards Committee until 1914, when Jewett joined the group. For the most part, AT&T leaders believed their competitive advantages flowed from keeping the company's patents and practices closed and secretive. Their insular attitude began to change in the years before the First World War. One indication of the change may be seen in Henry F. Albright's speech at the 1915 conference of Bell System engineering and manufacturing personnel. Albright, who was the leading proponent of scientific management at Western Electric, asked his colleagues to reconsider the potential benefits of professional activities outside the Bell System. He suggested that individual employees could gain "an enlarged circle of acquaintances" and learn about other engineering methods if they joined industrywide associations and societies. The company as a whole would benefit as well:

Through such associations the company obtains recognition for its principles and achievements; its worth and position in the community are better known; the quality of its scientific work and its efficiency in production becomes better known and our customers and friends learn to better appreciate our pioneer work in the development of the art of telephony.[44]

AT&T engineers quickly learned that "outside" cooperation had more than the social and civic benefits that Albright promoted. One good example of the technical benefits of cooperation was evident in AT&T's efforts to address a critical problem: inductive interference generated by the close proximity between telephone lines and the overhead electrical lines of power, lighting, and railroad companies.

Telephone engineers had long been familiar with interference, such as "crosstalk" (speech from one conversation was audible in another) and "babble" (unintelligible background noise), that resulted from placing telephone circuits in close proximity.[45] In the mid-1910s, however, Bell System

[44] H. F. Albright, "The Business Activities and Relations of Members of Engineering and Manufacturing Departments Outside the Western Electric Company," *Manufacturing and Engineering Conference* (1915). Stephen B. Adams and Orville R. Butler, *Manufacturing the Future: A History of Western Electric* (New York: Cambridge University Press, 1999), 79.

[45] Fagen, *Engineering and Science in the Bell System*, 324–337; John, *Network Nation*, 226.

engineers became increasingly concerned with new sources of electrical interference that originated not within their networks but rather from parallel and intersecting lines operated by power and light companies. This type of inductive interference was deeply problematic because it undercut one of the central technical objectives of Bell System engineers: to increase the efficiency, sensitivity, and reliability of transmission equipment. Bell System engineers were lowering their limits for acceptable levels of interference in an effort to improve call quality, but power company engineers were building more – and more powerful – lines and transmission facilities, which, infuriatingly for telephone engineers, generated more interference.

The first cooperative efforts to address the problem of inductive interference occurred in California, under the auspices of the California Railroad Commission. Between 1912 and 1917, the commission's Joint Committee on Inductive Interference – consisting of and funded by representatives from the telephone, power, and railroad industries – performed a number of field and laboratory tests and wrote dozens of technical reports, many of which were compiled in a 1919 final report. The report identified some "guiding principles" for preventing interference, including standards for minimum distance between power lines and communication lines as well as design and construction rules for apparatus, that were incorporated into commission rules. However, the report's authors also acknowledged the complexities of inductive interference and underscored the need to conduct further studies of the scientific and practical aspects of the problems at hand.[46]

In April 1919, Gherardi – who at this point still held the title of AT&T's acting chief engineer – distributed copies of the California report as an attachment to General Engineering Circular 901, which, like the other GECs, was mailed to the chief engineers of all Bell operating companies. After praising the careful work of the Joint Committee that drafted the California report, Gherardi cut to what he saw as the heart of the matter. "The greatest difficulty in obtaining adequate protective and remedial measures against inductive interference," he wrote, "consists in getting the power companies sufficiently interested in the problem and in convincing them that the protective measures asked for are necessary and reasonable." To accomplish this, Gherardi saw fit to begin a "campaign of education on this subject." He urged his colleagues in the Bell operating companies to make sure that "those of your engineers whose particular duty it is to deal with this subject familiarize themselves with the content

---

[46] Railroad Commission of the State of California, *Inductive Interference between Electric Power and Communication Circuits: Selected Technical Reports with Preliminary and Final Reports of the Joint Committee on Inductive Interference and Commission's General Order for Prevention or Mitigation of Such Interference* (Sacramento: California State Printing Office, 1919). See also A. H. Griswold and R. W. Mastick, "Inductive Interference as a Practical Problem," *Transactions of the American Institute of Electrical Engineers* 16 (1916): 1051–1094; and Frederick Bedell, "Characteristics of Admittance Type of Wave-Form Standard," *Transactions of the American Institute of Electrical Engineers* 16 (1916): 1155–1186.

of this report," so that they would be prepared to "discuss it convincingly with power company representatives."[47]

In his public statements, Gherardi did not frame his approach as a "campaign of education" as he did when writing to Bell System engineers. Instead, he emphasized that cooperation between power and telephone engineers could be most productive. Nevertheless, both telephone and power companies around the country turned to litigation throughout the 1920s in an effort to impose the costs of solving the problem onto their rivals. As early as 1920, a Bell System technical conference dedicated an entire session to working through AT&T's approach to the problem of inductive interference. Although existing laws and precedents seemed to indicate that the first party to construct facilities had the right to exclude other parties, AT&T's chief counsel N. T. Guernsey stressed that most interference cases were not so clear. Guernsey, in comments echoed by Gherardi, preferred to avoid litigation if possible. Guernsey found several advantages to this strategy: it would preempt the costs of litigation, it would avoid "controversy with our friends who are engaged in the power business," and it would help AT&T maintain a favorable image in the eyes of regulators and the general public. Both Gherardi and Guernsey had learned the lessons of the late-nineteenth century "robber barons" and did not wish to undermine the public image of AT&T as the service-oriented corporation that Vail had so carefully crafted.[48]

Gherardi personally directed AT&T's participation in the effort to settle through cooperation the problems generated by the interfering infrastructures. Beginning in 1921, Gherardi led the Bell System's involvement in two ad hoc Joint General Committees: one with the Association of American Railroads and the other with the National Electric Light Association (NELA). Because of his training as an electrical engineer and his long-standing participation in the AIEE, Gherardi was well suited for a task that was part diplomacy, part engineering. Because of his position in the Bell System managerial hierarchy, he was empowered to ask the presidents of all the Bell operating companies to ensure that "your people" would "continue the policy to reach, without sacrificing our interest, a harmonious understanding with the electric light and power people in any specific case that may come up."[49]

---

[47] Bancroft Gherardi, "Final Report of the Joint Committee on Inductive Interference and General Order No. 52 of the Railroad Commission of the State of California," GEC 901, April 29, 1919, Collection 6 AT&T Corp., RG 4 Corporate Functions, Box 4, AT&T Archives, San Antonio.

[48] Fagen, *Engineering and Science in the Bell System*, 336. For positions articulated at the AT&T conference, see Bancroft Gherardi, "Introductory Remarks on 'Our Legal Rights in Interference Cases'"; N. T. Guernsey, "Our Legal Rights in Interference Cases"; H. S. Warren, "Interference Problems"; Frederick L. Rhodes, "Remarks on 'Interference Problems'"; Harold S. Osborne, "Inductive Interference"; and D. H. Keyes, "Inductive Interference Problems – Method of Attack" – all in AT&T, *Plant and Engineering Conference of the Bell System*, Section IV.

[49] Bancroft Gherardi to George McFarland, April 27, 1921. Collection 3 Pacific Telesis Group, RG 5 Predecessors, Box 17, AT&T Archives, San Antonio. Gherardi became an AIEE Association

Cooperation continued in 1922, when NELA and the Bell Telephone System created a Joint Development and Research Subcommittee to investigate further the problems of inductive interference. They also created a Special Committee on the Joint Use of Poles in February 1923 to prepare "mutually acceptable principles and practices" for joint pole agreements between NELA's member companies and the operating companies of the Bell System. To complement this extension of interindustry cooperation, Gherardi continued to advance his strategy of intracompany education. In January, he arranged for engineers in the Department of Development and Research to conduct a six-week course of instruction on inductive interference, so that Bell employees in the "Engineering or Plant Departments directly responsible for the technical handling of inductive interference work" could "familiarize themselves with all the technical information" available to AT&T's most expert engineers. Gherardi attached a detailed six-page outline of "Notes for Inductive Interference School," with a characteristic reminder that he would "like to receive any suggestions" for the "proposed scope of the course."[50]

By 1924, the Bell-NELA Joint Committee joined with representatives from the electric and steam railroad industries to form a new group, the American Committee on Inductive Coordination. Gherardi was the group's chairman; Robert Pack, a respected power engineer and active member of NELA, was one of three vice chairmen. Together, the two men presented a report on the committee's work to the general session of the NELA convention in May 1926. In his remarks, Pack matter-of-factly noted three areas of effort. First, the committee had created "Principles and Practices for the Joint Use of Wood Poles" and distributed it to NELA member companies and AT&T associated companies. Second, he reported some progress toward a statement on procedures for dividing the costs of inductive coordination. Third, he noted the recent approval of funds for further development and research, which, to his regret, had not progressed as far as the first two areas.[51]

Gherardi departed from Pack's reporting style to relay his own personal reflections as a visitor to NELA. His performance was not only diplomatic

---

member when he graduated from Cornell in 1895, served as AIEE vice president from 1908 to 1910, was named an AIEE Fellow in 1912, served on a number of AIEE committees as well as the AIEE board of managers from 1905 to 1908 and from 1914 to 1917. He was later elected AIEE president for 1927–1928. Fagen, *Engineering and Science in the Bell System*, 336–337; Lewis Coe, *The Telephone and Its Several Inventors* (Jefferson, NC: McFarland & Company, 1995), 158–159; and "Bancroft Gherardi – Biographical Data," September, 1949, Box 1133, AT&T Archives, Warren.

[50] Bancroft Gherardi, "Progress of the Joint Committee on Relations of Supply and Signal Circuits," *Bell Telephone Quarterly* 1 (April 1922): 49–54. Bancroft Gherardi to J. C. Nowell, March 1, 1923 Collection 3 Pacific Telesis Group, RG 5 Predecessors, Box 17, AT&T Archives, San Antonio; Bancroft Gherardi to J. C. Nowell, January 20, 1923, Collection 3 Pacific Telesis Group, RG 5 Predecessors, Box 17, AT&T Archives, San Antonio.

[51] Bancroft Gherardi and Robert F. Pack, "Report on Joint General Committee, Bell System and N. E. L. A.," *National Electric Light Association Proceedings* 83 (1926): 191–193. Pack became NELA president for 1926–1927.

and cordial; it also indicates that Gherardi had imbibed the spirit of voluntary consensus standardization nurtured in the 1920s by people such as Paul Gough Agnew and Charles le Maistre and organizations such as the American Engineering Standards Committee (AESC). Gherardi, a telephone man among power and lighting engineers, noted the pleasure of "wearing the badge" of the group in his sixth consecutive appearance at their convention and speaking of "a change in my attitude toward the meeting, and a change in the meeting's attitude toward me." He continued,

I can feel that there has been a closer and closer bond between us.... We have put further and further behind us the proposition that inductive coordination was a problem to fight about, and we have more and more fully accepted the view that inductive coordination was a problem to work out together, quite a different attitude from fighting it out.[52]

Reports from NELA's Inductive Coordination Committee at the group's meetings in 1926 and 1927 further indicate that earlier tensions between the power and telephone companies had been reduced to a matter of cooperative research and routinized solutions. The 1926 report noted, "In contrast with the experience of previous years, the one now closing has been singularly free from controversy and threatened court actions." As we saw in Agnew's work described in Chapter 3, engineers were searching for ways to create their own norms and institutions that could preempt the involvement of lawyers and courts, in effect "turning from the courts with their numberless laws and intricate procedure to boards and reforms of [their] own choosing."[53]

The committee report for 1927 began by reflecting on two "outstanding facts" of the previous year. First, the committee had emerged "finally and completely" from its reputation as a body that handled a controversial problem with the telephone industry to a body that dealt with a "common electrical industry problem." Second, relations between the staff and engineers of NELA and AT&T had been further strengthened.[54] Gherardi, speaking from the audience at the 1928 AIEE meeting, confirmed that the group's turn from conflict to collegiality had borne fruit. Reflecting on the joint work between the Bell System and NELA over the past several years, Gherardi declared, "We came to the conclusion that 10 per cent of our problem was technical and 90 per cent was to bring about between the people on both sides of the question, a friendly and cooperative approach." Laurence Corbett, a power engineer who had been

52 Gherardi and Pack, "Report on Joint General Committee"; American Committee on Inductive Coordination, *Bibliography on Inductive Coordination* (New York: American Committee on Inductive Coordination, 1925).
53 P. G. Agnew, "How Business Is Policing Itself," *The Nation's Business* (December 1925): 41–43.
54 Howard S. Phelps, "Report on Inductive Coordination Committee," *National Electric Light Association Proceedings* 83 (1926): 851–852; J. C. Martin, "Report of Inductive Coordination Committee," *National Electric Light Association Proceedings* 84 (1927): 625–626.

involved with the study of inductive interference since the California Railroad Commission investigations, concurred in the preface of his 1936 book on inductive coordination: "A very minor amount of cooperative work in advance planning of facilities or in correction of existing unfavorable situations will in most cases enable both companies to serve the same customers – the public – at no greater cost."[55]

Although these ad hoc committees generated standards and recommended practices (such as recommendations for satisfactory distances between electrical wires connected to the same poles), they did not solve the underlying scientific and technical problems associated with inductive interference.[56] Nor did they need to. Cooperative organizations such as the Joint General Committees and American Committee on Inductive Coordination created organizational means for defusing a potentially costly confrontation between some of the most powerful corporations in American industry. Through this new approach – most visible in the rhetorical shift from "inductive interference" to "inductive coordination" – they redefined their confrontation as a problem that could be managed through collaborative research, good faith cooperation, and interindustry standardization.

Gherardi's enthusiasm for cooperative solutions to difficult technical and organizational problems reflected his growing commitment to the cause of industrial standardization and to the methods of voluntary consensus standardization. By the late 1920s, Gherardi's faith in engineering cooperation, combined with his long-standing interest in technical standardization, led him to become deeply involved with the activities of the AESC. One can imagine Gherardi nodding his head in approval the first time he came across P. G. Agnew's epigraph in the AESC *Year Book*: "Standardization is dynamic, not static; it means not to stand still but to move forward together."

At first, the Bell System participated in the AESC in a very limited way. It did not contribute to any AESC projects until 1921, when it sent an engineer to only one committee, Symbols for Electrical Equipment of Buildings and Ships.[57] Its involvement with the AESC began in earnest in 1922, when the Bell Telephone System formed the Telephone Group (together with its nominal partner, the United States Independent Telephone Association) and became

[55] Bancroft Gherardi, "Discussion at Pacific Coast Convention," *Transactions of the American Institute of Electrical Engineers* 47 (1928): 50; Laurence Jay Corbett, *Inductive Coordination of Electric Power and Communication Circuits* (San Francisco: The J. H. Neblett Pressroom, Ltd., 1936), xiii. See also "Symposium on Coordination of Power and Telephone Plant," *Transactions of the American Institute of Electrical Engineers* 50 (1931): 437–478.

[56] The engineering records of the Pacific Telephone and Telegraph Co. contain accounts of inductive interference problems much later in the century. Collection 3 Pacific Telesis Group, RG 5 Predecessors, Box 100, AT&T Archives, San Antonio.

[57] *Work of the American Engineering Standards Committee (Year Book)* (New York: American Engineering Standards Committee, 1921), 20, 25.

a dues-paying Member Body of the AESC.[58] By the end of 1927, dozens of Bell System engineers were involved in the work of twenty-one AESC sectional committees such as the National Electrical Safety Code Committee as well as committees that created standards for manhole frames and covers, tubular steel poles, methods for testing wood, direct-current rotating machines, induction motors and machines, and drafting room drawings.[59]

Each of these projects dealt with technologies that lay at the boundaries between the telephone business and other industries.[60] Each was important (or in some cases vital) for the operation of the Bell System, but, unlike standards for telephone transmission or equipment, they were not subject to AT&T's monopoly control. They all represented technologies that Bell System managers chose not to develop within the Bell System corporate hierarchy; rather, they decided to purchase these things (manholes, wood poles, electrical machines, and so on) from outside suppliers. Such managerial decisions created a need for modes of standardization that did not depend on hierarchical coercion.

Leaders of standardization activities in AT&T and the AESC initially kept a respectful if not wary distance. The AESC's leaders took care not to tread on AT&T's turf, as illustrated by the name of an AESC committee responsible for standards for insulated wires and cables: Wires and Cables, Insulated (Other Than Telephone and Telegraph). By the same token, I have found no evidence that AT&T submitted any of its specifications, D&R Bulletins, or Bell System standards to the AESC for approval as American standards, suggesting that the core functions of telephone transmission and equipment were safely contained within the boundaries of the monopoly Bell System.[61]

---

[58] *American Engineering Standards Committee Year Book* (New York: American Engineering Standards Committee, 1924), 17. AT&T engineers, consistent with their company's commanding technical and business position, were far more active: for example, in 1927, AT&T sent engineers to twenty-one committees; USITA engineers participated in nine. *American Engineering Standards Committee Year Book* (New York: American Engineering Standards Committee, 1928), 57, 64.

[59] *American Engineering Standards Committee Year Book* (New York: American Engineering Standards Committee, 1923).

[60] Thomas F. Gieryn, "Boundary-Work and the Demarcation of Science from Non-Science: Strains and Interests in Professional Ideologies of Scientists," *American Sociological Review* 48 (1983): 781–795; Susan Leigh Star and James R. Griesemer, "Institutional Ecology, 'Translations' and Boundary Objects: Amateurs and Professionals in Berkeley's Museum of Vertebrate Zoology, 1907–39," *Social Studies of Science* 19 (1989): 387–420; Etienne Wenger, *Communities of Practice: Learning, Meaning, and Identity* (New York: Cambridge University Press, 1998).

[61] *AESC Year Book* (1928), 39. For an example of AT&T Specifications that overlapped AESC work, see American Telephone and Telegraph Company, Department of Development and Research, "Specifications for Machine Bolts," Specifications No. 4423, March 31, 1925, Collection 6 AT&T Corp., RG 4 Corporate Functions, Box 10, AT&T Archives, San Antonio; and American Telephone and Telegraph Company, Department of Development and Research, "Specifications for Chestnut Poles," Specifications No. 4497, July 3, 1925, Collection 6 AT&T Corp., RG 4 Corporate Functions, Box 10, AT&T Archives, San Antonio.

Gherardi's personal involvement with the AESC grew as the organization reached a turning point in 1928. In response to increasing amounts of interest from all aspects of industry – not just engineers – the AESC made fundamental changes to its structure and process and reconstituted itself as the American Standards Association in July 1928. As we saw in Chapter 3, most of the organization's reforms made it more welcoming and efficient for industry representatives of all stripes – passing control, as the *New York Times* noted blandly, from engineers and scientists to "the executives of railroad, public utility companies and industrial concerns." In the reconstituted organization, engineers and scientists retained a smaller sphere of influence in the ASA Standards Council, while the industry executives formed a board of directors that assumed responsibility for the ASA's financial administration. Gherardi was a member of the board from 1929 to 1935 and also played a key role in the ASA Underwriters' Fund, which raised hundreds of thousands of dollars for ASA coffers by soliciting direct contributions from industrial firms.[62]

Gherardi's importance to the ASA – and the ASA's importance to Gherardi – was further underscored by his election as ASA president for the years 1931 and 1932.[63] Despite the potentially crippling effects of economic depression, Gherardi could boast by the end of his term in 1932 that the ASA consensus-driven standards process was alive and robust. During 1932, 2,700 individuals from 570 technical, trade, and government bodies were involved in ASA projects – more people than ever. Ten days after his term as ASA president was up, the AIEE awarded its prestigious Edison Medal to Gherardi, "for his contributions to the art of telephone engineering and the development of electrical communication."[64]

As Gherardi presided over the growth of the ASA, he also learned to use ASA committees to fulfill his duties as AT&T chief engineer. A close look at AT&T's extended efforts to revise a single, seemingly mundane standard for lock washers illustrates how the company's engineers used the industry standards process to leverage their status and power and extend their technical jurisdiction beyond the boundaries of the monopoly Bell System. One could just as easily examine other, more financially significant, technological and business decisions of the era – AT&T's entry into radio, television, or film, for example – that challenged Bell System managers to think across the boundaries

[62] "Executives to Direct Standards Body," *New York Times* (July 8, 1929), 36; *American Standards Year Book* (New York: American Standards Association, 1929), 7; "Plan to Enlarge Standards Work," *New York Times* (January 5, 1930), N21; William J. Serrill, "President's Report," *American Standards Year Book* (New York: American Standards Association, 1930), 9–10; "Milestones of the ASA," *Industrial Standardization* (1943): 330.

[63] "Gherardi Heads Standards Group," *New York Times* (December 12, 1930), 17.

[64] "Group Hears of Gain in Standards Work," *New York Times* (December 1, 1932), 38; "Industrial Standardization," *Wall Street Journal* (December 2, 1932), 2; "Bancroft Gherardi Wins Edison Medal," *New York Times* (December 12, 1932), 11; "Award of the Edison Medal to Bancroft Gherardi," *Science*, New Series, 76 (1932): 562.

FIGURE 4.1 Bancroft Gherardi, circa 1938. Gherardi's biographer, Oliver Buckley, wrote that those "close to him found beneath a somewhat austere exterior a warm, good-natured friend with a lively sense of humor and deep-seated human kindness." Buckley, "Bancroft Gherardi," 172.

Source: AT&T.

Courtesy of AT&T Archives and History Center.

of their monopoly.[65] By focusing on a mundane standard that was crucial for the system as it already existed, however, we can see how Gherardi's standardization strategy permeated the Bell System all the way down to what may appear to be the simplest matters of telephone design and everyday use.

To understand why AT&T engineers thought the standardization of lock washers could help solve a critical system problem, it is necessary to take a slight excursion and consider some of the history of coin-operated telephones. The first coin-operated telephone was invented in 1888, but Bell companies did not adopt them immediately on a large scale. When they first appeared in the 1890s, they were well suited for two different purposes: for convenient on-the-go calls in busy public areas, and for residential customers or shops – particularly in Chicago and San Francisco – who preferred the option to pay on a per-call basis instead of a more expensive monthly subscription. Chicago Telephone general manager Angus Hibbard installed nearly 40,000 "nickel-in-the-slot" phones between 1900 and 1906, as part of his strategy to cultivate the "telephone habit" among residents of the city.[66]

[65] See for example Reich, *The Making of American Industrial Research*; and Hugh R. Slotten, *Radio and Television Regulation: Broadcast Technology in the United States, 1920–1960* (Baltimore: The Johns Hopkins University Press, 2000).

[66] Fagen, *Engineering and Science in the Bell System*, 153–156, 160–162, 170–171; John, *Network Nation*, 290–298.

Chicagoans of modest means embraced the nickel-in-the-slots, but this innovation, like all innovations, also prompted complaints and resistance. Owners of drugstores – who charged customers ten cents to use the drug-store telephones – objected to the decline in foot traffic that came when their customers could make the same call for five cents less. Business subscribers complained that the increase in telephone traffic added congestion to the net-work, making it less useful as a tool for the commercial elite. Executives in upper echelons of the Bell System – including Carty in New York and Western Electric president Harry Thayer in Chicago – also disliked the irregularity and novelty of Hibbard's innovation: the five-cent rate was too low, the failure rate was too high, and the inconvenience to business subscribers was incompatible with their strategy to capture business customers and ignore the masses.[67]

Nevertheless, managers in Bell operating companies around the country followed Hibbard's lead during the next decades and installed nickel-in-the-slots in order to meet rising demand for low-cost public telephony. As Bell System executives grudgingly accepted coin-operated telephones, Bell System engineers slowly became aware of a new problem for the system: coin slots could be tricked. Instead of inserting nickels or dimes, some customers used metal objects – known as slugs – that were a similar size and weight to legal coins. Some local operating companies grew to tolerate and even encourage the practice. In 1909, Hibbard's company in Chicago published a brochure called "Nickel Prepayment Telephone Service" that implicitly approved "coins or tokens" to be used "instead of nickels," so long as the telephone subscriber paid the phone company collector five cents for every slug. Eventually, the Chicago Telephone Company (renamed Illinois Bell Telephone Company in 1920) contracted with Henry Goetz, a Chicago druggist, to manufacture slugs and escutcheons, devices that would block the coin slot and ensure that only the legitimate slugs would fit into the slot. Goetz's entrepreneurial venture, the Yale Slot & Slug Co. of Chicago, took out several patents on slug designs on its way to becoming a successful niche business for many years.[68]

Despite the flexible and entrepreneurial approach of Chicagoans Hibbard and Goetz, slugs attracted the scrutiny of Bell System officials because they hurt the bottom line. For example, one 1927 report suggested that in Detroit alone, more than 15,000 slugs were found in coin-operated phones each month, which translated to $750 in lost revenue.[69] Engineers from Western Electric, AT&T, and the operating companies studied the problem and quickly realized that any exclusively technical solution to the slug problem would be costly and

---

[67] John, *Network Nation*, 301–302.

[68] "The Slug Evil," *The Telephone Magazine* 21 (1901): 261–262; "Use of Telephone Slugs in Chicago," *The Telephone Magazine* 25 (1905): 27–28; Gerald Johnson, "Goetz Type Telephone Tokens," *TAMS Journal* (April 1978): 48–53; Chicago Telephone Company, "Nickel Prepayment Telephone Service," 1909, Illinois Bell Subject Files, Public Telephones – Slugs, 1944–1947, AT&T Archives, San Antonio.

[69] E. M. Gladden to L. B. Wilson, July 11, 1927, Location 482 07 03 08, "American Standards Association Committee on Washers, 1927–1934," AT&T Archives, Warren.

excessively difficult to engineer. One possibility they considered was to design coin boxes to use noncircular or octagonal tokens, but this solution would have triggered other substantial system problems, such as increased installation and maintenance costs.[70] Bell System engineers also considered making changes to the slots used to filter and collect nickels, dimes, and quarters, but these channels already were built to meet precise tolerances designed to allow legitimate coins to work. In both cases – the introduction of irregular tokens and the redesign of coin channels in existing telephones – the costs of fixing the slug problem within a system context were prohibitive, and both alternatives were rejected as short-term solutions.[71]

Unable to solve the slug problem through an internal technological fix, AT&T engineers chose to attack the problem by turning to organizations outside the Bell System. Between 1927 and 1938, AT&T cultivated relationships with two communities: private firms active in industry standards committees and government officials who took an interest either in the standardization process or in connections between "the slug racket" and other forms of organized crime. In their efforts with both communities, AT&T's strategy was based on a fascinating – and telling – assumption: it was easier to change the world than it was to change a technology embedded deep within the Bell System.

In 1927, the superintendent of the Michigan Bell Telephone Company alerted AT&T engineers that a significant portion of slugs discovered in coin boxes were in fact washers that were manufactured to conform with a particular industry standard. Many of the slugs that turned up in Bell coin boxes were, from a different perspective, simply standard iron washers (or commemorative coins or tokens) that coincidentally had similar dimensions to nickels, dimes, or quarters.[72] Two of the leading engineering societies in the country – the American Society of Mechanical Engineers and the Society of Automotive Engineers – had separately published these washer standards in the early 1920s. Beginning in 1926, these two groups combined efforts under the auspices of the American Standards Association, and formed ASA Sectional Committee B27, Standardization of Plain and Lock Washers. Gherardi saw the committee as an opportunity to eliminate the offending sizes of washers that were being used as slugs, and sent a senior equipment engineer, George K. Thompson, to participate on Committee B27 beginning in late 1927. Thompson was a good choice: he had worked with coin boxes

---

[70] George K. Thompson to C. J. Davidson, October 11, 1927, Location 482 07 03 08, "American Standards Association Committee on Washers, 1927–1934," AT&T Archives, Warren.

[71] [AT&T Outside Plant Development Engineer] to L. F. Morehouse, November 7, 1932, Location 482 07 03 08, "American Standards Association Committee on Washers, 1927–1934," AT&T Archives, Warren.

[72] E. M. Gladden to L. B. Wilson, July 11, 1927, Location 482 07 03 08, "American Standards Association Committee on Washers, 1927–1934," AT&T Archives, Warren.

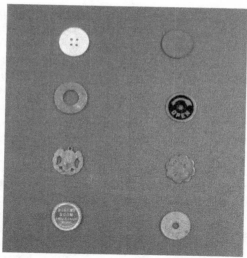

FIGURE 4.2 Slugs recovered from Bell System coin boxes.
Source: AT&T.
Photograph by the author.

for more than thirty years, ever since he filed the first Bell patent for coin telephones in 1895.[73]

The pace of work in the ASA committee was slow – so slow that when Thompson retired in 1930, it had not even published a draft of the revised washer standards. When he retired, Thompson left the AT&T washer standards campaign in the hands of Eliot W. Niles, an engineer in the Department of Development and Research. By early 1931, progress seemed imminent: the B27 Committee had prepared a tentative standard with revised dimensions for lock washers. In June 1931, however, the ASA Standards Council reviewed the committee's work and discovered a violation of ASA rules that caused further delay. The problem was that the ASA Procedure required sectional committees to have an even representation of producers and consumers – in this case, manufacturers and buyers of washers. With eighteen committee members designated as consumers and only eleven designated as producers, B27's membership failed to meet the ASA's procedural standard for balance. It took the committee another full year to canvass existing members for manufacturers who might be interested, convince six of these to join the committee, and obtain the ASA's

---

[73] George K. Thompson to C. J. Davidson, October 11, 1927, AT&T Archives, Warren; F. J. Schlink to George K. Thompson, November 19, 1927, Location 482 07 03 08, "American Standards Association Committee on Washers, 1927–1934," AT&T Archives, Warren; George K. Thompson to W. F. Hosford, December 20, 1928, Location 482 07 03 08, "American Standards Association Committee on Washers, 1927–1934," AT&T Archives, Warren.

approval for this change. After these new members were approved, they needed several additional months to review the proposed specifications.[74]

As the voluntary consensus standardization process plodded along, AT&T utilized a second, more aggressive tactic to recruit allies among other industrial firms. Thompson and Niles were eager to learn of companies that manufactured brass tags, commemorative coins, or nonstandard washers that could be used as slugs, and AT&T executives were not shy about dispatching company representatives to warn these companies about the damage their products were causing. Their approach worked well with small companies, but larger manufacturers or Bell System suppliers – such as Bethlehem Steel – were less easily persuaded (or intimidated) by letters, calls, or even visits from AT&T representatives.[75]

In 1933, a full six years after AT&T first identified the standard washers that were being used as slugs, AT&T officials finally found a strategy that helped them bring the work of Committee B27 to completion. On discovering that washer dimensions specified in an Air Corps standard contained the same specifications as some of the offending slugs, they pressured Harry H. Woodring, the assistant secretary of war, to support a new standard. Woodring, spurred to action by letters and meetings with Niles and A. E. Van Hagen (an AT&T official based in Washington), persuaded the Army-Navy Standards Board to back the changes favored by AT&T. Woodring's intervention seems to have sparked a final surge of support that culminated in the publication of the revised washer specification as an ASA-approved American Standard in 1934.[76]

Yet the long-awaited victory was bittersweet. By itself, the new standard – a significant technical, organizational, and political achievement that took seven years to complete – was not a wholesale solution to the slug problem. ASA standards were used only on a voluntary basis, and the ASA, by design, had no authority to enforce compliance with its standards. Even though AT&T had

74 C. B. LePage to E. W. Niles, June 9, 1931, Location 482 07 03 08, "American Standards Association Committee on Washers, 1927–1934," AT&T Archives, Warren; American Standards Association, "American Tentative Standard – Lock Washers," November 1931; C. B. LePage to P. G. Agnew, July 12, 1932, Location 482 07 03 08, "American Standards Association Committee on Washers, 1927–1934," AT&T Archives, Warren.

75 Correspondence in the AT&T Archives reveals at least three firms that cooperated with or submitted to AT&T's direct approach: the Rome Brass and Stamping Company of Rome, New York; the Dennison Manufacturing Company of Framingham, Massachusetts; and Patterson Brothers of Park Row, New York City. E. M. Gladden to L. B. Wilson, July 11, 1927, Location 482 07 03 08, "American Standards Association Committee on Washers, 1927–1934," AT&T Archives, Warren; "Fraudulent Use of Slugs in Coin Box Telephones (Confidential)," October 9, 1933, Location 482 07 03 08, "American Standards Association Committee on Washers, 1927–1934," AT&T Archives, Warren; [AT&T Outside Plant Development Engineer] to L. F. Morehouse, November 7, 1932, Location 482 07 03 08, "American Standards Association Committee on Washers, 1927–1934," AT&T Archives, Warren.

76 Harry H. Woodring to A. E. Van Hagen, October 13, 1933, Location 482 07 03 08, "American Standards Association Committee on Washers, 1927–1934," AT&T Archives, Warren; E. W. Niles to Z. Z. Hugus, December 22, 1933, Location 482 07 03 08, "American Standards Association Committee on Washers, 1927–1934," AT&T Archives, Warren.

spent the past seven years using the standardization process to build a strong network of partners through the standardization process, this alliance could not protect the Bell System from those elements of American industrial society that did not want to adhere to the consensus industry standard. The offending standard was eliminated, but the slug problem remained.

Exasperated AT&T executives appealed to regulators and law enforcement officials for their help in stopping the fraudulent manufacture and use of telephone slugs – a political strategy that began to pay dividends in 1936. In February, the New York district attorney arrested three men alleged to be responsible for manufacturing and selling a majority of slugs used to defraud coin-operated boxes used by telephone companies, public utility companies, and restaurants. As the arrests were announced, a representative from New York Telephone took advantage of the publicity to disclose the extent of the slug problem: he reported that, in 1935 alone, New York Telephone recovered 4,277,256 slugs, which amounted to $344,524 in lost revenue. The announcement was a shrewd public relations move, calculated to build a sense of indignation against the "slug racket." Twenty more suspects were arrested in an April 1936 sting, and sixteen of them (including their "spearhead") were convicted by the end of June.[77] Reflecting on these arrests, an outraged editorial in the *Washington Post* asked the public to rise above this "petty racket," and suggested that a cultural standard could succeed where a technical standard did not:

Petty rackets in which the public at large is able to participate with slight danger of detection are not so easy to control. They constantly crop up in one form or another. The ultimate hope of exterminating them lies in *elevating standards of personal conduct* through education in the home and schools.... For immediate relief from mass pilfering a great deal can be done by unrelenting pursuit of the individuals who earn a living by encouraging such practices.[78]

Buoyed by public support for police action against the slug racket, AT&T and the regional Bell associated companies pressed state regulators around the country to pass laws making the use of telephone slugs a crime punishable by fine, imprisonment, or both. In December 1937, the *Washington Post* reported the first arrest under the District of Columbia's new law prohibiting the use of telephone slugs. The article concluded by noting the financial benefits of such laws for the telephone company: "In 38 states where similar laws have been enforced, company officials said losses had 'dropped tremendously.'"[79] Of all the different tactics used by AT&T men since discovering

---

[77] "Third Man Seized in Sale of Slugs," *New York Times* (February 9, 1936), 24; "Merchant Is Guilty in Fake Coin Racket," *New York Times* (June 24, 1936), 19.
[78] "The Slug Racket," *Washington Post* (February 11, 1936), 8 (emphasis added).
[79] "D. C. Property, Telephone Slug Measures Pass," *Washington Post* (April 27, 1937), 15; "Police Accuse Two in Phone 'Slug Racket,'" *Washington Post* (December 3, 1937), 30; "Two Tried in First Phone Slug Case," *The Washington Post* (February 10, 1938), 18.

the slug problem in 1927, this lobbying offensive – a political solution to a technical problem – yielded the best results by far.[80]

"The cycle was finally broken," concluded an internal Illinois Bell history of slugs, "by the passage of an amendment to [federal] criminal code effective April 1, 1944" that outlawed the manufacture or sale of "any token, slug, disk, or other device similar in size and shape to lawful coins of the United States." The *Bell Telephone News* declared that the "Buy a Slug" signs, familiar to all Chicagoans, were coming down. "Before long," the story continued, "slugs which fit over shamrock-shaped escutcheons in coin slots will be museum pieces." Henry Goetz had passed away by 1944, leaving his son, Oscar, in charge of the family's now-illegal slug business. Oscar Goetz reportedly collected up to three million brass slugs, and shipped 23 tons of brass to New Jersey to aid in the American war effort for World War II.[81]

This episode in AT&T's anti-slug campaign illustrates some of the more general features of Gherardi's strategy for industrial standardization. Beginning in the 1920s, AT&T engineers joined dozens of consensus standards committees. Their experiences in these committees were as diverse as the standards they sought to influence. In many of these committees, such as those that set standards for wood poles and acoustic terminology, work proceeded harmoniously.[82] In other cases, such as the battles for control of radio transmission, the standards-setting process became a lightning rod for scientific, technical, and political controversy.[83] Sometimes AT&T participated in more targeted and specific institutions, such as the American Institute of Electrical Engineers, Institute of Radio Engineers, American Society for Testing Materials, and National Electric Light Association; other times it sent executives and engineers to larger and more bureaucratic bodies such as the ASA and the International Electrotechnical Commission.[84] AT&T's motives also varied along three strategic axes: the pursuit of reliable and efficient telephone transmission, the enhancement of individual reputation and professional status, and the drive to exercise power and bend American industrial society to its will.

Amid this variety, Gherardi's engineers effectively learned a valuable overarching lesson: they could use industry standards committees to solve critical

---

[80] Nevertheless, slugs – and their links to organized crime – remained common for several decades. See "7 Indicted in Slug Racket," *New York Times* (May 6, 1941), 23; and "Slug Dropped in Phone Box Leads to Mobster's Arrest," *Hartford Courant* (November 21, 1952).

[81] "Those 'Buy a Slug' Signs in Chicago Are Coming Down," *Bell Telephone News* (August/September 1944): 10; Illinois Bell Subject Files, Public Telephones – Slugs, 1901–1944, AT&T Archives, San Antonio; Johnson, "Goetz Type Telephone Tokens."

[82] Location 484 04 04 02, "ASA Sectional Committee on Wood Poles," AT&T Archives, Warren; and Location 419 01 02 16, "ASA Committee Z24 on Acoustic Terminology, 1932–1938," AT&T Archives, Warren.

[83] Reich, *The Making of American Industrial Research*, 170–238; and Slotten, *Radio and Television Regulation*.

[84] Harold S. Osborne, "Standardization in the Bell System – II," *Bell Telephone Quarterly* 8 (1929): 150–151.

problems with the telephone system that AT&T could not solve on its own. The standardization process could be painfully slow over the short term, but over the long term it had tremendous – albeit imprecise and somewhat unpredictable – strategic potential for dozens of firms like AT&T. Gherardi's engineers, like their colleagues throughout American industry, excelled in leveraging standards committees to defend their existing interests. Eventually, they also learned that standards – if they were published and maintained in a more open and public process – also could undermine proprietary interests and hierarchical control.

Conclusions

By design, the architects of the modern Bell System did not pursue a breakneck pace of development in the telephone art. AT&T managers since the 1890s tempered their company's utilization of radical or disruptive innovations in favor of a more incremental approach. The most influential engineers of the early Bell System, including Edward Hall, John Carty, Frank Jewett, and particularly Bancroft Gherardi, used standardization as the centerpiece of this incremental approach to maintaining and improving the Bell System. In this respect, the engineer-executives of the early Bell System applied lessons that they learned from nineteenth-century industries, such as railroads and telegraph, that generated and captured network effects. In each of these cases, engineers learned that standards were essential for ensuring that their contrivances would perform well – especially under conditions of stress and duress, when reliability and safety mattered most.

At the same time, AT&T engineer-executives appreciated that important technological developments would occur outside their jurisdiction, either in the shape of competition, regulation, or technological developments from ostensibly unrelated sectors. Under Gherardi's tenure as chief engineer, AT&T tried to keep up with outside developments by participating in industry standards committees, such as those that met under the auspices of the American Committee on Inductive Coordination and the American Standards Association. In these strategic decisions, we can see Gherardi and his colleagues wrestling with practical questions that we now, with the wisdom of hindsight, can see in theoretical terms: When faced with diverse and complex purchasing needs, what did managers decide to purchase through market relations, what did they decide to build within corporate hierarchies, and what arrangements – hybrids of markets and hierarchies – did they pursue?

Gherardi's organizational responses to the two central problems of the "engineering of the present" – standardization within the Bell System and standardization across the boundaries of the Bell System – were, over the short term and medium term, extraordinarily successful. They generated new economies of scale and scope for the Bell System, and helped engineers extend their control over aspects of the telephone business that had proved to be elusive. Over the long term, however, Gherardi's solutions worked too well and contributed to the Bell System's ultimate demise. In this sense, there is a deep irony that the thorough program of standardization in the Bell System attracted the attention of American regulators in the Department of Justice and Federal Communications Commission beginning in the 1930s. As we will see in Chapter 5, these regulators believed that Bell System executives and engineers had used their monopoly control to stifle innovation illegally and to inflate the cost of telephone service – the very outcome that Jewett anticipated in 1915. The Bell ideology of standardization had indeed sacrificed innovation and competition for stability and reliability – such were the opportunities of the particular economic and political contexts in which the Bell System was created, justified, and maintained. When these contexts changed over time, however – and when the technological foundations of communication networks also changed – Americans renewed their skepticism toward hierarchical control.

5

Critiques of Centralized Control, 1930s–1970s

> Standardization provided a stamp of approval and a level of acceptance and stability. The characteristics of the language would no longer be subject to the whims of a single organization. There would be an industry-wide voice in the definition of features and the timing of their introduction.
>
> – Martin Greenfield, "History of FORTRAN Standardization," 1982[1]

By the middle decades of the twentieth century, the architects and engineers of communication networks understood that standardization provided a powerful strategy and tool kit for control. In the early twentieth century, AT&T chief engineer Bancroft Gherardi and his colleagues at the top of the Bell System hierarchy nurtured a corporate culture – and an ideology of standardization – that privileged stability and caution over radical technological or organizational change. The question of control, as we have seen, was at the core of tensions between telephone engineers in local Bell operating companies and AT&T executives in New York. Even as Gherardi and his fellow executives struggled mightily to centralize control, they faced a wide range of recalcitrant obstacles, such as resistance from the Bell operating companies and competing firms, adversarial state and federal regulators, and rapid changes in the scientific and technological foundations of their industry.

From the 1930s to the 1970s, critiques of AT&T's style of centralized control arose from a variety of sources in American society. They are noteworthy as critiques – and not mere criticisms – because they were more elaborate than ordinary gripes about high rates or other aspects of telephone service. They did not only criticize the status quo; these critiques also began to build alternatives to the status quo that would take power out of the hands of Bell System

---

[1] Martin H. Greenfield, "History of FORTRAN Standardization," *Proceedings of the 1982 National Computer Conference* (Arlington, Virginia: AFIPS Press, 1982), 819–824.

employees and put it into the hands of a more diverse group of engineers, regulators, and users.

A foundational critique of centralized control – a regulatory critique – came of age during the 1930s, when the Federal Communications Commission (FCC) developed and began to exercise its capabilities to investigate AT&T's telephone monopoly. In the 1940s and 1950s, antitrust settlements by the U.S. Justice Department required AT&T to license crucial patents and prohibited AT&T from engaging in any business outside of its regulated telephone services. The outcome of these actions achieved the policy makers' intentions: they decentralized control over the development of new communication technologies and laid a path toward a future based on competition, innovation, and entrepreneurship. In the 1960s, policy entrepreneurs in the FCC, led by Common Carrier Bureau chief Bernard Strassburg, initiated proceedings that ultimately prevented AT&T from leveraging its monopoly power to dominate new markets. In the place of AT&T's traditional dominance, the FCC published new rules and new standards for "customer-premises equipment" such as computer modems and thus undermined AT&T's ability to influence the direction of the fledgling data communications industry. At the same time, the FCC signaled its reliance on the expert recommendations of AT&T's competitors and engineers in industry and government committees.

The critiques of AT&T's centralized control that arose from the 1940s to the 1960s had many common features with the discourses of control in another important industry – electronic computers. IBM created and consolidated its dominant position in domestic and international markets during this era. Like AT&T, its success raised suspicions and inspired critiques from engineers in competing firms and regulators in Washington. In both industries, antitrust proceedings constrained the actions of the firms and yielded to an increasingly strong entrepreneurial push to develop new digital and wireless technologies and provide new services.

Two features of the computer industry, however, meant that critiques of IBM's style of centralized control differed from critiques of AT&T. First, the American federal government lacked the same authority, desire, and capabilities to regulate the computer industry that it had for the telephone industry. The FCC in particular had no statutory authority to regulate the computer industry or to define technical specifications. Second, computer technology during this period inspired a committed user base to articulate and act on their desire to control the future of computing. Communities of engineers and programmers developed discourses of innovation, competition, liberation, and freedom that brought an individualistic and antihierarchical edge to their work. They envisioned new possibilities for opposing the "closed world" ideology of computers that aided centralized command and control. Their opposition prompted them to build new tools and new institutions to sustain their own shared visions of decentralized control and open competition.

During the 1970s, AT&T's critics found a natural affinity with the critiques of centralized control that were inspired by IBM. By the end of the 1970s, critiques of IBM and AT&T, despite their divergent origins and targets, converged on a common coordinating mechanism for industrywide standards: a loose network of technical committees that operated beyond the reach of any dominant company. It was these critiques, rather than any qualities of decentralization that are presumed to be inherent in digital networks, that structured the realm of ideological and practical possibilities for the convergence of telecommunications and computing technologies.

The Roles of Government in American Communications, 1930s–1970s

As AT&T engineers in the 1920s and 1930s learned to work within industry standards committees, AT&T executives learned to mobilize different sources of government authority – including military procurement officers, state regulators, and law enforcement officials – to defend the interests of the Bell System. Despite problems within the system (such as the protracted transition to machine switching we saw in Chapter 4) and beyond the system (such as the prolonged economic depression in the 1930s), by the end of the 1930s, the Bell System was in good health – "sitting pretty" in the assessment of historian John Brooks. At the same time, however, new challenges to the technological and legal foundations of centralized control emerged. The reemergence of antimonopoly sentiment in the 1930s prompted government officials to renew their efforts to regulate the Bell System. In the Communications Act of 1934, Congress created a new independent regulatory body – the Federal Communications Commission – with the authority to regulate commercial services that provided communication by radio and wire. Like other agencies born during the New Deal, the FCC was formed as a response to the lax oversight of business that New Dealers saw at the heart of the causes of the Great Depression.[2]

The FCC was an administrative solution to a technical problem: the scarcity of the radio spectrum. When too many users tried to send radio signals at the same frequencies in the same areas, all signals would be rendered useless by interference. The FCC, acting on the basis of the "scarcity rationale," developed a complex licensing scheme in which it allocated specific portions of the spectrum for specific uses and assigned licenses to private parties who could exclusively use those portions of spectrum. For example, the FCC assigned

[2] John Brooks, *Telephone: The First Hundred Years* (New York: Harper & Row, 1976), 206; Fred W. Henck and Bernard Strassburg, *A Slippery Slope: The Long Road to the Breakup of AT&T* (New York: Greenwood Press, 1988), 1–7; Alan Brinkley, "The New Deal and the Idea of the State," in Steve Fraser and Gary Gerstle, *The Rise and Fall of the New Deal Order, 1930–1980* (Princeton, NJ: Princeton University Press, 1990); Ellis W. Hawley, *The New Deal and the Problem of Monopoly: A Study in Economic Ambivalence* (Princeton, NJ: Princeton University Press, 1966).

licenses to television broadcasters to use the spectrum in the 512–608 MHz band only for broadcast television. FCC regulations prohibited the broadcasters (and everyone else) from using the 512–608 band for any other purpose, such as mobile telephony, radio broadcasts, or satellite transmission. This "command and control" model of licensing, in the apt summary of a later FCC chairman, had the flaw of being "an entirely reactive and too easily politicized process." As a result, the radio industry settled into a stable oligopolistic structure, with the FCC acting less as an adversarial regulator and more as partner in maintaining a corporatist status quo.[3]

Because it was constantly on the verge of being overwhelmed by the industry's political heft and technical prowess, the FCC developed a process for allocating spectrum through comparative hearings in which it would evaluate technical proposals from private firms. Although these hearings – commonly known as "beauty contests" – ostensibly were based on technical criteria, in practice they subjected the FCC to tremendous amounts of pressure from private firms, lobbyists, and elected officials (often acting on behalf of private interests). FCC commissioners and engineers in the 1940s learned they could not avoid the consideration of economic, social, and political factors and struggled to act decisively in the face of political pressure and technological uncertainty. The FCC, desiring to base its decisions on what it deemed to be widespread industry consensus, consulted a variety of industry groups – such as the Institute of Radio Engineers, the Radio Manufacturers Association, and the National Television System Committee – as well as its own staff and ad hoc committees for advice in its spectrum allocation and assignment proceedings. One memorable example of the FCC's difficulty in this role came in its establishment of color TV standards in the 1950s. At first, FCC commissioners decided to mandate a standard that would benefit CBS, only to reverse that decision and mandate a different standard in 1953 – a decision that CBS's chief rival, RCA president David Sarnoff, celebrated by literally jumping for joy in front of photographers.[4]

FCC commissioners encouraged private firms to develop new technologies that would use spectrum more efficiently, in part because such innovations

---

[3] Robert W. McChesney, *Telecommunications, Mass Media, and Democracy: The Battle for the Control of U.S. Broadcasting, 1928–1935* (New York: Oxford University Press, 1994), 151–225; Leonard S. Reich, *The Making of Industrial Research: Science and Business at GE and Bell, 1876–1926* (New York: Cambridge University Press, 1985), 218–238; Richard H. K. Vietor, *Contrived Competition: Regulation and Deregulation in America* (Cambridge, MA: Belknap Press, 1994), 176–178; Michael K. Powell, "Broadband Migration III: New Directions in Wireless Policy," October 30, 2002, http://www.fcc.gov/Speeches/Powell/2002/spmkp212.html (accessed September 25, 2013).

[4] Hugh R. Slotten, *Radio and Television Regulation: Broadcast Technology in the United States, 1920–1960* (Baltimore: The Johns Hopkins University Press, 2000), 168–231; Donald G. Fink, "Perspectives on Television: The Role Played by the Two NTSC's in Preparing Television Service for the American Public," *Proceedings of the IEEE* 64 (1976): 1322–1331.

would relieve pressure on the FCC and allow it to take a more passive – and less controversial – approach. To that end, the FCC experimented with limited forms of competition in new services, including mobile telephony. In 1949, it established two separate sets of land mobile frequencies in every geographical region, one for firms with existing wire line operations (such as AT&T) and a second for a group of hundreds of smaller firms classified as Radio Common Carriers. As the mobile industry grew during the next thirty years and its need for spectrum intensified, lobbyists tried to persuade the FCC to reallocate UHF spectrum from television broadcasters. Predictably, the broadcasters resisted and left FCC commissioners and staff with the unenviable task of mediating the demands of both industries.[5]

The FCC was more timid, at first, to intervene with the operations of the Bell System based on the authority granted by Title II of the 1934 act. Its initial approach toward the telephone monopoly was an odd mix of menace and deference – a striking contrast from its command and control approach to radio spectrum. The FCC's conflicted stance was especially evident in its investigation of standardization in the Bell System in the mid-1930s. The FCC began its Telephone Investigation in 1935 that culminated four years later with a 661-page *Report on the Investigation of the Telephone Industry*. The authors of the *Report* viewed standardization as a sort of necessary evil: "There is a natural tendency incident to any program of standardization to retard the utilization of technical improvements. On the other hand, the lack of standardization operates to impair uniformity of service when such uniformity is an essential factor in the furnishing of satisfactory service. Such uniformity is an essential of a Nationwide telephone service."[6]

Although it recognized the paramount need for standardization, FCC officials, like other New Dealers, were concerned that the giant monopoly would suppress innovations and harm the nation's economic recovery. The FCC's *Report* featured a sharp critique of control centralized in private hands, based on the assumption that the Bell System's motivations might not be in line with the FCC's own vague mandate to be a steward of the "public interest": "The concentration of authority over so large a utility makes possible the Nationwide dissemination of practices which may be against the public interest, especially where this interest may be at variance with that of the management and stockholders."[7]

---

[5] Dale N. Hatfield, "FCC Regulation of Land Mobile Radio – A Case History," in Leonard Lewin, ed., *Telecommunications: An Interdisciplinary Text* (Dedham, MA: Artech House, 1984), 105–132.

[6] United States Congress, *Report of the Federal Communications Commission on the Investigation of the Telephone Industry in the United States* (Washington, DC: Government Printing Office, 1939), 585. See also N. R. Danielian, *AT&T: The Story of Industrial Conquest* (New York: Vanguard Press, 1939).

[7] U.S. Congress, *Investigation of the Telephone Industry*, 252.

The authors of the *Report* did not define the public interest with any precision, but they carried this theme – the contrast between the public interest and the pursuit of profit – throughout the *Report*, including an entire chapter on Engineering and Standardization. The FCC singled out "centralized control over engineering, standardization, and manufacturing" as an area for concern because it could provide opportunities for the suppression of inventions, the failure to take advantage of outside improvements, and the sale and installation of outdated or inferior equipment by Western Electric to the regional operating companies.[8]

Despite its confrontational tone, the *Report* concurred with AT&T's assessment of three important issues: that AT&T had generated hundreds of millions of dollars in savings through standardization; that standardization had generated substantial social benefits such as the development of long-distance capabilities, service to rural areas, and increases in service quality; and that AT&T's organization of standards enabled both "flexibility in the interchange of equipment and trained personnel" and a "uniformly high quality of service." Consequently, the FCC recommended continued regulatory scrutiny of AT&T's standardization activities – especially those that required large capital investment – but did not contest AT&T's jurisdiction over standardization. "The overall results of standardization by the American Co.," the *Report* concluded, "are such as to justify a continuance of research and standardization."[9] Moreover, the FCC (and the public) did not see equipment and standardization as the most pressing problems with AT&T's telephone monopoly. In the coming years, the sensitive topics of prices and rates received the most attention from the press, from state and federal regulators, from the booming cottage industry of telecommunications economists and lawyers, and from historical accounts of the Bell System's demise.

On balance, AT&T protected itself from drastic regulatory reform in the 1930s and 1940s thanks to a strategy that combined advertising, public relations, and a reputation for high-quality service. This strategy created a positive public image for the giant telephone monopoly. It helped that the company responded heroically in times of national need: its engineers reacted bravely to a devastating hurricane in 1938 and, perhaps more important, AT&T, Bell Labs, and Western Electric made fundamental contributions to American military communications and weaponry during World War II. After the war, officials in the Defense Department depended on Bell executives to manage efforts to build nuclear weapons, providing further confirmation that the Bell

---

[8] U.S. Congress, *Investigation of the Telephone Industry*, 252.

[9] U.S. Congress, *Investigation of the Telephone Industry*, 247–283, 584–585. AT&T estimated savings of $99 million from cable development, $50 million from the use of less expensive metals, and $5 million per year from improvements in switchboard cords.

System was a vital asset for national security as well as for the infrastructure of American communications.[10]

Nevertheless, the good deeds of the Bell System could not protect it from the rekindling of antimonopoly sentiment after World War II. The timid complaints in the 1939 FCC *Report* fed a full-fledged antitrust suit filed in 1949. In the suit, the government repeated a familiar complaint: AT&T and Western Electric had stifled competition and inflated rates through their monopolization of the manufacturing and distribution of telephone equipment. The adversarial edge of the antitrust suit, however, was dulled by conflicts within the Justice Department, which eventually adopted a much more tolerant attitude toward big business after the election of Dwight D. Eisenhower in 1952. Justice officials and AT&T executives pursued an amicable settlement that would keep the relationship between Western Electric and AT&T intact. They ultimately agreed to a Consent Decree that settled the antitrust suit in January 1956.[11]

The 1956 Consent Decree fits comfortably within what David Hart has described as the "deconcentrationist" character of antitrust policy in the New Deal era between the 1930s and 1970s. A significant clause in the Consent Decree prevented both AT&T and Western Electric from engaging in business outside of common carrier telephone service, thus restricting the Bell System's ability to control the equipment, interfaces, and standards being developed in the nascent American electronics and data processing industries. Additionally, by forcing the Bell System to license its patents at reasonable rates, the Consent Decree limited the effectiveness of using patents and technical standards as strategic tools to marginalize competitors – a strategy that had been developed to great effect by earlier generations of Bell executives, most notably Edward Hall and Theodore Vail. AT&T executives claimed victory on the basis of their continued control over the American telephone industry, but the Consent Decree, combined with subsequent proceedings over the next three decades, provided the legal foundations for continued surveillance over the boundaries of AT&T's monopoly. The major thrust of these proceedings was to facilitate competition in markets for customer-premises equipment, including devices such as telephones, network switches, and computer modems that users could attach to the telephone network. As a result, a flood of entrepreneurial competitors entered niche markets that had been dominated by AT&T and Western Electric.[12]

---

[10] Roland Marchand, *Creating the Corporate Soul: The Rise of Public Relations and Corporate Imagery in American Big Business* (Berkeley: University of California Press, 1998), 48, 86; Brooks, *Telephone*, 199–214.

[11] Milton S. Goldberg, *The Consent Decree: Its Formulation and Use* (East Lansing: Michigan State University, 1962), 37–48.

[12] Brooks, *Telephone*, 233–256; Louis Galambos, "Looking for the Boundaries of Technological Determinism: A Brief History of the Telephone System," in Renate Mayntz and Thomas P. Hughes, eds., *The Development of Large Technical Systems* (Westview Press, 1988), 145–148; David M. Hart, "Antitrust and Technological Innovation in the US: Ideas, Institutions, Decisions, and

Entrepreneurs, who had been attempting to create new products and new markets at the boundaries of the Bell System since its formation, renewed their efforts after the end of World War II. An initial attack on AT&T's right to control attachments to the telephone network came in 1948, when the Hush-A-Phone Corporation filed a complaint against AT&T to establish the legal right to attach their cuplike device to telephone handsets. Two acoustic scientists – the Boston entrepreneur Leo Beranek and the Harvard researcher J. C. R. Licklider – testified that the Hush-A-Phone would not endanger the technical capabilities of the Bell System.[13] Despite the testimony of Beranek, Licklider, and other communications experts, the FCC remained steadfast in its support of AT&T's end-to-end control over the American communications system. Rulings by the FCC in 1955 and a federal appeals court in 1956 preserved the AT&T's ability to decide which "foreign attachments" could be permitted, but they added the caveat that such decisions be just, fair, and reasonable. A telephone subscriber, the court ruled, had the right to "use his telephone in ways which are privately beneficial without being publicly detrimental" – although it left the crucial question of "public detriment" to be decided by the FCC, which, in turn, deferred to the judgment of AT&T's experts.[14]

The entrepreneurial challenges that continued throughout the 1960s would not have fared so well without a group of sympathetic regulators in Washington, DC. Staff at the FCC – particularly Commissioner Nicholas Johnson and Common Carrier Bureau chief Bernard Strassburg – used their authority to create a more permissive and fluid context for entrepreneurs who were eager to build systems for computer communication. Strassburg joined the FCC in 1942 and embodied the liberal enthusiasm for regulatory experimentation that was so characteristic of New Deal agencies such as the FCC. In a 1988 interview, Strassburg recalled that "the first awareness that we had of the fact that computers and data processing had something in common with communications started to emerge in early '65." Strassburg recognized that cultural factors might inhibit innovation more than technical or legal factors: "The whole mentality of the industry, as well as the regulator, was predicated on almost a static

Impacts, 1890–2000," *Research Policy* 30 (2001): 923–936; and Vietor, *Contrived Competition*, 184–185; and Gerald W. Brock, *The Telecommunications Industry: the Dynamics of Market Structure* (Cambridge: Harvard University Press, 1981), 187–197. Congress objected to the modesty of AT&T's concessions, and used seventeen days of investigative hearings to "ridicule and embarrass" the Antitrust division for its weakness. Steve Coll, *The Deal of the Century: The Breakup of AT&T* (New York: Simon & Schuster, 1986), 59.

13 Both men would later figure in the development of network technologies: Beranek as one namesake of the firm, Bolt Beranek and Newman, that built equipment for the Arpanet; and Licklider as a critic of AT&T and an inspiration for a generation of packet-switching researchers.

14 *Hush-A-Phone Corp. v. United States*, 238 F.2d 266 (1956); Henck and Strassburg, *A Slippery Slope*, 32–67.

kind of service, with very measured change, controlled change. I don't say that critically, either, because everything worked so well."[15]

Strassburg's first tentative step to change these static mentalities came in 1966, when he proposed that the FCC issue a "Notice of Inquiry on the Regulatory and Policy Problems Presented by the Interdependence of Computer and Communications Services and Facilities." Strassburg hoped that such a public inquiry could "lay the foundation for regulatory policy and other actions, maybe even legislative recommendations that might be necessary." He was right; the proceeding would later be called *Computer I*, the first in a series of three Computer Inquiries that the FCC used to evaluate and regulate the convergence of telephone and computer technologies between the 1960s and the late 1980s.[16]

Strassburg's Notice of Inquiry invited comments from the general public. Predictably, the vast majority of comments came from industry leaders such as AT&T and IBM. Smaller firms in the data processing industry and industry associations such as the Business Equipment Manufacturers' Association also chimed in. Strassburg recalled that the FCC's primary goal was to protect competition in markets for data processing: "We didn't want AT&T or the other telephone companies to so intermix their operations and make a mess of both markets, both the monopoly [telephone] markets and the competitive [computer] markets." AT&T executives agreed, to the extent that they valued clarity and simplicity in the marketplace. Their concern was that the meddling bureaucrats in the FCC would be the ones to make a mess of both markets. They were understandably wary of Strassburg's ambition and of the instability that competition would introduce to AT&T's intricate rate structure and complex technical operations. AT&T vice president and general counsel Horace P. Moulton, speaking frankly in a May 1968 speech in Washington, reminded his audience that the stakes were high: "The accommodation of communications to computers is the subject of one of the most far-reaching inquiries ever undertaken by the FCC.... Here, as in the case of other technological advances, both we in the communications industry and our regulators must display a very high order of adaptability, or progress may kill us all."[17]

The FCC's final decision in *Computer I*, published in 1971, defined a standard of "maximum separation" under which no common carrier (such as

---

[15] Bernard Strassburg, interview by James Pelkey, Washington, D.C., May 3, 1988, courtesy of James Pelkey. See also Bernard Strassburg, "The Marriage of Computers and Communications – Some Regulatory Implications," 9 *Jurimetrics Journal* (September 1968): 12–18; and Henck and Strassburg, *A Slippery Slope*, 126–142.

[16] Strassburg interview. See also *Regulatory and Policy Problems Presented by the Interdependence of Computer and Communications Services and Facilities (Computer I)*, Docket No. 16979, NOI, 7 FCC 2d 11, 13 (1966); and James Pelkey, "Entrepreneurial Capitalism and Innovation: A History of Computer Communications, 1968–1988" (2007), chapter 1.

[17] Strassburg interview; Horace P. Moulton, "Monopoly and Competition Issues Facing the Communications Industries," *Antitrust Bulletin* 13 (1968): 889–897.

AT&T and the Bell operating companies) could enter data processing markets except through an entirely separate corporate entity, complete with different personnel, accounting, equipment, and facilities. In its 1973 decision to uphold the FCC's rules, the Second Circuit Court of Appeals admired the via media carved out by the FCC: "The Commission has avoided here the extreme of totally barring the carrier from data processing directly or indirectly on the one hand, and the regulation of the data processing industry on the other." Technological complexity belied this legalistic clarity, however, and the FCC soon began a second Computer Inquiry to investigate the rapid changes in computing technology and user demand.[18]

In the meantime, a flurry of entrants began to compete in new markets for telephone and data processing equipment as a direct consequence of the FCC's decision in *Computer I*. Dozens of new companies entered the market for modems, devices that enabled the transmission of computer data over telephone lines. Military procurement – here again a notable presence in the charting of new technological directions – had already nourished this small industry in the mid-1960s. AT&T deployed its first commercial modem in 1958, the 300 baud per second (bps) Bell System Data Set 103, but by 1962 Air Force officials requested faster speeds than the 2,400 bps that AT&T had reached in 1962. After looking elsewhere, the Air Force awarded contracts to specialized manufacturers such as Codex and Milgo, who were in better positions to meet the air force's request for modems as fast as 10,000 bps. As FCC decisions protected these firms from the dual threats of AT&T's dominance and federal regulation, they and dozens of other start-up firms raised capital, created new high-tech products, and struggled to learn the nuances of competing in the shadow of regulatory uncertainty, commercial risk, and rapid technological change. These trends were fueled by time-sharing systems in the academic computing research community, particularly at MIT.[19]

Subsequent challenges to AT&T's control over the interfaces of the telephone network followed the same pattern established by the Hush-A-Phone case in the mid-1950s and the Computer Inquiry begun in 1966: an entrepreneur would first introduce a new service or device, then encounter resistance from AT&T, and finally look to the FCC and the courts to challenge AT&T's

---

[18] *Tentative Decision of the Commission, In re Regulatory and Policy Problems Presented by the Interdependence of Computer and Communication Services and Facilities*, 28 F.C.C.2d 291 (1970) (Computer I Tentative Decision); *Final Decision and Order, In re Regulatory and Policy Problems Presented by the Interdependence of Computer and Communication Services and Facilities*, 28 F.C.C. 2d 267 (1971) (Computer I Final Decision), aff'd in part and rev'd in part sub nom. *GTE Service Corp. v. FCC*, 474 F.2d 724 (2d Cir.1973), decision on remand, 40 F.C.C.2d 293 (1973). See also Stuart L. Mathison and Philip M. Walker, "Regulatory and Economic Issues in Computer Communications," *Proceedings of the IEEE* 60 (1972): 1254–1272; and Robert Cannon, "The Legacy of the Federal Communication Commission's Computer Inquiries," *Federal Communications Law Journal* 55 (2003): 167–206.

[19] Pelkey, "Entrepreneurial Capitalism and Innovation," chapter 3.

authority. Lawsuits from Thomas Carter (filed 1965, decided 1968) and Microwave Communications, Inc. (filed 1967, decided 1969) convinced the FCC to sanction legal competitors to AT&T, albeit in small niche markets. The combined force of these small challenges dealt the first of a series of fatal blows to AT&T's dominant position in the American telecommunications industry.

Carter's case was especially important because it disputed AT&T's control over the equipment that customers could attach to the ends of the network, known to regulators as customer-premises equipment. Carter's company manufactured a device – the "Carterfone" – that connected a call between the telephone network and a mobile radio in an oil field or other remote location. AT&T, acting on its legal authority to approve or deny the use of devices connecting to the telephone network, cut off service to subscribers who were using a Carterfone. AT&T executives and engineers justified their actions with a technical rationale: the indiscriminate and uncontrolled use of foreign attachments, they insisted, would disrupt and harm the network. Strassburg and his colleagues in the FCC agreed that new equipment would introduce new risks and uncertainties, but they concluded that AT&T's restrictions were not "just and reasonable." Therefore, they requested in 1968 that AT&T propose new terms for the interconnection of outside devices and equipment. The economic and legal logic of the situation, in other words, outweighed AT&T's unconvincing claims of technical harm. AT&T moved quickly to accommodate the FCC's decision and responded by permitting the use of private mobile systems, data modems, and private exchanges for businesses and other large organizations. But there was a catch: AT&T required any attachments to interconnect through a "protective connecting arrangement" that it would provide. Such arrangements would preserve AT&T's exclusive control over the interfaces of the telephone network, even as it lost control over attachments at the end of the network. The boundaries of its monopoly were shrinking.[20]

In June 1972, the FCC commissioners, provoked by complaints from equipment entrepreneurs and Strassburg's continuing vigilance, created a joint board of state and federal regulators to consider alternatives to AT&T's exclusive control over protective connecting arrangements for foreign attachments. This board began to develop technical criteria and a regime for equipment certification that would not require a protective connecting agreement with AT&T. In a speech in September 1973, John DeButts, AT&T's combative new chairman, valiantly maintained his company's position that it should maintain strict control over the interfaces of the telephone network: "We cannot live with the deterioration of network performance that would be the inevitable consequence of

---

[20] Vietor, *Contrived Competition*, 190–193; Nicholas Johnson, "Carterfone: My Story," *Santa Clara Computer & High Tech Law Journal* 25 (2009): 677–700; Christopher H. Sterling, Phyllis W. Bernt, and Martin B. H. Weiss, *Shaping American Telecommunications: A History of Technology, Policy and Economics* (Mahwah, NJ: Lawrence Erlbaum Associates, 2006), 118–144.

'certification' and the proliferation of customer-provided terminals that would ensue from it."[21]

By this point, however, Strassburg and the FCC staff were in no mood to give in and continued to assert their authority and extend their jurisdiction over technical interfaces. A culmination of sorts came with the 1975 publication of Part 68 of the FCC's rules, which defined technical standards for customer-premises equipment that connected to the network. This slight revision to one of the FCC's rules could readily be mistaken for a mundane bureaucratic act; in retrospect, however, it was a significant institutional innovation that moved jurisdiction over network interfaces away from AT&T's centralized control.[22]

Subsequent antitrust lawsuits in the 1970s and 1980s further eroded AT&T's control over the technical interfaces and the sociotechnical boundaries of the telephone network. The critique of AT&T's centralized control that had been advanced by small competitors continued to gain support within the federal government. Innovation, which had been a peripheral concern of engineers and policy makers in the 1930s, became a central focus. Policy makers in Washington became convinced that new equipment and services were unlikely to emerge from the old system of centralized control.

The deconcentrationists persisted in their campaign to prevent the telephone monopoly from leveraging past successes into dominance in emerging markets. AT&T was named in two antitrust lawsuits in 1974, the first filed by MCI in March and the second filed by the Department of Justice in November. These lawsuits eventually reached a stunning conclusion in the 1982 settlement that broke up the Bell System. Despite the skill and persistence of public relations experts from AT&T and its allies, Washington elites on the political right and left developed a consensus ideology that elevated the virtues of competition and entrepreneurship as the best means for generating the innovations that would, in turn, uphold the public interest. Such an ideology would have been anathema to Theodore Vail, Bancroft Gherardi, and the other architects of the modern Bell System. The new enthusiasm for competition even would have sounded strange to AT&T's nominal adversaries in the FCC, who in 1939 reluctantly justified the continuance of AT&T's centralized control because they believed that "uniformity is an essential of a Nationwide telephone service."[23]

The combined effect of the Telephone Investigation, the Computer Inquiries, the Carterfone case, and numerous antitrust lawsuits between the late 1940s

[21] John DeButts, "An Unusual Obligation," reprinted in Alvin von Auw, *Heritage and Destiny: Reflections on the Bell System in Transition* (New York: Praeger, 1983), 426.

[22] Coll, *Deal of the Century*, 104–111; Henck and Strassburg, *A Slippery Slope*, 126–142; Peter Temin with Louis Galambos, *The Fall of the Bell System: A Study in Prices and Politics* (New York: Cambridge University Press, 1987), 41–47, 63–65.

[23] U.S. Congress, *Investigation of the Telephone Industry*, 585. For a closer look at the political, economic, cultural, and ideological aspects of the Bell System's decline, see Coll, *Deal of the Century*; Henck and Strassburg, *A Slippery Slope*; Temin, *The Fall of the Bell System*; and von Auw, *Heritage and Destiny*.

and the early 1980s – which were, it must be emphasized, reactions to challenges from entrepreneurs who were trying to capitalize on rapid changes in technology – was a sea change in jurisdiction over standards for American communication networks. Oversight of telephone standards and network interfaces moved out of AT&T's proprietary and unilateral control and into a more open and public process where the FCC set technical standards and allowed market entry for new equipment and services. The shift in regulatory focus between the 1930s and 1970s was gradual but, in retrospect, quite striking: where regulators in the 1930s investigated the level of efficiency *within* the Bell System, their successors in the 1960s and 1970s became increasingly concerned with the interfaces that facilitated voice and data communication *across the boundaries* of the Bell System.

Standards for the Computer Industry, 1950s–1970

In some ways, the critiques of AT&T's centralized control of the American telephone industry echoed IBM's dominant position in the international computer industry. During the 1950s and 1960s, IBM products became de facto standards for computer users around the world. American antitrust prosecution constrained IBM's business practices in some significant ways, but no federal agency existed that had the authority to regulate the computer industry in the same way that the FCC regulated the radio, television, and telephone industries. In the absence of an overbearing federal presence, IBM and the myriad other participants in the American and international computer industry competed to define basic standards for their complex and rapidly maturing products. Market competition spilled over into new industry committees that were created in the early 1960s in the hopes that they could define common standards through collaboration and consensus.

IBM and Its Critics

From the late 1940s through the 1950s, IBM manufactured different electronic computers – six models in total in 1960 – that could be programmed to meet rising demand from almost every type of user in scientific, government, military, and business settings. Contracts with the federal government included IBM's work to build the military's Semi-Automatic Ground Environment (SAGE) in the late 1950s and early 1960s.[24] However, incompatibilities among the six different models created tensions between divisions within the firm as well as excessive diversity in components. In the early 1960s, IBM executives responded to these problems with a strategy that replaced all existing models

[24] On SAGE, see Paul Edwards, *The Closed World: Computers and the Politics of Discourse in Cold War America* (Cambridge, MA: The MIT Press, 1997), 75–112.

of IBM machines by centralizing design and production around a single family of computers: System/360.[25]

The chief architectural innovation of System/360 was its interchangeable components, including peripherals (such as storage devices, printers, and terminals) and software that customers could mix and match in order to meet their individual needs. From the standpoint of IBM executives and managers, the modular design of System/360 streamlined the design and manufacturing processes, which generated new economies of scale and scope through the reduction of variety and incompatibility. Standardization within the firm led to unprecedented success for IBM in the marketplace: the company received more than a thousand orders within a month of the System/360 announcement in April 1964, and IBM's gross income more than doubled between 1965 and 1970. As a result, System/360 became more than a series of compatible computer components; it became a platform and a new international de facto standard that enabled IBM to reassert global dominance of the computer industry.[26]

When seen as episodes in the history of system architecture and standardization, there are several similarities between IBM's rise to dominance in the decades after World War II and AT&T's successful monopolization of the telephone industry. Just as Edward Hall and Theodore Vail devised AT&T's corporate strategy in order to control the terms of telephone interconnection, so too did IBM system architects Gene Amdahl and Fred Brooks design the System/360 so that IBM could define and control the standardization of interfaces among computer components. In both cases, system architects used standardization as a strategy when their respective companies faced organizational and technological complexity. In addition, the suspicious gaze of antitrust regulators shaped the decisions that system architects made about the creation of compatible interfaces.[27] Their ability to set standards generated a crucial set of organizational capabilities for both firms: managers used standards to simplify

---

[25] Alfred D. Chandler, *Inventing the Electronic Century: The Epic Story of the Consumer Electronics and Computer Industries* (New York: The Free Press, 2001), 7–9, 85–176; Emerson W. Pugh and William Aspray, "Creating the Computer Industry," *IEEE Annals of the History of Computing* 18 (1996): 7–17.

[26] Paul E. Ceruzzi, *A History of Modern Computing* (Cambridge, MA: The MIT Press, 1998), 145; Kenneth Flamm, *Creating the Computer: Government, Industry, and High Technology* (Washington: The Brookings Institutions, 1988), 96–102; Emerson W. Pugh, *Memories That Shaped an Industry: Decisions Leading to IBM System/360* (Cambridge, MA: The MIT Press, 1984); Bob O. Evans, "System/360: A Retrospective View," *IEEE Annals of the History of Computing* 8 (1986): 155–179; Steven W. Usselman, "Unbundling IBM: Antitrust and Incentives to Innovation in American Computing," in Sally H. Clarke, Naomi R. Lamoreaux, and Steven W. Usselman, eds., *The Challenge of Remaining Innovative: Insights from Twentieth-Century American Business* (Stanford, CA: Stanford University Press, 2009): 249–280; Andrew L. Russell, "Modularity: An Interdisciplinary History of an Ordering Concept," *Information & Culture: A Journal of History* 47 (2012): 257–287.

[27] Usselman, "Unbundling IBM."

coordination within the firm while simultaneously creating higher barriers to entry for competing firms. Centralized control over system architecture and interfaces reinforced and extended their dominant positions – and also earned continued scrutiny from antitrust regulators.

IBM's success in centralizing control also inspired opposition from a variety of sources. The first is what we might think of as a "competitive critique" that private firms articulated, primarily by building new products that differentiated them from IBM. Domestic competitors such as Sperry Rand, National Cash Register, Burroughs, Honeywell, RCA, and General Electric responded to IBM's dominance by introducing their own proprietary computer systems, but did not find substantial market success through imitation. Instead, IBM's competitors pursued a strategy of opening new market segments or looking to fill unmet demand at the margins of IBM's customer base. Two of the notable successes in this era, Control Data Corporation and Digital Electronics Corporation ("Digital"), survived by focusing on niche markets. Control Data found some success in the mid- to late 1960s with a strategy that focused on making high-performance supercomputers for scientific uses at the high end of the market. Digital, on the other hand, created a line of minicomputers in the early 1960s that appealed to users who wanted to modify and experiment with machines for their own specialized purposes. In stark contrast to IBM's policy of leasing computers to customers, Digital sold their computers outright and provided customers with detailed specifications for tinkering with the machine.[28]

Digital's machines, particularly the PDP series, fed a growing appetite among users who resented the IBM approach that kept its computers out of reach in sealed rooms. This "hacker critique" of IBM emphasized the importance of access to computers and a rejection of authority in favor of decentralization. In many ways, hackers were responding to the prevailing closed world discourse that was obsessed with geopolitical containment and driven by the military centralization of command and control. The hacker ideology of the 1960s had its roots in communities of programmers at MIT and in the San Francisco Bay area. According to Steven Levy's account, East Coast hackers emphasized their technical fascination with computers (the Hands-On Imperative) and disdain for any gatekeepers that interfered; West Coast hackers tended to situate

---

[28] Steven W. Usselman, "IBM and Its Imitators: Organizational Capabilities and the Emergence of the International Computer Industry," *Business and Economic History* 22 (1993): 1–35; Timothy F. Bresnahan and Shane Greenstein, "Technological Competition and the Structure of the Computer Industry," *The Journal of Industrial Economics* 47 (1999): 1–40; Timothy F. Bresnahan and Franco Malerba, "Industrial Dynamics and the Evolution of Firms' and Nations' Competitive Capabilities in the World Computer Industry," in David C. Mowery and Richard R. Nelson, *Sources of Industrial Leadership: Studies of Seven Industries* (New York: Cambridge University Press, 1999), 79–132; Chandler, *Inventing the Electronic Century*, 94–106; Ceruzzi, *A History of Modern Computing*, 128–143, 161–173; Gerald W. Brock, *The Second Information Revolution* (Boston: Harvard University Press, 2003), 106–111.

their tinkering within a broader countercultural critique of technology and authority in modern society. In all cases, hacker culture dovetailed nicely with the growing recognition of the importance of user groups for sales and continued innovation in the computer industry.[29]

A third response – a "regulatory critique" – to IBM's dominance was articulated by the same officials in the FCC and Justice Department who had set limits on AT&T's telephone monopoly in the 1950s and 1960s. A Justice Department investigation of IBM concluded in 1956 with a Consent Decree that ordered IBM to license its patents and allow its customers to buy system components from competing firms – similar to the remedies used in the 1956 Consent Decree with AT&T. In the words of historian Steven Usselman, government officials believed they were "taking these measures to break open the closed world of IBM" and thus facilitate competition in the industry.[30] This regulatory critique shared with the competitive critique a fundamental faith that market competition could only be effective if incumbents were not allowed to erect high barriers to entry from entrepreneurs and competing firms. By the mid-1960s, it was easy for critics to paint AT&T and IBM with the same brush: the power of regulation was needed to ensure that both of these dominant incumbents did not use their power as obstacles to competition and innovation.

Regulators watched carefully as IBM gained market share during the 1960s, prompting another antitrust investigation in 1967, which led to a lawsuit in 1969 that lasted until the early 1980s. Under pressure from the Justice Department, IBM executives in the late 1960s decided to "unbundle" their hardware and software products – a watershed event in computing history. Their decision to unbundle stimulated new markets for software and "plug-compatible" peripherals, undermined IBM's dominant position as the de facto guardian of computer interfaces, and ultimately disaggregated control in the industry.[31] By the time President Ronald Reagan's Justice Department dismissed the IBM antitrust suit in 1982, the combined force of critiques from

[29] Steven Levy, *Hackers: Heroes of the Computer Revolution* (New York: Doubleday, 1984); Edwards, *Closed World*; Atsushi Akera, "Voluntarism and the Fruits of Collaboration: The IBM User Group, Share," *Technology and Culture* 42 (2001): 710–736; Fred Turner, *From Counterculture to Cyberculture: Stewart Brand, the Whole Earth Network, and the Rise of Digital Utopianism* (Chicago: University of Chicago Press, 2006); John Markoff, *What the Dormouse Said: How the Sixties Counterculture Shaped the Personal Computer Industry* (New York: Penguin, 2006); Ted Friedman, *Electric Dreams: Computers in American Culture* (New York: New York University Press, 2005); Steven W. Usselman, "Comment: Mediating Innovation: Reflections on the Complex Relationships of User and Supplier," *Enterprise & Society* 7 (2006): 477–484.

[30] Steven W. Usselman, "Public Policies, Private Platforms: Antritrust and American Computing," in Richard Coopey, ed., *Information Technology Policy: An International History* (New York: Oxford University Press, 2004), 100.

[31] Usselman, "Public Policies, Private Platforms," 97–120; Steven W. Usselman, "Fostering a Capacity for Compromise: Business, Government, and the Stages of Innovation in American Computing, *IEEE Annals of the History of Computing* 18 (1996): 30–39; Usselman, "Unbundling IBM."

entrepreneurs, hackers, and regulators had radically changed the foundations of the computer industry. Innovative ideas, companies, and machines were everywhere, but an important question remained: If IBM would not be allowed to be the system architect for the entire computer industry, who or what would be responsible for industrywide compatibility? Who would set the standards?

The Jurisdictional Challenge from New Standards Committees

A fourth response to IBM – a "consensus critique" – came from a diverse group of computer professionals who were eager to steer the future trajectory of computer technologies. IBM gained market share and power during the late 1950s and 1960s, but it was by no means monolithic. During the 1960s, new institutions began to set standards for terminology, symbols, and programming languages – some basic building blocks that would facilitate cooperation among computer and data processing professionals on an international scale. Two of the most significant institutions in this new realm were the International Organization for Standardization (ISO) and the International Federation for Information Processing (IFIP).

ISO, which officially began operations in 1947, was the type of institution that standards pioneers such as Comfort Adams, Paul Gough Agnew, and Charles le Maistre had envisioned since the early decades of the twentieth century. Agnew, le Maistre, and other engineer-diplomats had created the International Federation of National Standardization Associations in 1926, but the group never established enough internal harmony to be as effective as its founders hoped. International cooperation was reinvigorated during World War II, mainly through the United Nations Standards Coordinating Committee dominated by Britain, the United States, and their allies. After a series of meetings and conferences in 1945 and 1946, ISO began operations in early 1947 as a nongovernmental federation of national standards bodies that would "coordinate, not promulgate standards" through the "consensus" sectional committee method discussed in Chapter 3. Under the ISO procedure, a national standards body could propose a project for international standardization; if approved, that standards body would become responsible for creating a secretariat to lead and coordinate the committee process.[32]

ISO first became involved with computer standards in 1960, when it acted on a Swedish proposal to form Technical Committee 97 (TC 97) for Computers and Information Processing. Participants from nine countries

---

[32] JoAnne Yates and Craig N. Murphy, *The International Organization for Standardization (ISO): Global Governance through Voluntary Consensus* (New York: Routledge, 2008); P. G. Agnew, "Standardization," *Encyclopedia Britannica*, 14th ed. (1940), reprinted as "Standards in Our Social Order," *Industrial Standardization and Commercial Standards Monthly* 11 (1940): 141–148; Chapter 3, "Ideological Origins of Consensus Standards II: American Standards, 1910s–1930s."

(Belgium, Sweden, Czechoslovakia, France, Italy, Germany, Poland, United Kingdom, and the United States) as well as observers from an additional fifteen countries took part in the first meeting of TC 97 in Geneva from May 16 to 19, 1961. At the meeting, the representatives chose the U.S. member body, the American Standards Association (ASA), to be the secretariat (or sponsor) of TC 97's work. TC 97 also created six working groups: Glossary, Character Sets and Coding, Character Recognition, Input and Output Media, Programming Languages, and Digital Data Transmission. American participation in all of these activities was coordinated by the American Standards Association's Sectional Committee X3. The end result of this extended exercise in bureaucratic and functional delegation was that ASA's X3 committee and subcommittees became a crucial venue where American computer professionals – including engineers from IBM and competing computer manufacturers – would attempt to translate their local customs and practices into international standards recognized by ISO.[33]

At the same time that TC 97 was born, a second group – IFIP – also sprung to life. IFIP's architect and first president, Isaac Auerbach, envisioned the group as an international forum for professional computer scientists and engineers to "exchange information about the state of the computer art." After meetings with leaders from a number of related professional societies, Auerbach convinced UNESCO to "fund, organize, and convene the First International Conference on Information Processing in 1959," which featured nearly 1,800 participants from thirty-eight countries – including, in a striking display of Cold War collaboration, the Soviet Union.[34]

With IFIP's official creation in 1960, Auerbach and others saw that the group could help meet the "need for greater standardization of terminology throughout the industry."[35] IFIP created a committee with a somewhat broad mandate that covered the "standardization of terminology of digital computers and data processing devices, equipments, media, and systems."[36] Before they began the detailed technical work required to complete a comprehensive and multilingual glossary of terms, members of IFIP's Committee for the Standardization of Terminology and Symbols tended to areas of overlap

---

[33] International Organization for Standardization, "Draft Report, First Meeting of ISO/TC 97, May 16–19, 1961, Geneva, Switzerland," July 20, 1961, Isaac L. Auerbach Papers, Charles Babbage Institute, Minneapolis, Minnesota, Box 5, Folder 24 [hereafter, Auerbach Papers]; ISO/TC 97 Org Chart, July 1968, Auerbach Papers, Box 5, Folder 4; J.A.N. Lee, "The 25th Anniversary of Committee X3," *IEEE Annals of the History of Computing* 9 (1988): 345–354.

[34] Isaac L. Auerbach, "The Start of IFIP – Personal Recollections," *IEEE Annals of the History of Computing* 8 (1986): 180–192; Ksenia Tatarchenko, "Cold War Origins of the International Federation for Information Processing," *IEEE Annals of the History of Computing* 26 (2010): 46–57.

[35] Auerbach, "The Start of IFIP," 192.

[36] Isaac L. Auerbach to Council Representatives and Members of IFIPS Standards Subcommittee on Terminology and Symbols, February 2, 1961, Auerbach Papers, Box 4, Folder 18.

and liaison relationships with a number of existing committees that were engaged in standardization work.

Auerbach's first step was to recruit the British computer scientist Geoffrey Tootill, a leader in the British Standards Institute's production of a draft glossary, to become chairman of IFIP's new committee. Auerbach was keenly aware of a great variety of work on computer terminology and symbols (hence his desire to create standards), and was eager for IFIP to become deeply involved in what was already a crowded organizational field. "It is through these interlocking memberships," he explained in a 1961 letter, "that I believe most effective coordination can be had at the present time."[37]

IFIP's work on a glossary standard carried out Auerbach's collaborative vision. In early 1962, IFIP forged an agreement with the International Computation Centre (ICC) in Rome, Italy, to create a Joint IFIP-ICC Terminology Committee that would prepare a multilingual glossary that they would propose for international standardization to ISO TC 97's glossary working group. Auerbach, Tootill, and other committee leaders assumed that their glossary's acceptance as an international standard could only occur through ISO; they also assumed that their contributions, the result of a collaborative effort of an international committee of professionals and experts, would be welcomed as a foundation of ISO's work.[38] They were eventually proved correct in their view of ISO's authority over international standards, but they soon discovered that jurisdictional jostling and personality clashes threatened the viability of any international consensus.

One problem surfaced in early 1962 between IBM engineer Robert Bemer and Westinghouse engineer W. Barkley Fritz. Both men had been working, independently, on glossaries of computer terms. Tensions were evident. Fritz was a prominent member of the community, having been named chairman of a joint committee of representatives from the Institute of Radio Engineers, the American Institute of Radio Engineers, and the Association for Computing Machinery (ACM) that was developing a glossary of computer terms. In addition, Fritz was the chairman of the ASA X3.5 subcommittee on data processing standards, as well as chairman of an ACM standards subcommittee on programming terminology. He was, in other words, well established in the nascent hierarchy of American computer standards-setting.

Bemer, IBM's director of systems programming, had already played a key role in the formation of the glossary working group of ISO TC 97. He introduced himself to the IFIP Standards Committee with a February 1961 memo that declared he was working on information processing standards "both in a professional capacity and through my work assignment with my employer."

---

[37] Isaac L. Auerbach to L. C. Hobbs, July 7, 1961, Auerbach Papers, Box 4, Folder 18.
[38] Isaac Auerbach to J. F. Traub, April 10, 1962, Auerbach Papers, Box 4, Folder 19; G. C. Tootill, "Report for the Period March to September 1961," September 15, 1961, Auerbach Papers, Box 4, Folder 19.

Bemer sought to preempt suspicions that he might try to steer the committee toward IBM's favor by adding, "I personally feel that this is a fortunate situation, since I have large resources at my disposal and yet I am not constrained to act for any proprietary interests.... I wish to state from the outset that my employer expects me to act with professional integrity in the long-range interests of the most people."[39]

The simmering conflict between IBM's Bemer and Westinghouse's Fritz came to a boil in the spring of 1962 evident in series of memoranda between Fritz, Bemer, Auerbach, and several other leaders in the computer standards community. In an April 19 memo to Auerbach, Bemer implied that Fritz, in his role as chairman of the ASA and ACM committees, was hindering communication between the American working groups operating under Fritz's direction and the international activities in ISO TC 97 (to which Bemer was an official American delegate). Bemer worried that Fritz's aggressive and provincial approach would "lead to hard feelings and disagreement in the international area." Bemer concluded with a call for drastic action: "The mechanism for representing U.S. interests in the standardization of terminology is not functioning and should be overhauled."[40]

Fritz, understandably offended, responded on May 1 that Bemer's memo was an "uninformed" summary of existing glossary work that contained "a considerable amount of misinformation." In an unusually testy and personal four-page letter ("Apparently you can't understand why the U.S. effort is centered around me and not you"), Fritz declared that his ASA X3.5 subcommittee was fully in control and "not ready to relinquish U.S. world leadership in the field of computers and information processing to the British or anyone else."[41] The argument-via-memo – an analog foreshadowing of the digital "flame wars" that raged over electronic mailing lists in the 1980s and 1990s – escalated in subsequent days until Auerbach weighed in at the end of May.

Auerbach's intervention came in the form of a letter to Charles Phillips, a respected Defense Department official who had recently left the government to become director of the powerful Business Equipment Manufacturers Association and the chairman of the ASA X3 Committee. Auerbach began by noting the recent flurry of activity: "Since everybody is writing to everybody else and circulating copies of their letters as if they were newspapers, I see no reason why you in your new capacity should be slighted." Auerbach, whose distribution list included carbon copies to Fritz, Bemer, and six others, continued:

[39] R. W. Bemer to All Members of the IFIPS Standards Subcommittee on Terminology and Symbols, February 20, 1961, Auerbach Papers, Box 4, Folder 18; R. W. Bemer to H. L. Mason, June 8, 1961, Auerbach Papers, Box 4, Folder 18.
[40] R. W. Bemer to Isaac L. Auerbach, April 19, 1962, Auerbach Papers, Box 4, Folder 19.
[41] W. Barkley Fritz to R. W. Bemer, May 1, 1962, Auerbach Papers, Box 4, Folder 19.

The recent exchange of letters concerning an international information processing glossary reflects poorly on the professional stature of our societies and personnel. I have waited for approximately one month for someone within an American technical society with more knowledge on the subject and with greater direct involvement to attempt to quench this fire and bring order to the situation.[42]

Anxious to resolve the conflict, and in the absence of any other intervening authority, Auerbach invited a small group of colleagues to meet and work out what he called the "Bemer-Fritz affair." Auerbach, an ally of Bemer, was motivated primarily by his vision of international standards developed through a scientific consensus, such as those he had been overseeing in IFIP. "The vituperative letters that have been exchanged," he complained, "divert energies away from the primary objective." He implored his colleagues to consider what was at stake: "A continuation or repetition of such activity will seriously impair the ability of the United States to provide the appropriate leadership in a field in which we are uniquely qualified and the recognized world leader."[43]

Two days later, Bemer telephoned Auerbach to explain his side of the story. Auerbach saved his notes from their frank conversation: "F [Fritz] thinks Bemer is holding up publication of his (F's) glossary. In fact Herb Bright is doing this because it is so bad."[44] Bright's opposition was a significant revelation, since he was the chairman of ACM's Standards Committee, a member of several ASA X3 subcommittees, and the only other official American delegate – alongside Bemer – to the international glossary working group in ISO TC 97. The combined diplomatic heft of Bemer and Bright, together with their damning technical appraisal of Fritz's project ("it is so bad"), suggested that Fritz's combative stance would be unsustainable.

Four weeks later, the six leading figures of American standards work for computer glossaries finally sat down for a tense conversation. Auerbach, Bemer, and Fritz were joined by Phillips (the new chairman of ASA Committee X3), Joe Traub (a Bell Labs engineer who was the ASA's official representative to the glossary working group in ISO TC 97), and Walter Carlson (a senior DuPont engineer who was active in the IFIP and ACM standards committees). The minutes of the summit noted, with considerable understatement, that "dual U.S. representation on the different International Committees has introduced difficult communication problems within the American standards effort."[45] Auerbach and Bemer stressed the professionalism of their approach to glossary standards, which they had pursued through an international alliance of professionals dedicated to scientific and educational goals. Furthermore, they had already laid the diplomatic path for international standardization by virtue of

---

[42] Isaac L. Auerbach to Charles A. Phillips, May 29, 1962, Auerbach Papers, Box 4, Folder 19.
[43] Isaac L. Auerbach to Charles A. Phillips, May 29, 1962, Auerbach Papers, Box 4, Folder 19.
[44] Isaac Auerbach, May 31, 1962, Auerbach Papers, Box 4, Folder 19.
[45] "Draft Minutes, Discussion of Glossary Standards," August 8, 1962, Auerbach Papers, Box 4, Folder 19.

their leading positions within the authoritative international committee in ISO TC 97.

The problem with their plan was that the official American delegation to ISO came through the ASA, and Fritz, as the leader of the ASA Glossary Committee, preferred his own glossary project to the IFIP glossary. Nevertheless, Fritz, outflanked and outnumbered, was forced to relent. By the end of the meeting, the group of six worked out a compromise that preserved Fritz's position at the head of the ASA Glossary Committee, but completely undermined his ability to promote his own glossary. All six men endorsed the authority of the "formal Standards Committees sponsored by ASA and ISO," but they stipulated that "the actual Standard would be prepared by less formal groups, such as the IFIP/ICC Joint Committee." The formal committees, including Fritz's ASA committee, were thus designated merely as "processing units to secure prompt adoption of the Glossary Standards now under development" by Auerbach and Bemer's international committee in IFIP. The group's explicit endorsement of the IFIP glossary must have come as a stinging rebuke to Fritz, who had to suffer the final indignity of being forced to accept Bemer as a new member of his ASA committee in order to "clear up the necessary communication channels."[46]

The Bemer-Fritz affair is not a terribly significant episode in the *technical* history of computing: the IFIP-ICC committee eventually published their *Vocabulary of Information Processing* in 1966, but the rapid, chaotic, and uncoordinated evolution of computer components and systems made it difficult to sustain a multilingual and international consensus around the field's central terms and concepts.[47] The significance of the Bemer-Fritz affair becomes more apparent when we see it as an episode in the *institutional* history of computing and in the contest for authority over computer standards. The affair illustrated the dominant strategic and organizational themes within the network of formal and informal American and international standards committees that sprung into life in the early 1960s. Alliances formed quickly as strong personalities and strong institutions sought to exert their authority and extend their jurisdiction. Experts such as IBM's Bemer were torn between their professional obligations to scientific integrity, their personal ambitions, and their corporate loyalties. They also worried deeply about the timeliness and relevance of their work – could committees reach consensus quickly, or would international competition and the rush of events prohibit cooperative standardization? The protagonists of computer standards recognized that speed was essential, that

---

[46] "Draft Minutes, Discussion of Glossary Standards," August 8, 1962, Auerbach Papers, Box 4, Folder 19.

[47] J. F. Traub, "USA Participation in an International Standard Glossary," *Communications of the ACM* 6 (1963): 658–659; J. F. Traub, "American Standard and IFIP/ICC Vocabularies Compared," *Communications of the ACM* 8 (1965): 361–363; I. H. Gould and G. C. Tootill, "The Terminology Work of IFIP and ICC," *The Computer Journal* 7 (1965): 264–270.

there were multiple and uncontrollable sources of technological change, and that the adoption of a single, clear standard had the potential to generate widespread benefits ("network effects," to use an anachronism) for producers and users alike.

Most important, the Bemer-Fritz affair reminds us that fierce personal rivalries and professional competition belied the professional rhetoric of consensus. We should assume that many other bitter conflicts existed, even (or especially) regarding issues for which the documentary record is thin. The rhetoric of consensus, the birthright and guiding ideal of all ASA committees, could easily be deployed as a smokescreen. As early as 1961, X3 chairman Herbert Bright issued a stern "X3 Policy Letter" that prohibited the "Use of Manufacturers' Names During Discussions." His plea is worth quoting at length:

Vigorous discussions during your technical meetings have been useful in helping those present to understand the problems which must be resolved. In the heat of discussion during several recent meetings, however, there has appeared a tendency toward the use of certain manufacturers' names for emphasis, particularly in connection with important economic considerations such as decisions which may result in the need for costly modifications to equipment. In an ASA-sponsored meeting such a tendency seems, in my opinion, to weaken valid arguments and to invite misinterpretation of the written record.[48]

Bright's request was typical of standards committees and engineering associations that, as a matter of tradition, sanitized their debates as they translated them from the "heat of discussion" to the "written record." It is difficult to imagine that Bright's allusion to "certain manufacturers' names" could be anything other than a reference to the elephant in the room, IBM. Yet technical meetings in X3 committees needed to be perceived as legitimate, so Bright reminded the group that it needed to convince anyone watching that their decisions focused only on valid technical grounds and thus transcended "important economic considerations." When faced with a disorderly and competitive situation, Bright found comfort in retreating to ASA norms and procedures for acceptable discourse. These norms and procedures, as we have seen, were developed in the late nineteenth and early twentieth centuries to signify a commitment to the professional ideologies of disinterested expertise, fairness, a concern for the public interest, and a genuine belief that cooperation was the key to technical and social progress. Bright's stern warning against "the use of certain manufacturers' names" illustrates the need for strong commitments to the professional values of fairness and expertise, given the conditions of persistent instability that computer scientists and engineers faced as they defined

---

[48] Herbert S. Bright to Chairmen and Secretaries of X3 Subcommittees and Task Groups, January 31, 1961, Herbert S. Bright Papers, Charles Babbage Institute, Minneapolis, Minnesota, Box 5, Folder 3 [hereafter, Bright Papers].

their collective professional identity and the technical foundations of their industry.[49]

As Auerbach, Bemer, Bright, and their colleagues struggled to impose order in an inherently disorderly situation, they learned to use standards committees to suit their scientific and strategic ambitions. The goals of all participants were never in perfect alignment, so the negotiations that took place within and between the cluster of computer standards committees were always highly charged affairs. Bright's ideal for conduct within these committees – that X3 could somehow keep its discussions and, more important, its documentary trail free of any unsavory connections with the interests of specific manufacturers – never had a chance.

IBM engineers assumed important positions in the new standards committees throughout the 1960s, prompting many observers to worry that they would use the standards process to consolidate its dominant position in the industry. For example, Bemer was an important proponent of a coding scheme developed in ASA Subcommittee X3.2 that would eventually be called ASCII (short for American Standard Code for Information Interchange). At the same time, however, a series of "no" votes from IBM made the consensus-building process work more slowly than it could have.[50] Even as IBM sent representatives to X3.2 (Bemer, the "father of ASCII," had moved to another company), it promoted an alternative code that would be less costly to implement.[51] In 1968, IBM's own management committee summarized its intentions: "IBM participation in the adoption of a[n ASCII] standard does not commit us to that standard."[52] Of course, it was to IBM's advantage, as the dominant supplier, to make its systems incompatible and thus irreplaceable. Such sophisticated tactics led one critic, the economist Gerald Brock, to conclude in 1975 that "IBM has generally opposed the development of standards in the computer industry that would result in an increase in computer compatibility.... In

[49] Nathan Ensmenger, *The Computer Boys Take Over: Computers, Programmers, and the Politics of Technical Expertise* (Cambridge, MA: The MIT Press, 2010); Brent K. Jesiek, *Between Discipline and Profession: A History of Persistent Instability in the Field of Computer Engineering, circa 1951–2006* (PhD dissertation, Virginia Polytechnic Institute and State University, 2006).

[50] See "X3.2 Documents 100–599," Honeywell, Inc., X3.2 Standards Subcommittee Records, Charles Babbage Institute, Minneapolis, Minnesota, Box 1, Folder 21 [hereafter X3.2 Standards Subcommittee Records] and "Materials on proposed American Standard Code for Information Interchange (ASCII), 1962," X3.2 Standards Subcommittee Records, Box 1, Folder 2. See also Gerald W. Brock, "Competition, Standards, and Self-Regulation in the Computer Industry," in Richard E. Caves and Marc J. Roberts, eds., *Regulating the Product: Quality and Variety* (Cambridge, MA: Ballinger, 1975), 85–91.

[51] Bob Bemer, "Thoughts on the Past and Future," http://www.bobbemer.com (accessed May 22, 2011); W. E. Andrus to Business Equipment Manufacturers Association, June 6, 1962, X3.2 Standards Subcommittee Records, Box 1, Folder 2. The alternative code was EBCDIC, Extended Binary Coded Decimal Interchange Code.

[52] Brock, "Competition, Standards and Self-Regulation in the Computer Industry," 88.

the absence of outside pressures, IBM can control the effective standards of the industry simply by virtue of its market position."[53]

Despite IBM's strategic position – or perhaps because of it – engineers in the ASA X3 committees deepened their commitment to the production of computer standards. The standardization of a programming language, FORTRAN, illustrates one way that computer engineers used the ASA consensus standardization process to coordinate the various outside pressures on the company, pry control away from IBM, and establish a viable alternative coordinating mechanism. FORTRAN, the first high-level programming language, was created between 1954 and 1957 by a team of IBM computer scientists led by John Backus. Users in the American military and scientific communities quickly adopted the language, thus providing the large user base that all successful standards need (as examples throughout this book demonstrate). Subsequent releases of FORTRAN II, III, and IV aimed to streamline the language, make it easier to learn, and enable machine independence and portability.

A significant turning point came in May 1962, when ASA Subcommittee X3.4 began a project to standardize FORTRAN. In keeping with the ASA requirement that committees represent a balance of interests, the committee leaders canvassed widely to bring major hardware vendors, user groups, software providers, and academic users into the committee membership. The committee circulated a draft of its work in 1964 and incorporated suggestions into its publication of USA Standard FORTRAN in March 1966. With this publication, which specified standard and machine-independent versions of FORTRAN, the committee succeeded in preserving the utility of the language for users who did not have IBM computers or who did not want to be IBM customers. Historians of computing celebrate this event – the standardization of FORTRAN by the ASA – as a major turning point in the history of programming languages: committee standards would deliver programmers from "the whims of a single organization" and replace it with "an industry-wide voice in the definition of features and the timing of their introduction.[54]

---

[53] Brock, "Competition, Standards and Self-Regulation in the Computer Industry," 91.

[54] FORTRAN history has been documented in great depth and breadth. See for example W. P. Heising, "History and Summary of FORTRAN Standardization Development for the ASA," *Communications of the ACM* 7 (1964): 590; S. Gorn, "FORTRAN vs. Basic FORTRAN," *Communications of the ACM* 7 (1964): 591–624; United States of America Standards Institute, *USA Standard FORTRAN*, X3.9-1966 (New York: American Standards Association, 1966); John Backus, "The History of FORTRAN I, II, and III," *IEEE Annals of the History of Computing* 1 (1979): 21–37; John Backus, "Early Days of FORTRAN," *IEEE Annals of the History of Computing* 6 (1984): 15–27; Jeanne Adams, "Institutionalization of FORTRAN," *IEEE Annals of the History of Computing* 6 (1984): 28–40; Martin H. Greenfield, "History of FORTRAN Standardization," *Proceedings of the 1982 National Computer Conference* (Arlington, VA: AFIPS Press, 1982), 819–824; Steve Lohr, *Go To: The Story of the Math Majors, Bridge Players, Engineers, Chess Wizards, Maverick Scientists and Iconoclasts – The Programmers Who Created the Software Revolution* (New York: Basic Books, 2001), 11–34;

This turning point was not universally acclaimed. Indeed, as two engineers noted in 1985, "Because the computer industry has been so dominated by IBM, consensus standards-setting procedures have been relatively unsuccessful."[55] Nevertheless, these procedures survived, contributed to stabilizing the industry, and eventually were essential for subsequent challenges to IBM's centralized control. During the 1960s, the X3 committee flourished in its new role as the "industry-wide voice" of professionals dedicated to the collaborative production of computer standards. In the process, it proved that the consensus critique of IBM could sustain industrywide cooperation in some select areas. X3 and its subcommittees consolidated their jurisdiction over computer standards domestically while also making substantial contributions to international standards development. The rapid growth of X3 indicated its ability to channel widespread interest in computer standards: by 1968, the X3 Facts Book listed fourteen member organizations classified as General Interest Groups, seventeen Consumer Groups, and twenty member companies in the Producer Group. The committee's work proceeded through eight subcommittees and working groups, which, by August 1968, had published twenty-six American standards on character codes and recognition, programming languages, terminology and symbols, data communications, and specifications for paper tape, magnetic tape, and punch cards equipment. It was, in short, a modest yet significant institutional innovation that helped guide the trajectory of a nascent industry. As in earlier episodes of standardization in engineering committees and trade associations, work that maintained close ties to business decisions made by buyers and sellers increased the likelihood that standards would be adopted widely.[56]

The consensus critique of IBM took shape through the actions of Auerbach, Bright, and their fellow pioneers in consensus standardization for the computer industry. They embodied their critique in formal standards committees within groups such as ASA X3 and ISO TC 97, as well as in informal committees in professional and scientific groups such as the ACM and IFIP. The growth of these committees marks a significant turning point in the history of the American and international computing industries: for the first time, an alliance of computer professionals created new organizational capabilities to challenge IBM's control – not in the production of hardware and software, but in the production of new techniques and technologies of *compatibility*, *interoperability*, and

---

Paul McJones, "History of FORTRAN and FORTRAN II," Computer History Museum, http://www.softwarepreservation.org/projects/FORTRAN/ (accessed February 20, 2011).

[55] Marvin A. Sirbu and Laurence E. Zwimpfer, "Standards Setting for Computer Communication: The Case of X.25," *IEEE Communications Magazine* 23 (1985): 37.

[56] Data Processing Group, *USASI X3 and ISO/TC 97 Facts Book* (New York: Business Equipment Manufacturers' Association, 1968), Bright Papers, Box 5, Folder 4; Business Equipment Manufacturers' Association, *Committee X3 Organization Manual* (New York: American National Standards Institute, 1970), Charles A. Phillips Papers, Charles Babbage Institute, Minneapolis, Minnseota, Box 1, Folder 2.

interconnection. Standards committees depended on the competitive energies of entrepreneurs, the implicit approval of regulators, and the cooperative ethos in the nascent computer professions; at the same time, they were consciously designed to build a future that was not dominated by IBM.

Conclusions

American attitudes toward advanced technologies crystallized into a broad-based ideological consensus during the 1970s. Many different groups – including hackers and university researchers, federal regulators, and engineers and scientists in the private sector – articulated new critiques of centralized control. Their consensus critique implicitly embodied three powerful convictions: the New Deal impulse to restrain corporate power; the entrepreneurial impulse to exploit new advances in electronics, materials, and information theory; and an economic impulse to seek efficiency and innovation through market competition and private ordering rather than government planning.

These impulses were specific to the telephone and computer industries, but they also shared expectations and assumptions toward technology, political economy, and control with broader swaths of American society. Historians such as Paul Edwards, Ted Friedman, and Fred Turner have analyzed, more than I have attempted to do here, the close links between counterculture ideals and skepticism toward unrestrained technological power. They point to films such as *Desk Set* (1957), *Dr. Strangelove* (1964), *2001: A Space Odyssey* (1968), *Blade Runner* (1982), and *The Terminator* (1984) as well as books such as Herbert Marcuse's *One Dimensional Man* (1964), E. F. Schumacher's *Small Is Beautiful* (1973), and Ted Nelson's *Computer Lib/Dream Machines* (1974) as indicators of an emerging critical approach to capitalist technology. These ideas took root in the freewheeling corporate cultures in Silicon Valley, which nurtured a fusion between the hacker critique of centralized control and a libertarian strain of individual freedom and empowerment.[57]

It would be oversimplifying matters, however, to reduce the critiques of centralized control that matured in the 1960s and 1970s to some sort of irresistible triumph of a populist or democratic control over technology. A close focus on standards has helped us see how a variety of uncoordinated attacks on AT&T and IBM facilitated *redistributions* of control within a dynamic network of overlapping bureaucracies.[58]

[57] Bryan Pfaffenberger, "The Social Meaning of the Personal Computer: Or, Why the Personal Computer Revolution Was No Revolution," *Anthropological Quarterly* 61 (1988): 39–47; Paulina Borsook, *Cyberselfish: A Critical Romp through the Terribly Libertarian Culture of High Tech* (New York: PublicAffairs, 2001); Markoff, *What the Dormouse Said*; Richard Barbrook and Andy Cameron, "The Californian Ideology," *Science as Culture* 6 (1996): 44–72.
[58] Coll, *Deal of the Century*, 373–4.

In the telephone industry, Strassburg had pushed the FCC to "cautiously and incrementally" experiment with competition in selected niches of telephone service. Strassburg, echoing the New Deal ethos of his early career, had no overarching plan or theory for how these experiments might play out. He simply acted on behalf of those who felt that monopoly control was harming the public interest and technological progress and sought to empower entrepreneurs and customers with direct economic interests.[59] The absence of any overarching strategy guiding the decades of attacks against the Bell System was noted by journalist Steve Coll, who concluded his *Deal of the Century* by putting the point more forcefully. "Precious little" in the antitrust case that ended with the Bell System's divestiture, he wrote,

> was the product of a single, coherent philosophy, or a genuine, reasoned consensus, or a farsighted public policy strategy. Rather, the crucial decisions made in the telecommunications industry in the 1970s and early 1980s were driven by opportunism, short-term politics, ego, desperation, miscalculation, happenstance, greed, conflicting ideologies and personalities, and finally, when Charlie Brown [AT&T's chairman at the time of divestiture] thought that there was nothing left, a perceived necessity.[60]

It is difficult to imagine a state of affairs more different than the stable, cautious, public-spirited approach to the "engineering of the present" that Bancroft Gherardi and his colleagues had devised in the early twentieth century. Nevertheless, divestiture suddenly ended AT&T's responsibility for the architecture of the national telephone system in 1984. The outstanding question, for engineers and regulators who worried about the institutional fragmentation of the American communications infrastructure, was: What came next? Who, or what, would be the new system architect, the new standard setter? What organizational alternatives did these critiques propose to build?

Regulators in the FCC quickly realized that they lacked the expertise and authority to define standards for these new networks. As AT&T's power faded, most interested parties – including engineers in private firms, university researchers, and regulators in the American federal government – looked to industry committees to fill the void. When the FCC in 1983 contemplated ways to ensure the technical coordination of the national network, it decided to turn to a new industry group called the Exchange Carriers Standards Association (ECSA). The ECSA's legitimacy was in part a function of its size: firms in the ECSA served 95 percent of American customers. In response to an FCC proceeding on standards in the postmonopoly world, the ECSA volunteered to sponsor a committee that would follow the guidelines of the American National Standards Institute that sought to ensure openness, due process, and balance of interests. With the approval of the FCC and the industry, the ECSA created Committee T1 in February 1984. The new T1 Committee proved itself capable

---

[59] Henck and Strassburg, *A Slippery Slope*, 255.
[60] Coll, *Deal of the Century*, 369.

of fulfilling the coordinating role previously assumed by the Bell System and thus provided a template for future standardization policy decisions.[61]

The computer industry took a different path toward its rendezvous with international standards committees: no federal authority dominated computer standards in the same way that the FCC, in the wake of the *Carterfone* decision, could dictate interface standards to the Bell System. Instead, engineers and scientists in the nascent computer professions struggled mightily to forge alliances and develop international agreement around basic computer standards. IBM's dominance was a primary concern for the leaders of the new standards committees in IFIP and ASA, just as much as it was a concern for the competitors, hackers, and regulators who had articulated their own critiques of IBM. Despite their divergent regulatory settings, the strategic issues in the telecommunications industry and the computer industry gravitated to the crucial question of who could attach new devices to existing systems. The consensus critique succeeded through standards committees hosted by trade associations, scientific societies, and international organizations. These committees, crucially, developed the organizational capabilities to sustain international agreements over complex and contested technical problems that emerged at the boundaries and interfaces of technological systems.

By the early 1980s, the threat of monopoly power and centralized control in both industries had been tempered by the forces of regulation and technological change. In the absence of monopoly power, voluntary standards committees became the key mechanisms for coordinating industrywide technological change. Standards committees would help knit together the modular characteristics of the computer industry and the artifacts that it produced in the new era of "alliance capitalism." Engineers and corporate managers became increasingly familiar with the structures and procedures of standards committees. As we will see, they gained a richer appreciation of the new strategic opportunities afforded by standards committees.[62]

During the post–World War II era there was a period of rapid innovation in two distinct areas: applications of information theory (digital techniques) and the convergence of computer and telephone industries. Many forward-thinking professionals – including MIT's J. C. R. Licklider and the FCC's Bernie Strassburg – recognized that the technological convergence of telephony and computing could only be sustained through innovations in the social and organizational foundations of these heretofore distinct industries. Through the late 1980s, Strassburg continued to insist that the commercial applications of

[61] Ian M. Lifchus, "Standards Committee T1 – Telecommunications," *IEEE Communications Magazine* 23 (1985): 34–37; Arthur K. Reilly, "Defining the U.S. Telecommunications Network of the Future," in Brian Kahin and Janet Abbate, eds., *Standards Policy for Information Infrastructure* (Cambridge, MA: The MIT Press, 1995), 579–593.

[62] Louis Galambos and Eric John Abrahamson, *Anytime, Anywhere: Entrepreneurship and the Creation of a Wireless World* (New York: Cambridge University Press, 2002).

new technologies – "packet switching, satellite communications, fiber optics, microwave, cellular radio, and other technologies having a variety of new service potentials" – could only be realized through "coordinated planning and action involving the entire telecommunications infrastructure." He continued, "This in turn requires the formulation of national, if not global, standards to facilitate optimum design and performance of networks with technically compatible transmission, switching, and terminal components.... Until now there have been no encouraging signs that either government or industry is disposed to launch the necessary initiatives to deal with the problem."[63]

Perhaps Strassburg did not appreciate (or did not yet know about) the robust coordination mechanisms in hybrid committees such as IFIP, ASA, and ISO that had been working since the 1960s. These committees, along with some new alliances, were taking shape in order to sustain new packet-switched networking experiments in the United States, Britain, and France. They had been creating exactly the sort of standards that, as Strassburg correctly noted, were not being created by government or industry. The new standards committees nurtured partnerships among private firms, academic and scientific researchers, professional societies, and government officials. No single institution had unquestioned authority, and no company or committee found itself in a position of centralized control. Grave uncertainties faced telephone and computer professionals in these committees as the era of regulated monopoly passed: Would these new alliances be effective? Could they keep up with the rapid pace of change in the telephone and computer industries? Could they keep IBM at bay? What strategies could they use to manage the explosive tensions unleashed by convergence, competition, innovation, digitization, and globalization?

---

[63] Strassburg, *A Slippery Slope*, 256–257. See also J. C. R. Licklider and Robert W. Taylor, "The Computer as a Communication Device," *Science and Technology* (April 1968): 20–41; and Strassburg, "The Marriage of Computers and Communications."

6

International Standards for the Convergence of Computers and Communications, 1960s–1970s

> It is often argued that the computer industry lacks internal discipline as it exhibits a conspicuous inability to produce standards. Usual counter arguments are the youth of this industry, an extremely fast pace of evolution, and the dominance of one manufacturer. On the other hand the telecommunications industry has produced an impressive lot of standards over its long existence.... Both industries are now bound to become so intertwined that they can no longer look at each other from a respectful distance.... Orderly diversity is the only way to foster progress and keep a hand on one's own destiny.
>
> – Louis Pouzin,
> "Standards in Data Communications and Computer Networks," 1975[1]

The formative technological innovations of the "open" digital age – electronic computers that exchange data through digital communication networks – did not, by themselves, drive the decentralization of authority in late-twentieth century global society. These technologies, the offspring of the stormy marriage of computing and communications, came into being as components of broader and more fundamental critiques of centralized control. Critiques came first; then came technical development, standardization, and, finally, social consequences.

In the United States and abroad, the computing industry was younger and more volatile than the communications industry. Telephone systems in all industrialized countries operated under the clear jurisdiction of national governments. The American arrangement, which featured the private administration of a national monopoly with the grudging approval of regulators, was abnormal (if not exceptional). American federal regulators in the 1950s,

---

[1] Louis Pouzin, "Standards in Data Communications and Computer Networks," March 1975, INWG 79, Alexander McKenzie Collection of Computer Networking Development Records, Charles Babbage Institute, University of Minnesota, Minneapolis [hereafter, McKenzie Collection].

1960s, and early 1970s used their authority to prevent AT&T from entering the computer business. These restrictions were motivated, in part, by the hope that competition would flourish and bring with it a new era of innovation that would fuel the continuation of American technological leadership.

Beyond the boundaries of the United States, an example of a more typical arrangement existed in France, where a government agency was the sole administration responsible for the provision of postal, telegraph, and telephone (PTT) service.[2] In some obvious ways, the directors of the French PTT shared the vision of orderly, centralized control pursued by Bell System architects such as Theodore Vail and Bancroft Gherardi. But where American political and legal traditions established clear limits to the boundaries of the Bell System by defining structural barriers at its institutional boundaries, no such barriers existed in France. Consequently, no legal restrictions stopped civil servants in the French PTT from planning to extend their jurisdiction into the terminal equipment that users would attach to the ends of the telephone network.

It may be too simplistic to characterize these national regimes as Tocquevillian voluntarism in America and Colbertist centralization in France, but it is fair to conclude that national communication policies reflected and projected the values of the politicians and bureaucrats who designed and administered them. In many ways, national responses to the convergence of computing and communications held true to national stereotypes. The American response, as we have seen, featured a pragmatic and ideologically inconsistent mix of antimonopoly regulation, direct and indirect government subsidies for research in private corporations, and a faith that entrepreneurship and competition would harness the creative energies of Americans and stimulate technological innovation. There was no coherent state policy for the American entry into the information age but rather a patchwork of overlapping strategies.

The French response, in contrast, featured greater ideological clarity: IBM's dominance threatened French technology and therefore French sovereignty. The French response, articulated in the Plan Calcul that Charles de Gaulle approved in 1966, would be a national response administered by the state. French elites were determined to create a strong domestic computer industry and to take every possible measure to ensure that IBM would not encroach on the French PTT's jurisdiction over communication networks.[3]

The divergent responses of national governments reflected both the prevailing ideologies of the bureaucrats in power in the 1970s as well as national varieties of insecurity and uncertainty about the high-tech global economy. The clarity that we bring to this recent history from our vantage point in the twenty-first century often obscures the conditions of uncertainty over the outcome of

[2] Eli Noam, *Telecommunications in Europe* (New York: Oxford University Press, 1992), 7–25.

[3] Simon Nora and Alain Minc, *The Computerization of Society: A Report to the President of France* (Cambridge, MA: The MIT Press, 1981); Valérie Schafer, *La France en Reseaux* (Paris: Nuvis, 2012).

the Cold War, the fragility of the global economic system in the 1970s, and the contested boundaries between public and private control over the global communications infrastructure. The neologisms coined during this era – such as "compunications" and "*télématique*" – remind us of the novelty and contingency of discourses, categories, and technologies that we today assume to be stable.[4] They help us remember that the celebrated innovators who worked across the boundaries of computers and communications – people such as J. C. R. Licklider, Vinton Cerf, and Louis Pouzin – worked in conditions of persistent instability and always needed to describe and justify their designs with respect to the technological and jurisdictional status quo. They tell us that the "history of the Internet" is but one thread in a much bigger and more complex tapestry of the history of networking in the 1970s.

The politics of patronage, rather than the deterministic arrow of technological progress, structured the realm of technological possibility. In the United States, the mix of strong leadership from a wealthy patron – the Department of Defense – and an informal, research-oriented culture were key enablers of innovation in the Arpanet community in the early and mid-1970s. In France, computer researchers worked in a less permissive environment. The leader of the French computer research community in the early 1970s was Louis Pouzin (b. 1931), a charming and sagacious man who worked for the French national research lab IRIA (Institut de Recherche en Informatique et en Automatique). Pouzin developed strong relationships with a number of his American colleagues: he had been a visitor at MIT from 1963 to 1965 and visited several Arpanet sites while he was designing the French packet-switched network, Cyclades, in the early 1970s. In the process, Pouzin learned a strategy of resistance – "to behave a little bit more like an American," as he put it in a recent interview. "When I was in my first month at MIT," Pouzin recalled, "sometimes there were technical points to discuss. Like the French, I was criticizing things. They said 'well, if you have another idea, then do it.' And I said 'Well, OK.' So I started doing things without criticizing."[5]

Pouzin quickly learned that researchers sponsored by the Advanced Research Projects Agency (ARPA)[6] had a key advantage over the French. In the United States, the bureaucrats in the FCC and Justice Department had no reason to notice, let alone care, about ARPA's experiments with packet switching. In France, however, Pouzin and his colleagues worked under the watchful

[4] Daniel Bell, "Introduction," in Nora and Minc, *Computerization of Society*, vii; Anthony Oettinger, "Compunications in the National Decision Making Process," in Martin Greenberger, ed., *Computers, Communications and the Public Interest* (Baltimore: The Johns Hopkins University Press, 1971).

[5] Louis Pouzin, oral history interview by Andrew L. Russell, April 2, 2012, Paris, France. Charles Babbage Institute, University of Minnesota, Minneapolis.

[6] ARPA was renamed DARPA (for Defense Advanced Research Projects Agency) in 1972; then ARPA again in 1993, before changing back to DARPA once again in 1996. For consistency, I will use "ARPA."

eyes of Parisian bureaucrats who expected that Cyclades would be part of the French fortifications against IBM's relentless expansion in international computer and communication markets. Pouzin was especially forthcoming with his objections to French policies to centralize power within a "national champion" computer company and the monopoly PTT. His critique of the French national policy – and his tireless advocacy of his vision of a decentralized and "heterogenius" network that could sustain an entrepreneurial and international computer industry – ultimately made him a martyr in the war against centralized control.

J. C. R. Licklider and the American Cold War Consensus

The Arpanet[7], conceived and funded by managers in the U.S. Department of Defense, was the most significant early experiment that erased the boundaries of computer and communication networks. The visionary behind the creation of the Arpanet was J. C. R. Licklider (1915–1990). Licklider's interest in computers began when he used them in the course of his research on psychoacoustics in the 1940s and 1950s. Over time, he became convinced that computers could help humans think and work in more sophisticated ways, resulting in a partnership of humans and electronic computers that, by the early 1960s, he called "interactive computing." Licklider's vision of interactive computing – inspired by his encounter with one of the first PDP-1 minicomputers made by Digital Equipment Corporation – differed dramatically from the existing methods of human-computer interaction, known as batch processing. As every user of an IBM mainframe knew, batch processing required users to run programs by submitting stacks of punched cards to a uniformed operator who was the only person authorized to run programs on the computer. Licklider, in his 1960 article "Man-Computer Symbiosis," argued with great conviction that interactive computing would unleash the social and intellectual potential of computers by allowing individual users to interact directly with the machine.[8]

The politics of Cold War national security provided Licklider with a chance to pursue his vision on a large scale. In the immediate wake of the "Sputnik shock" of 1957, President Dwight D. Eisenhower created ARPA within the U.S. Defense Department, with the expectation that it would help the American military keep up with the technological capabilities of the Soviets. This basic course of action – heavy government investment in science and technology, in the hopes that American scientists would be able to win another war – was one

---

[7] The proper rendering of the network's name is ARPANET, but I have chosen to use "Arpanet" for aesthetic purposes.

[8] J. C. R. Licklider, "Man-Computer Symbiosis," *IRE Transactions on Human Factors in Electronics* Vol. HFE-1 (March 1960), 4–11; M. Mitchell Waldrop, *The Dream Machine: J. C. R. Licklider and the Revolution That Made Computing Personal* (New York: Viking, 2001); Howard Rheingold, *Tools for Thought: The History and Future of Mind-Expanding Technology* (Cambridge, MA: The MIT Press, 2000 [1985]), 132–151.

of the core assumptions of the science policy consensus that Americans forged in the early Cold War.[9]

In 1961, Defense Department officials Jack Ruina and Eugene Fubini decided to centralize their research on computing technology into one program and asked Licklider (known to Ruina as a "man of some distinction") to lead it.[10] Licklider seized the opportunity to direct the Department of Defense's considerable resources to advance what he often referred to as an evangelical mission, his "religious conversion to interactive computing."[11] His primary strategy was to link his scientific vision of interactive computing to ARPA's strategic desire to find new ways to use computers in military communications (known at the time as command and control). "Every time I had the chance to talk," Licklider recalled of his first days at ARPA, "I said that the mission is interactive computing." He soon convinced his colleagues that

the problems of command and control were essentially problems of man computer interaction.... Fubini essentially agreed 100% with that and so did Ruina.... Why didn't we develop an interactive computing? If the Defense Department's need for that was to provide an underpinning for command and control, fine. But it was probably necessary in intelligence and other parts of the military too. So, we essentially found that there was a great consonance of interest here, despite the fact that we were using different terms we were talking about the same thing.[12]

Once he convinced his Pentagon bosses of the utility of his work, Licklider operated freely under conditions that he referred to as "benign neglect." With Ruina and Fubini concentrating on the agency's much larger programs for ballistic missile defense and nuclear test detection, Licklider was able to direct his program with a great deal of autonomy. He skillfully channelled military contracts to his friends and colleagues at universities and data processing firms. Throughout the mid-1960s, Licklider and his successors at ARPA's Information Processing Techniques Office (IPTO) funded research projects and computer

---

[9] David Hart, *Forged Consensus: Science, Technology, and Public Policy in the United States, 1921–1953* (Princeton, NJ: Princeton University Press, 1998); Bruce L. R. Smith, *American Science Policy since World War II* (Washington, DC: The Brookings Institution, 1990); Andrew L. Russell, "Ideological and Policy Origins of the Internet, 1957–1969," paper presented to TPRC 2001, the 29th Research Conference on Communication, Information, and Internet Policy, Washington, DC, October 28, 2001.

[10] Fubini, as director of defense research and engineering, was the only step in the chain of command between the director of ARPA and the secretary of defense. Ruina was ARPA director between February 1961 and September 1963. Jack Ruina, oral history interview by William Aspray, April 20, 1989, Cambridge, Massachusetts. Charles Babbage Institute, University of Minnesota, Minneapolis.

[11] Howard Rheingold, *Tools for Thought*, 138. The evocative language of "religious conversion" also appears in J. C. R. Licklider, oral history interview by William Aspray and Arthur L. Norberg, October 28, 1988, Cambridge, Massachusetts, Charles Babbage Institute, University of Minnesota, Minneapolis; and John A. N. Lee and Robert Rosin, "The Project MAC Interviews," *IEEE Annals of the History of Computing* 14 (1992): 16–17.

[12] Licklider interview, Charles Babbage Institute.

science departments that would help realize their shared vision of interactive computing. Licklider is remembered fondly for nurturing a diffuse community of researchers whom he called in a famous 1963 memo the "Members and Affiliates of the Intergalactic Computer Network." These researchers, in turn, developed technologies such as time-sharing, graphics, networking, and databases that were key components of Licklider's vision. By using American military patronage to create and sustain "centers of excellence" in computing research at American universities such as Berkeley, MIT, Carnegie-Mellon, and Stanford in the mid-1960s, Licklider successfully mobilized the resources of the American Cold War state to create the foundations of American global leadership in computer science.[13]

Licklider left IPTO in 1964 and was succeeded at first by computer graphics prodigy Ivan Sutherland (1964–1966) and then by the NASA psychoacoustics researcher and computer enthusiast Robert W. Taylor. In 1966 and 1967, Taylor and his deputy Lawrence Roberts initiated a series of conversations among IPTO-supported researchers about the possibility of building a network to connect over a dozen ARPA-funded computer centers around the country. Such a network – dubbed the Arpanet – could simultaneously address IPTO's major administrative and research goals. It would be able to conserve its computing budget by facilitating remote access to expensive computers, researchers at different sites could swap data and programs developed remotely and on incompatible systems (such as the IBM System/360 and Digital's PDP-10), and the network itself would be an exciting new research focus for the time-sharing community. Since every major computer manufacturer was following a proprietary strategy designed to prevent connections between dissimilar systems, the Arpanet represented a government-funded project that sought to compensate for this evident market failure.[14]

---

[13] J. C. R. Licklider and Robert Taylor, "The Computer as a Communication Device," *Science and Technology* 76 (1968): 21–31; Licklider interview, Charles Babbage Institute; Arthur L. Norberg and Judy E. O'Neill, *Transforming Computer Technology: Information Processing for the Pentagon, 1962–1986* (Baltimore: The Johns Hopkins University Press, 1996); Paul E. Ceruzzi, *A History of Modern Computing* (Cambridge, MA: The MIT Press, 1998), 154–156; David Walden and Tom Van Vleck, eds., *The Compatible Time Sharing System (1961–1973): Fiftieth Anniversary Commemorative Overview* (Washington, DC: IEEE Computer Society, 2011); Martin Campbell-Kelly and Daniel D. Garcia-Swartz, "Economic Perspectives on the History of the Computer Time-Sharing Industry, 1965–1985," *IEEE Annals of the History of Computing* 30 (2008): 16–36; Martin Campbell-Kelly and Daniel D. Garcia-Swartz, "The History of the Internet: The Missing Narratives," *Journal of Information Technology* 28 (2013): 18–33.

[14] The Arpanet's history is well documented in Janet Abbate, *Inventing the Internet* (Cambridge, MA: The MIT Press, 1999); Norberg and O'Neill, *Transforming Computer Technology*; Katie Hafner and Matthew Lyon, *Where Wizards Stay Up Late: The Origins of the Internet* (New York: Simon & Schuster, 1996); and Peter H. Salus, *Casting the Net: From ARPANET to INTERNET and Beyond...* (New York: Addison-Wesley Publishing Company, 1995).

The most distinctive attribute of the Arpanet was its reliance on packet-switching, a new method of transmitting data across telephone lines that had been invented independently in the mid-1960s by Paul Baran at RAND Corporation and Donald Davies at Britain's National Physical Laboratory (NPL). Where traditional circuit-switched telephone networks required a direct, dedicated connection between users, packet-switched networks broke data into discrete blocks, which Davies dubbed "packets," that contained basic information about their places of origin and destination. Packets could be transmitted indirectly and asynchronously throughout nodes in the network before eventually arriving at their destination, where they would be reassembled into their original form.

Despite the common technical features of their inventions, Davies and Baran worked in radically different political economies. Davies, a research scientist in Britain's prestigious national laboratory, emphasized the scientific and technical virtues of packet switching as a more efficient mode of data communication. Baran, in contrast, worked for RAND, the American Cold War think tank. His publications famously emphasized the strategic virtues of decentralized and distributed networks: networks with multiple nodes were more robust and more likely than a circuit-switched, centralized network to survive a catastrophic event such as a nuclear attack.[15]

AT&T's role in the creation of the packet-switched Arpanet was negligible, other than its provision of the leased telephone lines that the Arpanet used as its physical infrastructure. Baran approached AT&T executives with his ideas in the mid-1960s, but they dismissed the project because it suggested that their analog network was inadequate. Despite some support within Bell Labs for Baran's ideas, Baran and AT&T could not get past what he called a "cultural impasse": AT&T engineers were deeply committed to their legacy network, and could not understand or admit how a digital, packet-switched network could be superior to the venerable Bell System. Baran, bemused by what he perceived to be AT&T's arrogance, recalled in a 1999 interview that one AT&T official told him, "It isn't going to work, and even if it did, damned if we are going to put anybody in competition to ourselves."[16] Baran continued, "I suspect the major reason for the difficulty in accommodating packet switching at the digital transmission level was that it would violate a basic ground rule

[15] Abbate, *Inventing the Internet*, 7–42; Paul Baran, "On Distributed Communications Networks," *IEEE Transactions on Communications* 12 (1964): 1–9; Donald Davies, "An Historical Study of the Beginnings of Packet Switching," *The Computer Journal* 44 (2001): 152–162; Lawrence G. Roberts, "The Evolution of Packet Switching," *Proceedings of the IEEE* 66 (1978): 1307–1313; Stephen J. Lukasik, "Why the Arpanet Was Built," *IEEE Annals of the History of Computing* 33 (2011): 4–21.

[16] Paul Baran, electrical engineer, oral history interview by David Hochfelder, 1999, IEEE History Center, Rutgers University, New Brunswick, NJ, USA. See also Paul Baran, oral history interview by Judy O'Neill, March 5, 1990, Menlo Park, California. Charles Babbage Institute, University of Minnesota, Minneapolis.

of the Bell System – everything added to the telephone system had to work with all previous equipment presently installed. Everything had to fit into the existing plan."[17] Many scholars have interpreted Baran's remarks as indications of the Bell System's bias against outside innovations. A more charitable interpretation can be advanced if we take into account the ideology of the Bell System's cautious approach to "engineering of the present" described in Chapter 4: there was no place for equipment or procedures that might unbalance the intricate technical and bureaucratic arrangements that AT&T executives had put in place to maintain high-quality and low-cost telephone service for the American public. AT&T's executives may look silly in hindsight for turning down the chance to take over the Arpanet, but if we look at it from their perspective in the 1960s, their decision was undoubtedly correct.

It was clear from the start of the Arpanet in 1969 that a digital, packet-switched communications system would require new institutions to coordinate the informal and geographically dispersed community of researchers who were building it. Several ARPA researchers began to meet regularly and named themselves the Network Working Group (NWG). At first, the NWG was led by Steve Crocker, a graduate student at UCLA, and consisted mostly of graduate students at other ARPA-funded computer research projects at UCLA, the University of California at Santa Barbara, the University of Utah, and MIT. The group also included representatives from ARPA contractors in Bolt, Beranek and Newman (BBN); RAND; and the Stanford Research Institute.

The NWG's chief technical objective was to create technical specifications to connect mainframes – known as "host" computers – to the network. Each mainframe would be connected to a small computer, designed by engineers from BBN, called an IMP (interface message processor).[18] Because IMPs were the common gateways to the Arpanet, their software for communicating with hosts – known as the Network Control Program (NCP) – would be the defining point of interconnection. ARPA managers and its contractors at BBN and the universities were most concerned with the operation of computer hardware. Software and protocols, which were new, were left to graduate students.[19]

The split between hardware and software was a recent development in computing, and perhaps just as significant for the history of standardization as for the history of computing.[20] Indeed, most instances of compatibility standards to this point – with the minor yet telling exceptions of programming languages

---

[17] Baran interview, IEEE History Center, 1999.

[18] In this era, a "small" computer was about the size of a refrigerator.

[19] Abbate, *Inventing the Internet*, 59–69.

[20] See Nathan Ensmenger, *The Computer Boys Take Over: Computers, Programmers, and the Politics of Technical Expertise* (Cambridge, MA: The MIT Press, 2010); Thomas Haigh, "Software in the 1960s as Concept, Service, and Product," *IEEE Annals of the History of Computing* 24 (2002): 5–13; Martin Campbell-Kelly, "The History of the History of Software," *IEEE Annals of the History of Computing* 29 (2007): 40–51.

such as FORTRAN discussed in Chapter 5 – concerned interfaces between tangible objects such as nuts and bolts or electrical plugs. In the computer age, the boundaries between technologies and organizations increasingly became problems that could be solved by software. We will see how the speed with which software could be written and deployed introduced new dynamics into negotiations over standards that had previously implied changes in the manufacture of artifacts.

Uncertainty over software's status was evident in the attitudes of NWG members, who initially adopted a tentative stance toward promulgating protocols and practices that all Arpanet hosts should follow. Because NWG emerged from a closed research community and had no relevance to contemporary production networks, it had no reason to be connected to any recognized standards body, such as the International Federation for Information Processing (IFIP) or X3. In the absence of any obvious external authority, Crocker initiated a document series in 1969 – the Request for Comments (RFCs) – whose name and structure perfectly captured NWG's informal and experimental ethos. Crocker later recalled, "Most of us were graduate students ... we kept expecting that an official protocol design team would announce itself." Crocker slowly realized that no such team was going to arrive. Much to his surprise, the team of graduate students and contractors in NWG had uncontested jurisdiction over protocol development for the Arpanet. Nevertheless, Crocker proceeded with caution. "In my mind," he recalled, "I was inciting the wrath of some prestigious professor at some phantom East Coast establishment." He remembered being overcome by a "great fear that we would offend whomever the official protocol designers were, and I spent a sleepless night composing humble words for our notes. The basic ground rules were that anyone could say anything and that nothing was official. And to emphasize the point, I labeled the notes 'Request for Comments.'"[21]

The RFCs soon became the vehicle for the Network Working Group to publish consensus statements and technical standards for the Arpanet – even though they were specifically intended *not* to become standards.[22] The Arpanet, fueled by ample funding, a collaborative ethos, and absence of external authority, grew quickly: it began with only four host computers in 1969, grew to twenty-three in 1971, and then to almost fifty hosts by the end of 1973 – including one in England. An active NWG sustained this early growth.

[21] Stephen D. Crocker, "The Origins of RFCs," in Joyce Reynolds and Jon Postel, eds. (1987), "The Request for Comments Reference Guide," RFC 1000, http://tools.ietf.org/rfc/rfc1000 (accessed September 25, 2013).

[22] Steve Crocker, oral history interview by Judy E. O'Neill, October 24, 1991, Glenwood, Maryland. Charles Babbage Institute, University of Minnesota, Minneapolis. See also Reynolds and Postel, eds., RFC 1000; RFC Editor, et al. (1999), "30 Years of RFCs, RFC 2555, http://tools.ietf.org/rfc/rfc2555 (accessed September 25, 2013); and Stephen D. Crocker, "How the Internet Got Its Rules," *The New York Times* (April 7, 2009), A27.

Its members developed and implemented network applications on a number of host computers, including programs for file transfer, remote job entry, and electronic mail.[23]

As the Arpanet grew, IPTO managers and engineers in NWG recognized that they needed to cultivate contacts with network researchers in the United States and abroad. The opening act of this collaboration was the First International Conference of Computer Communications held in Washington, D.C., in October 1972. After months of preparations orchestrated by new IPTO director and former MIT and BBN engineer Robert Kahn, the Arpanet went on public display. The Arpanet's first live demonstration stunned the audience with demonstrations of applications such as interactive graphics, methods of reading Associated Press reports over the network, and several chess programs. It was a resounding success – even if utterly baffling to the ten AT&T executives who attended and reportedly declined ARPA's offer to operate the Arpanet or even purchase it outright.[24]

The meeting in Washington did more than host the Arpanet's coming-out party; it also provided an ideal opportunity to invigorate international cooperation. A growing number of networking experiments outside the United States had already begun to take shape, including two significant projects in France and Great Britain. Louis Pouzin, the computer scientist working for IRIA, was designing a packet-switched network called Cyclades; Donald Davies, the computer scientist at Great Britain's NPL, had begun his packet-switching experiments in the mid-1960s. Additionally, several PTT national monopolies in Europe were evaluating packet-switching technology, and the European Common Market had asked Derek Barber from Britain's NPL to direct the creation of a European Informatics Network. These researchers came together to pursue their shared objective: to design new network standards for a new era of digital, packet-switched communication.

The network researchers assembled in Washington, D.C. in October 1972 – including Pouzin, Davies, Barber, and leaders of the Arpanet community – took advantage of the unprecedented occasion to form the International Network Working Group (INWG). An Arpanet researcher, Vinton Cerf, became the first chairman of INWG. Cerf was among the dozens of graduate students in the late 1960s whom ARPA paid to work on the packet-switched Arpanet. After Cerf earned his PhD in 1972 from the University of California, Los Angeles, he became an assistant professor at Stanford's Digital Systems Laboratory

[23] Abbate, *Inventing the Internet*, 59–69.

[24] Robert Kahn (1972), "Demonstration at International Computer Communications Conference," RFC 371, http://www.ietf.org/rfc/rfc371.txt (accessed September 25, 2013); Hafner and Lyon, *Where Wizards Stay Up Late*, 176–186; Abbate, *Inventing the Internet*, 123–127. *Computer Networks: Heralds of Resource Sharing* (1972) was a documentary prepared for ICCC '72 but not completed in time. It is now available from http://www.archive.org/details/ComputerNetworks_TheHeraldsOfResourceSharing (accessed August 23, 2011).

and continued to work on ARPA-sponsored projects. He was, at the time, a relatively unknown member of the international networking community.[25]

The International Network Working Group, 1972–1975

In the early 1970s, Cerf collaborated with Robert Kahn to develop a new communications protocol for delivering packets between different types of packet-switched networks. Cerf and Kahn, like Licklider, mobilized ARPA's resources as leverage to advance their research agenda and to bring the diffuse research community closer together: Kahn had the power to issue ARPA contracts to network researchers, and Cerf soon moved to Washington, D.C., to join him and champion their approach to internetworking.

Their ambition, however, did not fare well within INWG. Much to his dismay, Cerf soon encountered political, technical, and bureaucratic obstacles that frustrated his efforts to promote ARPA network protocols as international standards. Researchers in Europe did not have access to the ample funding and political cover that Cerf enjoyed with ARPA. Consequently, many Europeans came to INWG with a graver and more sophisticated understanding of the political and economic pressures that surrounded them.

The founders of INWG hoped the group would become an international forum that mimicked the informal organizational structure and culture that had been taking place through the Arpanet's NWG. For example, researchers at INWG's first meeting in October 1972 established a document series of General Notes modeled on the tentative and research-oriented spirit of the Request for Comments used to communicate research questions and findings within the Arpanet community. The first INWG General Notes (numbers 0, 1, and 2) carry no founding charter or any grandiose claims of authority. Instead, they pragmatically and matter-of-factly document the creation of two subgroups to explore the technical aspects of network interconnection. Subgroup 1, on Communication System Requirements, set out to study the use of interfaces and gateways that could "allow convenient communications between computers and terminals" across more than one packet-switched network. Subgroup 2, whose members included Cerf, Barber, and Pouzin, prepared to examine more specific details of the Host-Host Protocol Requirements. Their objectives, articulated in the notes of their first meeting on October 24, 1972, include commitments to collaboration such as "select a design philosophy," "design the Host-Host protocol," and, most ambitiously, "agree to implement the protocol design by a certain date."[26]

---

[25] Alexander McKenzie, "INWG and the Conception of the Internet: An Eyewitness Account," *IEEE Annals of the History of Computing* 33 (2011): 66–71; Vinton Cerf, "International Packet Network Working Group Mailing List," November 7, 1972, INWG 3, McKenzie Collection; Abbate, *Inventing the Internet*, 123–125.

[26] Vinton Cerf, "International Network Working Group, Subgroup: Communication System Requirements," October 23, 1972, INWG 0, McKenzie Collection; Vinton Cerf, "International

Both subgroups acknowledged they would have to frame their technical work in a way that could inform the ongoing efforts of the "official protocol designers" that Steve Crocker anticipated when he created the Arpanet Request for Comments series in 1969. Their desire to design protocols that would connect private and public networks – that is, telecommunication networks that, in most countries, were operated by national monopolies – represented a major departure from the NWG's interest in protocols for one private network, the Arpanet.[27]

Private and experimental networks could afford to develop standards informally, but formal standards for public and international networks fell under the jurisdiction of the International Telecommunications Union (ITU). The ITU, founded in 1865 as the International Telegraph Union, became affiliated with the United Nations in 1947 and operated in a similar fashion: it consisted of representatives that had been chosen by sovereign national governments.[28] In 1956, the ITU created the International Telegraph and Telephone Consultative Committee (CCITT) to harmonize the operation of international telecommunications systems, to issue "Recommendations" on technical features of these systems, and to study emerging issues and plan for future developments. Voting members of CCITT were drawn from sovereign nations, and the group also invited (on a nonvoting basis) contributions from approved professional, scientific, or industrial organizations. The CCITT worked on an orderly schedule: Plenary Assemblies were held every four years, separated by four-year Study Periods during which Study Groups (and other small groups) could examine various issues and prepare agenda items for the next Plenary Assembly.[29]

At its 1968 Plenary Assembly, the CCITT signaled the intentions of the telecommunications companies to extend their jurisdiction over new, computer-based digital networks by creating a Joint Working Party on New Data Networks. At the 1972 Plenary Assembly, the CCITT took three significant actions. First, it published eleven Recommendations with the new, soon-to-be-famous "X" prefix (including Recommendations X.1, X.20, and X.30 that specified basic interfaces between circuit-switched public and private data networks).

Packet Network Working Group: Report of Subgroup 1 on Communication System Requirements," October 24, 1972, INWG 1, McKenzie Collection; Vinton Cerf, "International Packet Network Working Group: Report of Subgroup 2 on HOST-HOST Protocol Requirements," October 25, 1972, INWG 2, McKenzie Collection.

[27] Cerf, INWG 0.

[28] Most national representatives worked for the state-owned monopoly PTT. As the United States had no such monopoly, the Department of State chose the American delegates.

[29] D. M. Cerni, *The CCITT: Organization, U.S. Participation, and Studies toward the ISDN* (Washington, DC: U.S. Department of Commerce, 1982); Gerald C. Schutz and George E. Clark, "Data Communication Standards," *Computer* (February 1974): 32–37; Susanne K. Schmidt and Raymund Werle, *Coordinating Technology: Studies in the International Standardization of Telecommunications* (Cambridge, MA: The MIT Press, 1998), 43–84. In 1993, CCITT changed its name to ITU-T.

Second, it formalized and extended the mission of the Joint Working Party into Study Group VII on New Networks for Data Transmission. Study Group VII defined a twenty-three-point program of work for the 1972–1976 study period, an agenda that clearly illustrated its comfort with traditional circuit-switched networks as well as its unfamiliarity with packet-switched networks. Only one of the twenty-three points mentioned packet switching, and that was only in the vaguest of terms: "Should the packet-mode of operation be provided on public data networks, and if so, how should it be implemented?"[30]

CCITT's somewhat inexpert foray into the world of packet switching weighed heavily on the minds of the experts who gathered into INWG in October 1972. From the start, a strong consensus existed within INWG that the group should be contributing to CCITT's process for creating formal international standards. The report of the first meeting of INWG subgroup 1 on Communication System Requirements – circulated as INWG 1 – defined two objectives for further work. The first, discussed earlier, was technical: to consider interfaces and gateways for connecting private and public networks. The second objective was strategic: "to consider what recommendations to make on packet-switched networks, and how to provide for acceptance of these recommendations throughout CCITT and other international organizations."[31] INWG members understood that their group was a means for coordinating research as a basis for standardization that would occur elsewhere. Accordingly, INWG General Notes exhibited an increasing preoccupation with the CCITT process.

In early 1973, INWG researchers created an institutional arrangement that would allow them to contribute to CCITT: an affiliation with IFIP. INWG would become chartered as IFIP Working Group 6.1, under the umbrella of IFIP Technical Committee 6 on Data Processing. IFIP, as discussed in Chapter 5, was a federation of national professional societies designed to stimulate, coordinate, and disseminate research in the new field of information processing. By the early 1970s, IFIP had gained sufficient stature that INWG researchers saw it as a "catalyst and focal point for conceptual and technological developments" – just as Isaac Auerbach had envisioned when he created the group in 1960.[32] Pouzin was an active member, and he, Cerf, and Alex Curran, the chairman of IFIP Technical Committee 6, agreed to bring INWG under the IFIP

---

[30] Marvin A. Sirbu and Laurence E. Zwimpfer, "Standards Setting for Computer Communication: The Case of X.25," *IEEE Communications Magazine* 23 (1985): 38.

[31] Cerf, INWG 1.

[32] Alex Curran and Vinton Cerf, "The Work of IFIP Group 6.1," *ACM SIGCOMM Computer Communication Review* 6 (1975): 18–27; Vinton Cerf, "Affiliation of INWG with IFIPS," April 1973, INWG 24, McKenzie Collection; Vinton Cerf, "INWG Plenary Meeting, June 7 and June 8," June 7, 1973, INWG 28, McKenzie Collection; James Pelkey, "Entrepreneurial Capitalism and Innovation: A History of Computer Communications, 1968–1988" (2007), http://www.historyofcomputercommunications.info/Book/6/6.0-Overview.html (accessed September 25, 2013).

umbrella. Hubert Zimmermann, one of Pouzin's collaborators on the Cyclades project, later noted that INWG's affiliation was a natural progression for the group's collaborative effort: "IFIP was the right place for standards to mature in international arena before they were passed on to standardization bodies. It was a meeting place for the scientific community."[33]

The affiliation with IFIP further energized and emboldened the researchers in INWG. They sensed a real opportunity to influence international standards for packet switching because, as the British computer scientist Donald Davies noted to INWG members in late 1972, "almost all the [CCITT] work was, in fact, concerned with circuit switching."[34] As we will see, however, leading members of CCITT's packet-switching projects were suspicious of INWG's affiliation with IFIP. The French telecommunication engineer Rémi Déspres, for example, later noted that IFIP "was a largely academic group, and which had basically no interest in operators' objectives and was even an anti-PTT front. And Louis Pouzin was very active in that, and that was the place where coming from the PTT you would not be welcome anyway." The cultural divide between telecom and computer engineers thus was made visible in the different organizations they used to develop standards.[35]

The General Notes circulated among INWG members in 1973 and 1974 reflect the extent to which the group's protagonists – especially the leaders of the Arpanet, Cyclades, and NPL experimental networks – worried about how they could influence the formal international standards process. In early 1974, Canadian researcher Dave McLimont prepared a "Thinkpiece" that he planned to submit to CCITT to stimulate discussion on "the procedure for the interworking of packet switching networks." McLimont used a recent paper by Cerf and Kahn (circulated in September 1973 as INWG 39, and subsequently published in the *IEEE Transactions on Communications* in May 1974)[36] as a basis for defining "a number of problem areas" that CCITT should study in order to "reduce the risk that packet switching networks being developed will be virtually incompatible."[37] Cerf and other INWG leaders also referred more regularly and explicitly to the International Organization for Standardization (ISO) when they talked about international standards. For example, Cerf's April 1974 outline of the "Objectives and Organization of TC6.1" began:

This working group has several long and short term objectives. On the short term, we are committed to performing experiments, both real and on paper, which will improve

33 Hubert Zimmermann, interview by James Pelkey, May 25, 1988, courtesy of James Pelkey.
34 Donald Davies, "CCITT Documents – APV No. 21, 22, 23, 24," December 1972, INWG 11, McKenzie Collection.
35 Rémi Déspres, oral history interview by Valérie Schafer, May 16, 2012, Paris, France. Charles Babbage Institute, University of Minnesota, Minneapolis.
36 Vinton G. Cerf and Robert E. Kahn, "A Protocol for Packet Network Intercommunication," *IEEE Transactions on Communications* Com-22 (1974): 637–648.
37 Dave McLimont, "A CCITT Thinkpiece," January 15, 1974, INWG 45, McKenzie Collection; Donald Davies, "CCITT Contribution by IFIP WG6.1," August 1974, INWG 69, McKenzie Collection.

our understanding of technical issues in packet network interconnection. The results of this effort are to be supplied as source material to standards-making bodies such as CCITT and ISO. TC6.1 will, through TC6, maintain liaison with CCITT Study Group VII and ISO TC97.

In the course of planning and executing these experiments, it is to be expected that we will also uncover important social, economic, and political issues. Over the long term, our objective is to seek resolution of these issues by calling attention to their existence, and perhaps, by recommending courses of action.[38]

Because the number of researchers in INWG was growing steadily – almost 100 names appeared on the group's mailing list by mid-1974 – Cerf proposed to restructure the group into four working parties: Internetwork Experiment Committee (chaired by Peter Kirstein from the UK), Protocol Design Committee (chaired by Pouzin), Social Issues Committee (chaired by Terry Shepherd from Canada), and Legal and Political Issues Committee (chaired by Franklin Kuo at the University of Hawaii).[39]

Kuo, the only ARPA contractor out of the four chairmen of INWG's new working parties, hosted a meeting of the Legal and Political Issues Committee as part of the general INWG meeting in January 1974. Kuo's group included researchers from the United States, Germany, Japan, and New Zealand who had "extensive discussions on political, social, and economic issues that must be addressed before network interconnections can truly be feasible." INWG 52 summarized their frank and wide-ranging conversations on a number of topics, including the variety of motivations for interconnection, the types of interconnection arrangements that would be "easier politically," the restrictions and regulatory issues that PTT administrations might impose, network accounting issues such as costs and tariffs, privacy and security considerations of databases and files, and customs and licensing issues inherent in the communication of data across national boundaries. (It is interesting to note, as an aside, that many issues the six researchers discussed in Honolulu, such as pricing for network access, privacy and security of data, uncertainty in regulatory regimes, and "possible 'smuggling' of software programs," remain as central problems within information policy debates almost forty years later.)[40]

In his brief summary of the group's discussion of standards, Kuo argued that INWG researchers should not rush into promoting their experiments for international standardization:

The issue of standards is probably the most difficult and important problem that networks people must face.... It was generally agreed that we do not really know what standards for network interconnection are necessary and even desirable, and that standards

---

[38] Vint Cerf, "INWG Meeting in Stockholm, August 10–11, 1974," April 1, 1974, INWG 53, McKenzie Collection.
[39] Cerf, INWG 53; Vint Cerf, "Minutes of the Stockholm Meeting of IFIP WG6.1, August 10–11, 1974 (Aboard the good ship *BORE I*)," December 20, 1974, INWG 73, McKenzie Collection.
[40] Franklin F. Kuo, "INWG Workshop Report: University of Hawaii, 6–7 January 1974," March 19, 1974, INWG 52, McKenzie Collection.

must evolve after a number of successful interconnections. Therefore we agreed that INWG should not attempt to define a set of standards for network interconnection for ISO and CCITT, but rather *work to prevent the issuance of early standards* to avoid ending up with restrictive or unusable protocols.[41]

In the spring of 1974, Kuo revised the group's summary into a paper published in the Proceedings of the 1974 International Conference on Computer Communication in Sweden. In his revised passage on standards, he expanded on his cautionary logic: "It is the author's view that it is important to prevent the issuance of premature standards by these groups," which he named as CCITT, ISO, and, conspicuously, INWG. "Packet communication networks," he warned, "are still in an evolutionary stage. Early standards could only freeze the development process."[42]

Kuo's plea to continue research and development before premature standardization was intellectually sound, but his advice quickly became lost in the rush of events. The drive toward international standardization did not slow down; rather, competition and its cultural complements – urgency, cunning, and tribalism – only increased. In 1974 and 1975, new packet-switched networking architectures were in abundance, inspired and informed by developments in the Arpanet in the United States, Cyclades in France, and NPLnet in Britain. Several of these new architectures came from industry leaders, such as IBM's System Network Architecture (1974), Digital's DECnet (1975), and Xerox's PARC Universal Packet (1975). IBM's entry, predictably, captivated industry analysts who were well aware of the company's market power, its proven ability to impose standards unilaterally, and its ambition to use its new System Network Architecture as a means to, as one customer put it, "lock in users to IBM products and make them subject to IBM's whims."[43] In addition to the proprietary packet-switching products being marketed by IBM, Digital, Xerox, and other manufacturers, plans for public packet network services were popping up around the world. These networks included Transpac in France, Datapac in Canada, EPSS in the United Kingdom, DDX-1 in Japan, and networks in the United States operated by start-up companies such as Packet Communications, Tymnet, and Telenet.[44]

[41] Kuo, INWG 52 (emphasis added).

[42] Franklin F. Kuo, "Political and Economic Issues for Internetwork Connections," May 2, 1974, INWG 58, McKenzie Collection. Kuo revised INWG 58 for publication as Franklin F. Kuo, "Political and Economic Issues for Internetwork Connections," *ACM SIGCOMM Computer Communication Review* 5 (1975): 32–34.

[43] David Loehwing, "Computer Networks: Data Communications Have Spread Out From the Center," *Barron's National Business and Financial Weekly* (February 16, 1976), 8.

[44] Sirbu and Zwimpfer, "The Case of X.25," 36–37; Abbate, *Inventing the Internet*, 149; Tony Rybczynski, "Commercialization of Packet Switching (1975–1985): A Canadian Perspective," *IEEE Communications Magazine* (December 2009): 26–32; Rémi Déspres, "X.25 Virtual Circuits – Transpac in France – Pre-Internet Data Networking," *IEEE Communications Magazine* (November 2010): 40–46.

As expected, CCITT moved decisively to set standards for these emerging data networks. Throughout 1975, a small group of engineers who had worked with these public networks met under the auspices of CCITT Study Group VII to define X.25, a standard interface for public and private packet-switched networks. The designers of X.25 faced a fundamental technical decision that would have tremendous economic and political consequences: whether to use virtual circuits or datagrams. Virtual circuit service, as its name suggests, closely mimicked the circuit-switched networks that telephone companies had operated for decades. Virtual circuits created, in effect, a *connection* between two terminals at the edges of the network through which all packets would pass. Complex transmission tasks – such as ensuring that every packet arrived at its destination in the correct sequence – would be managed within the network, not the computer terminals at the ends of the network.

Datagram service, on the other hand, embedded transmission information such as packet address and sequence data within each individual packet. The datagram concept emerged from the work of Pouzin and his Cyclades team in France. Datagrams, as the Cyclades engineer Jean-Louis Grangé recalled, "really simplified the mechanism of switching packets because it made all packets totally independent from the others." In Pouzin's description, "The essence of datagram is connectionless. That means you have no relationship established between sender and receiver. Just things go separately. One by one. Like photons." As a result, it was up to the computers (or hosts), and not to the network itself, to reassemble packets, which may have arrived out of order, and to verify that the transmission was complete. It was this style of network architecture – known as *connectionless* service – that had been pioneered in Cyclades, and was now the primary topic of conversation within INWG. This is the essential point: INWG was created to unite datagram researchers and present a unified front to the international arena. Computer manufacturers, along with Cerf, Pouzin, and their collaborators, believed that datagrams had many technical advantages over virtual circuits, on the grounds that they were more robust, more flexible, and could provide faster service as the computers at the edges of the network became faster and more powerful. Network users would benefit, Pouzin insisted, because "they don't have to worry about opening a circuit, how long it's going to last, how much does it cost, and so on. In a virtual circuit system," Pouzin continued, "the virtual circuit is a resource that requires management and optimization and all kinds of things."[45]

The debate over virtual circuits versus datagrams, at one level, was a debate over the efficient use of resources that were in short supply at the time – bandwidth and processing power. Advocates of datagrams, however, more frequently viewed the debate in terms of *control*. With datagrams, they could push more

[45] Jean-Louis Grangé, oral history interview by Andrew L. Russell, April 3, 2012, Paris, France. Charles Babbage Institute, University of Minnesota, Minneapolis; Pouzin interview, Charles Babbage Institute.

tasks out of the network and to the computers at the end points – computers that were rapidly becoming more powerful and more flexible. The potential economic consequences of the technical decision between virtual circuits and datagrams were profound: where virtual circuits would allow the staid telephone monopolies to control all activity over the network and to guarantee a quality of service for its existing customers, datagrams would splinter control by making it possible to distinguish between data transport across the network and data applications that resided on host computers. With virtual circuits, telephone companies would be able to keep all applications inside the network; with datagrams, computer companies and users could take control over end-to-end transmission and thus provide services and applications that the telephone companies could not control.

The small group that drafted X.25 – with exclusive support for virtual circuits – included the Canadian telecom engineer Tony Rybczynski, the French PTT researcher Rémi Déspres, and the American computer scientists Larry Roberts and Barry Wessler. Rybczynski and Déspres worked on behalf of officials in the Canadian and French governments who wanted to quickly roll out their Datapac and Transpac networks. Roberts and Wessler, however, had strong connections to the Arpanet and its community. Roberts, as director of ARPA's IPTO from 1969 to 1973, presided over the team that created the Arpanet (Déspres called him "Arpanet's father"). Wessler completed his undergraduate and master's degrees at MIT before moving to Washington, D.C., to be a program manager under Roberts at ARPA from 1967 to 1970. Wessler them moved to Utah to complete his PhD at the University of Utah's ARPA-funded Center of Excellence over the next three years. In 1973, Roberts and Wessler became entrepreneurs and viewed the CCITT process as the best way to ensure that their start-up company, Telenet, would profit from government investments in packet-switching research. Neither Roberts nor Wessler were unshakable advocates of either virtual circuit or datagram service; their interest, rather, was to get a standard in place so they could take advantage of the limited amount of start-up capital at their disposal and thus be among the first movers in the commercial development that would inevitably follow.[46]

The rigid schedule for CCITT proposals – it would only approve Recommendations at its Plenary Assemblies every four years – meant that the small design group had to move quickly – too quickly to resolve the many technical issues that were the subject of ongoing research and experimentation. In the absence of technical clarity, the X.25 design team instead created the illusion of compromise by "disguising some of the differences of opinion in the wording of the draft recommendation." Critics correctly suspected that ambiguous writing was a sign that the design group had struggled to reconcile competing agendas – particularly between the economic interests of the

---

[46] Déspres, "X.25 Virtual Circuits."

telecom monopolies and the computer researchers' desire to move quickly and create the foundations for a new breed of communication networks.[47]

One notable decision of the X.25 design group was to favor virtual circuits over datagrams. The decision was a pragmatic one that followed logically from the power dynamics of CCITT. As Rybczynski later noted, telephone carriers unanimously supported virtual circuits because they "better fitted their service models, existing applications and user expectations." X.25 also fit well with the networking approach of a dominant user, IBM. IBM's System Network Architecture, according to IBM France engineer Marc Levilion, transmitted data between computers first by considering a "map of the routes available," and then by sending "each new communication between applications" along "one particular physical route across a network." This emphasis on fixed routes was, Levilion recalled, "very close to the philosophy of X.25." Such a consonance of interest was important for Déspres, who knew that the French telecom monopoly's users were "companies using IBM computers.... IBM had at that time more than half the market." Virtual circuits, from the vantage point of the PTTs and dominant computer firms, would provide an incremental and predictable transition into the world of digital data communications – a satisfactory solution for the leading providers and users of standards at the time. Datagrams, however, promised disruption. Datagram designs seemed likely to usher in a new era of unpredictable and radical technological change in networking – the very thing that both the monopoly telephone carriers and start-up services such as Telenet wished to avoid.[48]

With their agendas in alignment behind a specification that they agreed was good enough, CCITT approved X.25 as a Recommendation at its September 1976 Plenary Assembly. The strategic implications were striking, in the summary of historian Janet Abbate: "The CCITT's protocols were deliberately designed to put control of the network in the hands of the PTTs by locating most of the functionality within the network rather than in the subscribers' host computers." The technical and economic logic of the PTT position was clear: they could either force all of their potential customers to build their own interfaces for multiplexing, flow control, error control, and so on between their own operating system and the data network; or they could offer a single, standardized interface to entice those customers.[49] Nevertheless, opponents and critics of X.25 in the computer industry and research community assailed

[47] Rybczynski, "Commercialization of Packet Switching," 26–31; Déspres, "X.25 Virtual Circuits"; Sirbu and Zwimpfer, "The Case of X.25," 37–41; Abbate, *Inventing the Internet*, 152–161; Valérie Schafer, "Circuits Virtuels et Datagrammes: Une Concurrence à Plusieurs Échelles," *Histoire, Économie & Société* 26 (2007): 29–48.

[48] Rybczynski, "Commercialization of Packet Switching," 26; Déspres interview, Charles Babbage Institute; Marc E. Levilion, oral history interview by Andrew L. Russell, April 2, 2012, Paris, France. Charles Babbage Institute, University of Minnesota, Minneapolis.

[49] Alex McKenzie helped me appreciate the nuances of the PTT's difficult position. Personal communication, October 9, 2011.

the Recommendation as hastily designed, unstable, and motivated more by economic and strategic concerns than by technical merit or market acceptance, but their objections mattered little. The European PTTs had cleverly mobilized the standards process to advance their goal to monopolize new markets for data communications. All that was left for them to do was use X.25 to preserve their end-to-end control and extend their jurisdiction into the computer terminals attached to the edges of the network.[50]

Pouzin, the architect of the French Cyclades network and the world's leading theorist and engineer of datagram networks, emerged as the most outspoken critic of X.25 and the motivations of its supporters. By 1975, he was already a vocal advocate of connectionless datagrams and the simplicity he had designed into the Cyclades network. On a strictly technical level Pouzin was not impressed with X.25. He argued that X.25 would impose new problems on network users by introducing new incompatibilities that "cannot be resolved without additional complexity and cost to the user."[51] He also, somewhat unusually for a protocol designer of such technical acuity, identified some of the broader trends of political economy and industrial policy that framed the debate between virtual circuits and datagrams. The conclusion of his January 1976 essay, "Virtual Circuits vs. Datagrams: Technical and Political Problems," indicated that he understood with great lucidity the forces swirling around him.

The controversy DG [datagrams] vs. VC [virtual circuits] in public packet networks should be placed in its proper context.

First, it is a technical issue, where each side has arguments. It is hard to tell objectively what a balanced opinion should be, since there is no unbiased expert.... Second, the political significance of the controversy is much more fundamental, as it signals initial ambushes in a power struggle between carriers and the computer industry. Everyone knows in the end, it means IBM vs. Telecommunications, through mercenaries. It may be tempting for some governments to let their carrier monopolize the data processing market, as a way to control IBM. What may happen, is that they fail in checking IBM, but succeed in destroying smaller industries. Another possible outcome is underdevelopment.... It looks as if we need some kind of peacemaker to draw up boundary lines before we all get in trouble.[52]

Pouzin's concern for "smaller industries" should not be interpreted as a cynical or self-interested statement, since he was not an entrepreneur himself but rather a scientist employed in a French government research institute. Rather, his comments indicate his ideological stance against centralized control of all

---

50 Sirbu and Zwimpfer, "The Case of X.25," 40–44; Abbate, *Inventing the Internet*, 160.
51 Louis Pouzin, "The Case for a Revision of X 25," *ACM SIGCOMM Computer Communication Review* 6 (1976): 17–20.
52 Louis Pouzin, "Virtual Circuits vs. Datagrams: Technical and Political Issues," January 1976, INWG 106, McKenzie Collection. Subsequently published in the *1976 Proceedings of the National Computer Conference* (New York: AFIPS, 1976), 483–494.

forms – particularly within state monopolies and IBM – and in favor of a more entrepreneurial and distributed structure for the data communications industry. Such beliefs made Pouzin something of an iconoclast among French policy makers, who believed that heavily subsidized "national champion" firms, rather than market competition, would ensure the viability of the struggling French computer industry.

It is tempting to see Pouzin, in some ways, as a French version of Bernard Strassburg, chief of the Common Carrier Bureau at the American FCC. In 1971, thanks to Strassburg's initiative, the FCC defined a standard of "maximum separation" under which no common carrier (such as AT&T and the Bell operating companies) could enter data processing markets except through an entirely separate corporate entity. The FCC and federal courts agreed with Strassburg's logic that structural separation between the monopolized telephone industry and the chaotic computer industry would not allow "AT&T or the other telephone companies to so intermix their operations and make a mess of both markets."[53]

Pouzin was a more articulate critic of centralized control but lacked Strassburg's regulatory standing and political support. He was, nevertheless, fearless in warning his colleagues in the computing research community about the treacherous political and economic landscape in front of them. Perhaps his bravery was a function of the political sensibilities he developed in the 1960s, an era when "intellectually I was in favor of the insurrection."[54] In an August 1976 conference paper titled "The Network Business – Monopolies and Entrepreneurs," Pouzin warned that "protectionism tends to soften competitiveness" and that "following their familiar business practices, [the telephone monopolies] dislike the computer jungle, and mean to introduce some discipline: standards and monopoly." He continued by outlining the problems the telephone monopolies faced in their attempts to set standards for data networks: "Interface standards are no longer limited to technical matters, since they determine boundaries between two competing domains." Finally, he turned his attention once again to the X.25 Recommendation that would be approved by CCITT in its Plenary Assembly one month later:

This draft standard, called X25, is the first in a bag of tricks devised to give carriers a captive market of terminal access and services.... What the squadron of carriers tends to overlook is that computer specialists are old hands at masking out interfaces they do not like. They already have a few well proven schemes at hand, which will bury X25 under a protective shield. The game is similar to anti-missile missiles. What counts is to be ahead by one trick.[55]

[53] Bernard Strassburg, interview by James Pelkey, Washington, D.C., May 3, 1988, courtesy of James Pelkey.

[54] Pouzin interview, Charles Babbage Institute.

[55] Louis Pouzin, "The Network Business – Monopolies and Entrepreneurs," (n.d. 1976), INWG Legal/Political Note 6, McKenzie Collection.

Pouzin's verve – one can almost see the twinkle in his eye when he referred to the "well proven schemes" that the "old hands" in the computer world were ready to deploy – can be explained, in part, by returning to the technical debates that took place within INWG in 1974 and 1975. There were, as we have seen, a great variety of developments outside of INWG that added urgency to their work, including the proliferation of commercial networks to meet growing customer demand, IBM's ominous entry into the crowded field, and the rapid schedule for the approval of X.25. The combined force of these developments in 1974 and 1975 exacerbated tensions within the group, which was struggling to achieve the goal they had declared in their first meeting in October 1972: to design a common host-to-host protocol for packet-switched datagram networks and implement that design. As the history of standardization so often demonstrates, common standards had the potential to generate network externalities, avoid technological segmentation, and create new opportunities for market competition and scientific collaboration. In this case, such common standards continued to elude the researchers in INWG.

Collaboration, Competition, and a Meaningless Consensus

An energetic mix of collaboration and competition emerged with INWG researchers during 1973, as those who had been working independently on packet-switching experiments in the United States, France, and England exchanged ideas, met in person, refined their designs, and continued to build implementations.[56] At the second INWG meeting in June 1973, a small group of researchers including engineers from the Arpanet (Cerf, Kahn, McKenzie, and David Walden), Cyclades (Hubert Zimmermann), and NPL (Roger Scantlebury), drafted an International Transmission Protocol that Cerf edited and circulated as INWG General Note 28. During the summer of 1973, Cerf and Kahn incorporated French and British ideas into their specification for a Transmission Control Program (TCP)[57] that was the foundation for "Internetwork Communication" – a term they used to describe the "interconnection of individual packet switching networks." In the cover letter Cerf wrote as a preface to the Cerf/Kahn paper that was circulated as INWG 39, Cerf took a collegial tone and emphasized that "the document is in DRAFT form and we

[56] Abbate, *Inventing the Internet*, 124–127; Valérie Schafer, "Appropriating Packet Switching Networks, Making Cyclades Network: The Threat of American Dominance in the 1970s and the French Answer through Cyclades," courtesy of the author.

[57] At first, the acronym "TCP" referred to the Transmission Control Program that Cerf and Kahn drafted in 1973. By 1977, however, Cerf, Postel, and other ARPA contractors used the acronym "TCP" to refer to the Transmission Control *Protocol* (rather than Program). The shift from "Program" to "Protocol" indicated that the Transmission Control Program had been revised to include two distinct functions, which Cerf and his colleagues referred to as the Transmission Control Protocol and the Internet Protocol.

hope that, by presenting this material in its unpublished form, we will stimulate critical remarks and perhaps even counter-proposals from INWG."[58]

As Cerf and Kahn continued to refine and implement TCP on the Arpanet in late 1973, a counterproposal was already in the works. In December 1973, Hubert Zimmermann and Michel Elie circulated INWG 43, an outline of an alternative Host-Host protocol that was being developed in Cyclades. The next March, Pouzin published INWG 60, "A Proposal for Interconnecting Packet Switching Networks," in which he introduced the term "Catenet" to describe "an aggregate of networks [that] behave as a single logical network."[59] One month later, Zimmermann and Elie distributed a complementary document, INWG 61, "Transport Protocol: Standard Host-Host Protocol for Heterogeneous Networks." Their design, as Alex McKenzie described in his excellent technical history of INWG, aimed both to incorporate and improve on the proposals that had been circulated within INWG during the past year. The proposal in INWG 61 shared many technical features with INWG 39 (Cerf and Kahn's TCP draft) but described a different way to handle packet fragmentation and reassembly that, Zimmermann and Elie argued, would "greatly simplify implementation" across different types of packet-switched networks.[60]

For the remainder of 1974, the Arpanet and Cyclades design teams settled into two opposing camps, neither of which could establish decisive technical superiority or first-mover advantages. Both teams had strong leadership, but the Arpanet community used its advantages in size and funding to move its drafts and proposals closer to operational protocols. At the INWG meeting in August, Pouzin reported that the protocol described in INWG 61 was being tested in Cyclades and in links with other European networks. In the meantime, TCP built momentum within the large Arpanet community, and the protocol described in INWG 39 was quickly refined and had already advanced to the next stage: specification. Cerf continued to implement, test, and publish improvements to TCP with Kahn, who, as IPTO's director, was personally invested in the success of the Arpanet. Kahn also was able to increase financial support for TCP research by issuing contracts to Stanford, BBN, and University College London to implement TCP.[61] In May, Cerf and Kahn published a description of TCP in the *IEEE Transactions on Communications* (a paper destined to be a landmark in the history of the Internet). In December, Cerf and

[58] Vint Cerf and Robert Kahn, "HOST and PROCESS Level Protocols for Internetwork Communication," September 13, 1973, INWG 39, McKenzie Collection.

[59] Louis Pouzin, "A Proposal for Interconnecting Packet Switching Networks," March 1974, INWG 60, McKenzie Collection. On "Catenet," see also Vint Cerf, "The Catenet Model for Internetworking," July 1978, Internet Engineering Note 48, http://www.rfc-editor.org/ien/ien48.txt (accessed October 24, 2012).

[60] Hubert Zimmermann and Michel Elie, ""Standard Host/Host Protocol for Heterogeneous Computer Networks," April 1974, INWG 61, McKenzie Collection; McKenzie, "INWG and the Conception of the Internet," 67–69.

[61] Robert E. Kahn, oral history interview by Judy O'Neill, April 24, 1990, Reston, Virginia. Charles Babbage Institute, University of Minnesota, Minneapolis.

two of his students at Stanford, Yogen Dalal and Carl Sunshine, published the "Specification of Internet Transmission Control Program" as contributions to both the Arpanet Request for Comments series (where it was RFC 675) and as INWG General Note 72.[62]

Arpanet researcher Alex McKenzie, deeply concerned that INWG was splintering into two rival camps, proposed an "internetwork Host-to-Host protocol" in 1974 to try to move the community forward together. McKenzie conceived of his proposal as "a synthesis of the two main streams of protocol thinking currently being discussed in INWG": the Arpanet TCP approach described by Cerf and Kahn, and the Cyclades approach described by Zimmermann and Elie. McKenzie acknowledged that it was correct for INWG, a research-oriented group, to "foster as many different protocol experiments as possible," but he argued that "there are almost no significant differences between the current formulations of these two ideas. Thus," he continued ominously, "independent experiments would produce little 'light,' although possibly much heat."[63] McKenzie's proposal was intellectually sensible but strategically untenable. Both camps within INWG continued to pursue their own variations of host protocol design – despite McKenzie's best efforts at technological diplomacy (which Pouzin reviewed favorably as "a further step towards a reasonable basis for a common standard").[64]

In the early months of 1975, external pressure – the formation of the X.25 Committee in preparation for the CCITT Plenary in 1976 – forced the proponents of competing proposals within INWG to try to resolve their differences and renew their effort to reach a consensus behind a single protocol design. Twenty-seven researchers, including Cerf, McKenzie, Pouzin, and Zimmermann, attended an INWG meeting in Anaheim on May 23.[65] Minutes from the Anaheim meeting report spirited and detailed technical discussions in response to presentations by McKenzie and Pouzin, as well as the meetings of two subgroups to discuss INWG contributions to CCITT and the "consolidation of the two proposed end-to-end protocols into a single, coherent proposal." The minutes concluded, somewhat dramatically, with an action item that laid bare the group's need to make a decision, even if such a decision was likely to be hotly contested and, in all likelihood, technically premature.

---

[62] Vint Cerf, Yogen Dalal, and Carl Sunshine (1974), "Specification of Internet Transmission Control Program," RFC 675, http://tools.ietf.org/html/rfc675 (accessed September 25, 2013); Vint Cerf, Yogen Dalal, and Carl Sunshine, "Specification of Internet Transmission Control Program," December 1974, INWG 72, McKenzie Collection.

[63] Alex McKenzie, "Internetwork Host-Host Protocol," December 1974, INWG 74, McKenzie Collection; McKenzie, "INWG and the Conception of the Internet," 69–70.

[64] Louis Pouzin, "Standards in Data Communications and Computer Networks," March 1975, INWG 79, McKenzie Collection.

[65] Vint Cerf, "Minutes of IFIP WG 6.1 Meeting," May 23, 1975, INWG 88, McKenzie Collection. Larry Roberts, the Arpanet pioneer who was now deeply involved in the effort to use virtual circuits as a foundation for the design of X.25, also attended the meeting.

CCITT SG VII plans to meeting [*sic*] September 15–19, 1975 in Geneva to prepare final submissions and recommendations to CCITT on packet mode operation for 1976–1980 time period. It is imperative that WG6.1 make its proposals for end-to-end protocols at at [*sic*] meeting. A voting policy has been prepared (see INWG General Note #85) for this purpose.[66]

The voting policy, jointly signed by Cerf and Pouzin[67] and distributed as INWG 85, began blissfully: "Discussions in the Anaheim meeting confirm that there is no major difficulty in reaching now an agreement on a common E-E [end-to-end] protocol." However, Cerf and Pouzin continued, it would not be possible to come to such an agreement at an INWG meeting as attendance was variable and "new arguments are raised on the spot, and cannot be studied carefully." Because the group was running out of time – and because meetings evidently were too heated to allow detached and objective evaluations of competing ideas – Cerf and Pouzin agreed to stop discussion and simply put the matter to a vote. Their timetable was brisk: after a period for initial comments, proposals, and criticisms, Cerf and Pouzin required "all complete proposals for a common protocol" to be sent to them by September 1. Cerf and Pouzin would then distribute all complete proposals to INWG members, who had until October 1 to mail their votes to them.[68]

Cerf added to the sense of drama and urgency in July, when he notified INWG members that their timetable for submissions would be accelerated to match the schedule that the CCITT Study Group VII had established. Complete proposals, under the new schedule, should be submitted by August 1 (not September 1 as originally planned). Pouzin needed to receive all votes before September 15 so that he could bring the consensus INWG protocol to the Study Group VII meeting that was scheduled for September 15–19.[69] In a desperate effort to create a workable consensus, four of INWG's leaders decided to meet in London in July. According to Alex McKenzie, the four men agreed "to stay in session for as long as it took to present a single proposal ... or to reach the conclusion that we could not agree." These four leaders represented the most powerful nodes within the INWG network: Cerf (from the Arpanet), Zimmermann (from Cyclades), Roger Scantlebury (from the British NPL), and McKenzie (the BBN engineer who had the unenviable task of being "designated as output editor").[70]

[66] Cerf, INWG 88.

[67] Cerf was the chairman of INWG in May 1975 and had appointed Pouzin to be chairman of the INWG Protocol Design Committee responsible for "technical issues in the design of an internetwork communication protocol."

[68] Louis Pouzin and Vint Cerf, "Vote for a Basic End-to-End Protocol," May 1975, INWG 85, McKenzie Collection.

[69] Vint Cerf, "GENERAL NOTE #85 on Voting Procedures," July 1975, INWG 91, McKenzie Collection.

[70] Cerf, INWG 91; McKenzie, "INWG and the Conception of the Internet," 69.

The gang of four came to an agreement and created a document "in less than a week," but it is not at all clear that anybody was happy with it from a technical point of view.[71] Moreover, from a bureaucratic point of view, the document was not completed in time to go through the INWG voting process. In a note to INWG on September 4, the four men announced that they had sent a copy of their "Proposal for an Internetwork End to End Protocol" (subsequently referred to as INWG 96) to CCITT as a "delayed contribution." Such a designation was sure to weaken the status of the document in the unfriendly gaze of the CCITT. INWG 96 was further weakened by an accompanying note, written by Cerf, stating that the proposal did not have the consensus support of INWG. The voting procedure to gain INWG support would go on nevertheless: other proposals would be accepted until October 1, at which point the group would finally vote on what proposal – if any – to send to CCITT as the consensus submission of INWG (and its institutional sponsor, IFIP).[72]

In the meantime, Pouzin had attended two interim meetings of CCITT Study Group VII (in May and September 1975) and did not have good news to report. Delegates at the May meeting, he wrote in INWG 97, "were cool about packet switching ... they do not object to packet switching, as long as it looks just like circuit switching." His conclusion was equally blunt: "It does not appear realistic to expect comprehensive CCITT recommendations on packet switching by the 1976 plenary." Pouzin's report from the September meeting was even more disheartening. He explained that CCITT bureaucrats had been given ample reasons to reject the proposal that Cerf, McKenzie, Scantlebury, and Zimmermann had scrambled together at the end of July: "INWG Note 96 (end-to-end protocol) was also waiting and Mr OKABE [the secretary of the meeting] objected to distribute it, since the cover letter sent by Vint CERF mentioned that this document had not yet been submitted to the agreement of W.G.6.1." Pouzin, clearly annoyed, concluded pessimistically. "The going is getting rough. The only carriers that count are Bell Canada, France, UK and Telenet," who were, as we have seen, the principal authors of the X.25 proposal that favored virtual circuits over datagrams. "Since there is little time left before things are frozen for the 1976 plenary," Pouzin concluded, "they are trying to sort out their differences, without care for outsiders' views."[73]

---

[71] John Day, "35 Years Ago Internetworking Was Headed in the Right Direction, Then Someone Got the Bright Idea for IP and Botched the Whole Thing," August 2011, courtesy of the author; McKenzie, "INWG and the Conception of the Internet," 69.

[72] Vint Cerf, Alex McKenzie, Roger Scantlebury, and Hubert Zimmermann, "End-to-End Protocol," September 4, 1975, INWG 95, McKenzie Collection; Vint Cerf, Alex McKenzie, Roger Scantlebury, and Hubert Zimmermann, "Proposal for an Internetwork End-to-End Protocol," July 29, 1975, INWG 96, McKenzie Collection.

[73] Louis Pouzin, "Meeting of the CCITT Rapporteur's Group on Packet Switching, Geneva, May 26–27, 1975," August 1975, INWG 97, McKenzie Collection; Louis Pouzin, "Meeting of the CCITT Rapporteur Group on Packet Switching – Geneva, 16–19 September, 1975," INWG 101, McKenzie Collection; Sirbu and Zwimpfer, "The Case of X.25," 40.

Pouzin's report of CCITT's intransigence was not altogether unexpected, but it was nevertheless a deeply demoralizing setback for the researchers in INWG. They were articulate, diverse, well funded, and completely powerless to set international standards. On December 1, Cerf finally circulated the long-awaited (and now practically useless) ballot for INWG members to approve or disapprove a motion: "the End-to-End Protocol Design described in INWG General Note No. 96 should be submitted to ISO and CCITT as an official opinion of IFIP W.G. 6.1." Since no other INWG members had advanced any proposals, the long-awaited vote was simply an up-or-down referendum on the "synthesis" that Cerf, McKenzie, Scantlebury, and Zimmermann created back in July.[74]

The results of the vote, distributed to INWG in March 1976, were 25.8 in favor, 7.5 against, and 8.7 abstentions.[75] The split within the group indicates the diverging strategic agendas of INWG participants. Support for INWG 96 included Pouzin and Zimmermann from Cyclades, Cerf from Stanford, McKenzie from BBN, Scantelbury from NPL, and Steve Crocker from the University of Southern California. Notable votes against were cast by Rémi Déspres, the representative of the French PTT who was drafting the X.25 proposal in CCITT; Jon Postel, the ARPA-funded researcher at the Stanford Research Institute; and Jerry Burchfiel and Ray Tomlinson, two researchers at BBN who were working on an ARPA contract for packet radio networks. More compelling, perhaps, were the nonvotes of Robert Kahn, the director of ARPA IPTO who had funded and coauthored with Cerf the first TCP specification; and Larry Roberts, the "Arpanet's father" who was, by 1975, working with Déspres on the X.25 design so that his company could grow more quickly from the establishment of an international standard. The fact that these two towering figures in the community did not bother to vote on the INWG proposal suggests they had already abandoned hope that a consensus within INWG would generate meaningful results.[76]

Despite the absence of unanimity – and a few conspicuous nonvotes – a large majority of INWG researchers sensed that they had reached a milestone within their community by establishing strong consensus behind the proposal in INWG 96. In March 1976, INWG forwarded the document formally to the relevant committees in CCITT and ISO. Cyclades, NPL, and the European Informatics Network all announced plans to convert their networks to the specifications in INWG 96, but it soon became evident that the Arpanet community would not

---

[74] Vint Cerf, "End-to-End Protocol Voting Procedure," December 1, 1975, INWG 102, McKenzie Collection.

[75] The fractions occurred because voting privileges were extended only to institutions, not individuals. Two institutions – the University of Tokyo and BBN – had multiple individuals active in INWG, with evident disagreements on INWG 96. Two out of three from Tokyo abstained, with one in favor; the BBN vote split evenly, with two in favor and two against.

[76] Cerf, INWG 102; "Result of Vote on End-to-End Protocol," March 1976, INWG 109.

follow suit. Cerf, in a 1988 interview with James Pelkey, lamented that despite his best efforts, he could not "persuade the TCP community to adopt the compromise, given the state of implementation experience of TCP at the time and the untested nature of the IFIP document." McKenzie, who seemed to have been more committed to an international consensus than to pleasing his patrons at ARPA, remembers the events differently. In a 2011 article, McKenzie reported that "we were all shocked and amazed" when the INWG community was told that ARPA researchers were too invested in the implementation of TCP to abandon that effort. Cerf and Kahn simply were unwilling to require all ARPA contractors to abandon their work on TCP and instead introduce the end-to-end protocol defined in INWG 96 into the Arpanet.[77]

Cerf's participation and standing in INWG diminished rapidly as he rededicated his energies to TCP and other protocols that he and the Arpanet community had been defining and implementing. One can only imagine Cerf's frustration with his first experience in international standards setting. He and his colleagues had failed to convince the bureaucrats in the monopoly-dominated CCITT to consider a datagram option for their packet-switching standards. Even worse, Cerf had been unable to convince his peers in INWG that the Arpanet's protocol design was superior to the Cyclades-inspired synthesis presented as INWG 96. Surely this failure rubbed extra salt into the wounds they had all suffered at the hands of the CCITT.

In the meantime, Kahn and the ARPA management presented an alternative that Cerf found much more appealing: they asked Cerf to leave his tenure-track job at Stanford and join Kahn in Washington, D.C., as the director of a new Internet Program in IPTO. In a 1990 oral history interview, Cerf reflected at length on his choice to move to ARPA. He recalled that he "had been asked to go more than once," but preferred to stay in Palo Alto. When Kahn offered again in 1975, Cerf reconsidered.

This time I was more interested, in part because I was finding my time very fragmented at the University. A lot of people needed to see me, I had seventy students in the master's degree program and about a dozen graduate students in Ph.D. programs. I was finding it hard to get any research done because I was so busy doing that, or writing proposals, or writing reports telling people what I would do if I wasn't spending all my time telling them what I was going to do. It started to appeal to me that maybe I could get more done, if I had more resources available. At DARPA I would be able to call on the resources of some of the best places in the country to actually do substantive work.[78]

---

[77] McKenzie, "INWG and the Conception of the Internet," 70; Derek Barber, "INWG Affairs," March 1976, INWG 111, McKenzie Collection; Louis Pouzin, "TC 6 Contributions to ISO," June 15, 1976, INWG 122, McKenzie Collection; Pelkey, "Entrepreneurial Capitalism and Innovation," chapter 6.

[78] Vinton Cerf, oral history interview by Judy O'Neill, April 24, 1990, Reston, Virginia. Charles Babbage Institute, University of Minnesota, Minneapolis.

Cerf also remembered more personal reasons – including his wife's enthusiasm and concerns over his professional reputation – that weighed heavily on his mind.

And my wife finally said that she was willing to go to Washington because she was curious. She'd grown up in Kansas and lived in Los Angeles and San Francisco. It was really her interest and enthusiasm for moving to Washington that sort of tipped me over the edge. I was very reluctant to go.... I also was a little worried about taking the job because I thought, "Gee, it's very visible. If you screw up, you will screw up in a very visible way and everybody will know." "Ah, that guy really loused it up." That gave me a few sleepless nights. But I had the same trouble deciding to go to Stanford when I was offered a job right out of graduate school at UCLA. I thought, "I don't have anything to teach anybody. What am I going to say?" That made me nervous. So I wasn't sure about going to DARPA, but I finally got my arm twisted and they said, "Well, you'll never know unless you go.'" So I said, "Well, all right, I'll try."[79]

Strangely, Cerf's account of his decision to leave Stanford and join the ARPA staff omits entirely the somewhat tragic role that he had been playing in the international standardization drama. Perhaps the frustrations of international committees, much like the daily buzz of graduate students and deadlines for grant proposals, were pests that he happily exterminated when he moved to Washington. In any case, the international consensus behind INWG 96 (a document Cerf helped write), expressed in the vote in December 1975, did not seem to concern him once he became the director of the Internet Program. As the Cyclades engineer Michel Gien recalled, when Cerf moved to ARPA he became "the money-giver as opposed to the guy doing the work. So that's why there might have been also some differences." Cerf's departure for ARPA, combined with Kahn's evident opposition to a consensus international protocol, meant that Cerf would be free to focus all of his energies on Arpanet experiments for the military and the development of the TCP protocol that could facilitate communication between three different packet-switched networks being developed by ARPA. In the 1990 interview, Cerf recalled his new mission with great clarity: "The single-minded goal was to get the Internet system up and running."[80]

IFIP and INWG moved quickly to restructure their leadership after Cerf announced his departure at the IFIP Technical Committee 6 (Data Processing) in Sao Paulo, Brazil, from October 27 to 31, 1975. Cerf offered his resignation at the same time that Pouzin was named chairman of IFIP TC6. Pouzin appointed Derek Barber from England's NPL to be the new chairman of WG 6.1, effective January 1, 1976. He also appointed Hubert Zimmermann to assume the chair of the INWG Protocols Committee, a position that Pouzin vacated when he moved up IFIP's organization chart.[81] Zimmermann, as we

---

[79] Cerf interview, Charles Babbage Institute, 1990.

[80] Cerf interview, Charles Babbage Institute, 1990.

[81] Vint Cerf, "Report on TC-6 Meeting in Sao Paulo, October 1975," December 1, 1975, INWG 103, McKenzie Collection; Louis Pouzin, "Chairmanship of WG 6.1," March 16, 1976, INWG 110, McKenzie Collection.

will see in Chapter 8, continued his emergence as a highly effective technical and political leader in the international standards community.

The administrative reshuffling within INWG marked a fundamental and irreparable split among the one-time collaborators: Cerf, as we have seen, assumed a new leadership role in ARPA and, before long, stopped attending INWG meetings. He, along with Kahn, would eventually be singled out and celebrated for their pioneering roles in the creation of the Internet. The pair's fame stems from a fateful strategic decision: to abandon the international standards process in order to build a network for their wealthy and powerful client, the American military. Although there were many different developments beyond the control of Cerf and Kahn that contributed to the eventual growth of the Internet, it is tempting to wonder what might have happened if the leaders of INWG had been able to build and sustain a meaningful international consensus design for datagram networks. The Cyclades engineer Michel Gien believed that "technically – purely technically speaking – it would have been a better solution for Arpanet to actually take [INWG] 96.1 and then implement that. And that could have been a more general standard to fight CCITT rather than to being separate between Europe and the U.S." Alex McKenzie concluded a recent article on INWG's history with his own speculation: "Perhaps the only historical difference that would have occurred if DARPA had switched to the INWG 96 protocol is that rather than Cerf and Kahn being routinely cited as the 'fathers of the Internet,' maybe Cerf, Scantlebury, Zimmermann, and I would have been."[82]

What about Louis Pouzin and his team that had built Cyclades? Once Pouzin emerged as the clear victor from INWG's internal technical and political debates, he continued his polemics against the overbearing French PTT, their co-conspirators in the CCITT, the American giant IBM, and the ill-conceived French "national champion" strategy for its computing industry. In a critical review of X.25, circulated as INWG 106 in January 1976, Pouzin pointed out the political dangers of antagonizing the established powers: "It is not considered as a wise stand to antagonize the European PTT's when they are known to be ticklish."[83] Nevertheless, Pouzin championed the distribution of control to users, small manufacturers, and entrepreneurs. After all, he had designed the Cyclades datagram network to be a "heterogenius [*sic*] network" (see Figure 6.1) that would permit the decentralization of control. By design, Cyclades thus provided the technical component to Pouzin's social and political critique of the IBM and PTT ideology of centralized control.

Pouzin's writings in 1975 and 1976 turned increasingly caustic as he recognized that his designs for a different political and technological future were

---

[82] Michel Gien, oral history interview by Andrew L. Russell, April 3, 2012, Paris, France. Charles Babbage Institute, University of Minnesota, Minneapolis; McKenzie, "INWG and the Conception of the Internet," 70.

[83] Pouzin, INWG 106.

Heterogenius

computer

network

FIGURE 6.1 Pouzin's "heterogenius" network. The play on words is very much intentional, since such a network would keep users from being captured by one of the vendors pictured.

Source: Louis Pouzin.

Courtesy of the Charles Babbage Institute, University of Minnesota, Minneapolis. INWG Note #49, Louis Pouzin, "Network Architectures and Components," Box 1, Alexander McKenzie Collection of Computer Networking Development Records.

doomed. The only hope of avoiding the dystopic scenario that Pouzin fearlessly referred to as "IBM vs. Telecommunications, through mercenaries" was to increase the level of cooperation with users, small manufacturers, and the community of expert computer researchers. But Pouzin was pessimistic that such cooperation would be allowed to occur.

Standards are a weapon used by the carriers to expand their territory.... National monopolies belittle the opinion of customers and minor manufacturers, since they can exercise enough arm-twisting to keep the latter in line.... Good standards would require more cooperation with the data processing field. For the time being a standard war is predictable.... There is little chance to expect a truce soon.[84]

Pouzin knew that his increasing belligerence did not endear him to the French government officials who controlled the budget for his experiments with Cyclades and his international collaborations with datagram advocates. Dépres, who was leading the French PTT networking project that used X.25's virtual circuits, explained that "the PTT was in a position where they were subsidizing this project with Louis Pouzin, a great showman in international conferences explaining that we were doing everything wrong." John Day recalled in a 1988 interview that "Cyclades was, to some extent, a big embarrassment for the PTT, because they had built a packet switch network, and the PTT had failed, to that point." Similarly, the Cyclades engineer Najah Naffah emphasized that Pouzin had "a lot of enemies" who "didn't like the way he approached the networking game.... And this idea of having a free datagram going from A to B in any route they want, it's not something acceptable. So he was excluded from some of the circles."[85]

By the early months of 1976, the technical and political differences between Pouzin and the French PTT had become irreconcilable. Rather than depend on Pouzin's design, the French PTT stood behind X.25 and, in April 1976, announced that its new Transpac data network would utilize X.25's virtual circuit design that Pouzin had assailed as technically inferior and politically inept. Administrators in the PTT and in the Ministry of Industry decided that datagram service was too unstable and that the plans for Transpac made Cyclades unnecessary. Furthermore, changes in the administration of President Valery Giscard d'Estaing meant that IRIA would no longer exist as an independent national research facility. Instead, it would build closer ties with the newest incarnation of the "national champion" computer firm, CII-Honeywell-Bull, to help in the French strategy to develop a powerful domestic computer manufacturing firm. The continuation of Cyclades, therefore, was undesirable because it would only undermine the French national policy to subsidize CII-Honeywell-Bull and use the Transpac X.25 network as the foundation of a new Maginot Line defense against IBM. Pouzin, shorn of his pet

---

[84] Pouzin, "The Network Business – Monopolies and Entrepreneurs."

[85] Dépres interview, Charles Babbage Institute; John Day, interview with James Pelkey, July 11, 1988, Canton, Massachusetts, courtesy of James Pelkey; Najah Naffah, oral history interview by Andrew L. Russell, April 2, 2012, Paris, France. Charles Babbage Institute, University of Minnesota, Minneapolis.

project, received new instructions from Paris: the time for research was over, and he should now turn to the more practical work of developing applications for France's emerging "computerized society."[86]

Administrative fiat thus ended the innovative research agenda of the man that Cerf later admired as the "datagram guru."[87] Pouzin had been cast as Socrates, and the French PTT and Ministry of Industry administered the hemlock.

Conclusions

The departure of Vint Cerf from INWG, together with the political termination of the Cyclades project, meant that INWG's prominent role as a forum for networking discussions ground to a halt by late 1975. The group continued to exist for another few years, but its rush to standardize protocols had the unintended consequence of preempting research projects that examined theoretical and experimental approaches to network pricing, flow and congestion control, security, and new approaches to networking architectures. This consequence may have been unintended, but it certainly was not unpredictable; after all, prominent INWG members including Pouzin and Franklin Kuo had warned their colleagues that premature standardization "could only freeze the development process" and lead to underdevelopment of research on datagrams and packet-switched networks.[88]

INWG researchers voted in early 1976 to support the "synthesis" host-to-host protocol described in INWG 96, but their rough consensus, ironically, might have done more damage than anything else. The vote and subsequent personnel changes within INWG and IFIP marked a turning point in what had once been a promising international collaboration. The split between the Arpanet researchers and the European researchers, with very few exceptions, soon became permanent. National telephone monopolies in Europe continued to assert jurisdiction over data networks. The French authorities responded to

[86] Derek Barber, "Meeting in Toronto, August 1976," (n.d., 1976), INWG 128, McKenzie Collection; Louis Pouzin, "Cyclades ou Comment Perdre un Marche," *Recherche* 328 (Fevrier 2000), 32–33; Stéphane Foucart, "Louis Pouzin, l'homme qui n'a pas inventé Internet," *Le Monde*, 5 août 2006; Alain Beltran and Pascal Griset, "Le Projet Cyclades Sacrifié," *Codesource* 11 (2007): 1; Nora and Minc, *The Computerization of Society*; Valérie Schafer, "Appropriating Packet Switching Networks, Making Cyclades Network"; Schafer, *La France en Reseaux*; Després interview, Charles Babbage Institute; Pouzin interview, Charles Babbage Institute.

[87] Vint Cerf, quoted in Ian Peter, "Separating TCP and IP," September 30, 2004, available from http://mailman.postel.org/pipermail/internet-history/2004-September/000431.html (accessed August 24, 2011). A dénouement of sorts occurred in 2013, when Pouzin shared the first Queen Elizabeth Prize for Engineering with Cerf, Kahn, and World Wide Web pioneers Sir Tim Berners-Lee and Marc Andreesen. Royal Academy of Engineering, "Queen Elizabeth Prize for Engineering," June 25, 2013, http://qeprize.org/ (accessed June 30, 2013).

[88] Pouzin, INWG 106; Kuo, "Political and Economic Issues for Internetwork Connections"; McKenzie, "INWG and the Conception of the Internet"; John Day, *Patterns in Network Architecture: A Return to Fundamentals* (Upper Saddle River, NJ: Prentice Hall-PTR, 2007).

Pouzin's provocations by backing Déspres's design of the X.25-based Transpac network and cutting off the support Pouzin needed to continue his research with Cyclades.

Despite the best intentions of the INWG members, incompatibilities reigned at nearly every level. The technical incompatibilities that were the initial catalyst for INWG – including the proliferation of different types of packet-switched networks and the debates between proponents of datagrams and virtual circuits – had not only persisted, they had intensified. Technical incompatibilities, as we have seen, both reflected and further exacerbated the incompatible worldviews of the various technological communities that were working at the boundaries of computing and telecommunications. These worldviews stemmed in large part from the divergent institutional motivations, constraints, patronage strategies, and political economies of researchers and engineers. Just as Arpanet researchers benefited from the generous support of ARPA, Pouzin and other European researchers struggled to reconcile the conflicting agendas of their scientific curiosity and the strategic ambitions of national policy makers. The fundamental question was one of control: telephone engineers believed they had a right – indeed, an obligation – to exercise end-to-end control over all communications; computer engineers, for the most part, favored a regime that limited telephone companies to the tasks of transmission and freed users to configure and manipulate computer terminals in any way they chose.

Another type of incompatibility that doomed INWG was the mix of individual motivations and ambitions. Most leaders of INWG – including Cerf and Pouzin – proclaimed to have cordial, even friendly relations with their colleagues, but hints of personal rivalries and ego clashes were difficult to keep under the surface. INWG members tried to follow the long-cherished tradition of engineering communities that sanitize personal disputes from official records.[89] They succeeded for the most part, but in the process they left scant evidence to suggest that INWG members as a whole shared an esprit de corps that existed in the early 1970s within the distinct communities that developed the Arpanet in the United States and Cyclades in France. The persistent tension between cooperation and competition obscures an important point: for engineers who designed new networks, participation in standards committees was a necessary – if painful – step. For this and subsequent generations of network architects, active and energetic representation in standards committees became routine.

Above all, the discussions within INWG and negotiations within CCITT failed to resolve the fundamental incompatibility that arose with the convergence of computing and communications: the question of jurisdiction. Who

---

[89] As historian Edwin Layton noted and the examples in Chapters 2 and 3 demonstrate, "It is considered bad form to publicize the inner workings of engineering societies." Edwin T. Layton, *The Revolt of the Engineers: Social Responsibility and the American Engineering Profession* (Cleveland, OH: Press of Case Western Reserve University, 1971), 15.

was in control of the new digital data networks? By the late 1970s, three competing communities of researchers – Arpanet engineers, telecommunications professionals in the ITU, and a loose alliance of American and European computer professionals – were seeking to establish their own designs as the definitive architecture for packet-switched networks. They would press on with the benefit of Pouzin's insights but without the presence of the sage of datagrams himself. INWG member John Day summarized the significance of Pouzin's technical work, suggesting that the conventional wisdom about the "invention" of packet-switching and internetworking is incomplete without reference to Pouzin and the young cohort of computer researchers whom he inspired:

The real breakthrough in networking is not packet switching (Baran and Davies independently), but datagram packet switching (Pouzin). I have always found it somewhat interesting that every project Baran and Larry Roberts have been involved in since the ARPANet have been connection-oriented networks, not connectionless ones. Interestingly enough, what is seen as revolutionary seems to depend on age cohort. For the older group where communication was based on telecom, packet switching is revolutionary. Sending data in small packets rather than as a continuous stream. But for the slightly younger group (and the shift is sudden) schooled in computing, packet switching is obvious (data is in buffers, you want to communicate, pick up a buffer and send it. How else would you do it?). What isn't obvious is datagram packet switching. The idea that the communication can be probabilistic, that we can build reliable communication from unreliable parts. That was the insight of Louis Pouzin. That is what captured the imagination and what the Internet was built on.[90]

External events continued to dictate the research and standardization strategies of the computer networking community. Pouzin was not the only computer researcher to realize that contributions to CCITT were fruitless, and that they would need a different strategy if they wanted to create international standards for packet-switched networks. The best alternative left for the INWG researchers – since, in Pouzin's words, "IFIP power is only intellectual" – was to bring their work to yet another standards-setting committee: the International Organization for Standardization.[91]

By 1975, ISO had already proved to be a viable forum for the creation of international computer standards. Moreover, it offered computer researchers a respite from powerful political forces of the telephone monopolies that dominated the international standards process. The problem with moving the datagram standards effort to ISO was that it was likely to force the computer researchers into a head-to-head confrontation with IBM – and, eventually, a renewed confrontation with telecommunications engineers as well. Matters would be complicated as it became clear that Cerf and Kahn were

[90] John Day, "What Went Wrong? How the Internet Stagnated, or the 30 Years War of the Bellheads and the Bitheads," University of Minnesota, October 19, 2010.

[91] Louis Pouzin, "The Communications Network Snarl," *Datamation* 21 (1975): 70–72; Abbate, *Inventing the Internet*, 152–155.

more interested in developing a network for their sponsor, the U.S. Department of Defense, than they were in working toward an international standard. There would be no significant ARPA presence within ISO's project on Open Systems Interconnection (OSI) that began in 1977. As Zimmermann recalled in a 2005 interview,

The sad thing for the standardization of computer network protocols is that the Arpanet community, the core of the Arpanet community, did not join the ISO discussions. I'm not sure why, probably the Arpanet guys would be able to tell you. There were representatives from the U.S. computer industry, from the U.S. telco industry, but there was no significant representation or strong representation from the Arpanet TCP/IP community.... We had, within IFIP, started to discuss and elaborate a common protocol which was somehow trying to take the best from TCP/IP and Cyclades, and this was then contributed to ISO, but without giving a supporting implementation while Cyclades and TCP/IP had their own supporting implementations. But that's something which was not just an adjustment from an existing implementation, but was something new, built from existing things.... When experts discuss and meet and design something, they don't start completely from scratch.[92]

Cerf later justified his decision – omitting any mention of the bitter strategic battle within INWG – by arguing that "the DARPA effort had been under way since 1973. By the time the first papers on OSI architecture emerged, 1978, we were 5 years into the development and testing of TCP/IP and it was time to implement and deploy. As the DARPA program manager for the effort, I was not willing to spend another 5 years in negotiations on a new suite of protocols when we had something that looked pretty solid." The subsequent trajectories of the two networking projects diverged: where the Internet stayed protected under ARPA's singular sponsorship, OSI opened itself up to the forces of international politics.[93]

[92] Hubert Zimmermann, interview with Mariann Unterluggauer, July 14, 2005, courtesy of Mariann Unterluggauer.
[93] Vint Cerf, quoted by Mariann Unterluggauer, "denktage," December 10, 2007, available from http://motz.antville.org/stories/1734089/ (accessed November 22, 2012).

7

Open Systems and the Limits of Democratic Design, 1970s–1980s

> Networks are political issues at government or corporate levels. In hush statements, communications and computer organizations leak their feelings that the other guys had better mind their own business. Since they are big, and more or less monopolies, each scrambles at shaping the world into specific designs before the others do it. Lost in the turmoil is the user. He has no strategy, no power. He thinks the big brothers know what they're doing. No doubt, there are some who do. Take for example our beloved friend IBM.[1]
>
> – Louis Pouzin, "The Communications Network Snarl," 1975

The promise of data networking was evident to everyone who worked near the intersections of digital computer and communication technologies. The Bell Labs economist Jeffrey Rohlfs articulated the new conventional wisdom in a pioneering 1974 article, in which he based his "theory of interdependent demand for a communications service" on a forceful assertion: "The utility that a subscriber derives from a communication service increases as others join the system."[2] Rohlfs theorized and mathematized the lessons that the packet-switching researchers we met in Chapter 6 – such as Louis Pouzin, Vinton Cerf, Robert Kahn, Rémi Déspres, and Marc Levilion – understood intuitively: the value of a digital computer-communication network would reside primarily in its ability to facilitate and sustain interconnections on an ever-increasing scale.

As we have seen, these packet-switching researchers gained firsthand experience with the obstacles that prevented the universal adoption of new standards for data networking. As the "virtual circuit versus datagram" controversy described in Chapter 6 demonstrated, many technical disagreements mapped

---

[1] Louis Pouzin, "The Communications Network Snarl," *Datamation* (December 1975): 70–72.
[2] Jeffrey Rohlfs, "A Theory of Interdependent Demand for a Communications Service," *Bell Journal of Economics and Management Science* 5 (1974): 16.

directly to conflicting strategic priorities and conflicting visions of control: telecommunications engineers used virtual circuits to keep more control inside the network, but computer engineers favored connectionless datagrams to shift control to the computers at the edges of the network. As positions hardened in the mid-1970s and it became clear that the argument was unlikely to be settled on technical merits alone, the network engineers we will meet in this chapter sought to create effective alliances that could constrain and direct the creation of international networking standards.

By the mid-1970s, the most powerful organizations in computer networking had already charted their respective strategic directions and invested heavily to advance them. The national monopoly telephone companies assembled in the International Telegraph and Telephone Consultative Committee (CCITT) had approved a virtual circuit design for their X.25 protocol in 1975, computer manufacturers such as IBM and Digital introduced a competing set of proprietary network architectures, and managers at the Advanced Research Projects Agency (ARPA) sponsored the development of the Arpanet to meet the American military's demand for enhanced command and control capabilities.

Pouzin, the director and inspiration of the French Cyclades datagram network, emerged during the early 1970s as an advocate for the relatively powerless and voiceless who had become "lost in the turmoil" – a group that included computer users, researchers, entrepreneurs, and engineers in small companies around the world. Pouzin, along with a few researchers from the United States, France, and Great Britain, created the International Network Working Group (INWG) in 1972. They hoped that INWG could influence international standards and thus counteract the ambitions of the powerful incumbents of the global computer and telecommunications industries. But as we saw in Chapter 6, INWG's double failure – its inability to unite all datagram researchers behind the protocol described in INWG 96 and its impotence in the international discussions at CCITT – led, by 1975, to the disintegration of its alliance.

Undeterred, many members of INWG continued to push for international standards. Some of these researchers – including veterans of the French, British, and American packet networking experiments – turned to the International Organization for Standardization (ISO) to renew their attempts to create international standards for packet-switched networks. Officials in ISO were accommodating and, in 1977, agreed to create a new subcommittee – known as SC16 – on Open Systems Interconnection (OSI) to investigate "the need for standardization in the area of open systems, as it relates to system interconnection."[3]

---

[3] "Resolution 11. Establishing Subcommittee 16 – Open System Interconnection," Box 19, Folder 1, Charles W. Bachman Papers, 1951–2007, CBI 125, Charles Babbage Institute, University of Minnesota, Minneapolis [hereafter, Bachman Papers].

The researchers working on OSI made rapid progress during the next five years. Thanks to their diligent and well-publicized work, the OSI project became the most visible example of a new vision of open systems that allowed users to mix and match any software or hardware component that adhered to nonproprietary standards and interfaces. The ideology of openness proposed to eliminate boundaries among the wide variety of existing proprietary systems. User empowerment would follow naturally from a design process that forged a consensus among all interested parties rather than being imposed by the will of a single powerful actor.

As much as the open systems concept appealed to users, it was even more appealing to regulators, engineers, and entrepreneurs who still were looking for ways to "get them out from under IBM's thumb."[4] For these diverse constituencies, the ideology of openness allowed IBM's many critics to come together and advocate an alternative to the status quo. The technical details of this alternative were left to the future, but its fundamental character was clear: it would be *open*. Open systems thus existed as an ideological concept – a critique of the existing order – before they existed as technological systems.

INWG's fate had already demonstrated an enduring lesson: organizational innovation did not guarantee successful technological standardization. ISO's vast size and international status created new possibilities that INWG could not, but it also forced computer engineers to navigate new bureaucratic and diplomatic obstacles. Because the utility and legitimacy of ISO standards depended on widespread consensus and voluntary adoption, matters of structure and process were paramount: Who controlled the committee? Who could make technical contributions? Who could vote? How could the committee ensure that it represented a "balance of interests" that ISO rules required? How could it forge a meaningful consensus among participants with diverging political and market strategies?[5]

The work in ISO SC16, like the earlier work in INWG, ultimately failed to generate global standards for computer networking. Accounts of these projects are valuable nevertheless, in part because they remind us that the history of networking is much more than a chronology of "inventions" or technical accomplishments. The story of the growth of computer networks necessarily must be much broader: it is a story of the interplay of ideas and institutions, of technologies and ideologies, and of strategies to harness technological change for economic and social gain.

Chapters 7 and 8 investigate the history of two efforts to develop international standards for computer networks between the 1970s and the 1990s. The first effort, ISO's Open Systems Interconnection, most closely resembled a

---

[4] Martin Libicki, *Information Technology Standards: Quest for the Common Byte* (Boston: Digital Press, 1995), 11.

[5] Louis Galambos, "Recasting the Organizational Synthesis: Structure and Process in the Twentieth and Twenty-First Centuries," *Business History Review* 79 (2005): 1–37.

form of *democratic* control over innovation and standardization.[6] Because OSI was organized under ISO's auspices, its structure and process followed well-established rules for membership, meetings, and decision making. ISO was an "industrial legislature" on an international scale – a direct descendant of the efforts of Paul Gough Agnew and Charles Le Maistre, two pioneers of consensus standardization in the early twentieth century.[7] Indeed, the organization that Agnew had championed for so long – the American Standards Association, renamed the American National Standards Institute (ANSI) in 1969 – was responsible for coordinating American participation in OSI committees.

Like the United Nations, voting members within ISO's international democracy came from national delegations. However, whereas delegates to the United Nations were named by national governments, delegates to ISO were named by national standardization bodies, such as the ANSI X3 Committee in the United States. Engineers could not simply walk off the street and attend these international meetings; they had to be nominated by their respective national standards body. These formal bureaucratic structures were designed to ensure that no single powerful party could dominate the standardization process. As SC16 chairman Charles Bachman summarized in 1978, "The adjective 'open' means to imply that all participants come to the system as equal partners."[8]

As an open institution, ISO SC16 mimicked other features of democratic governance: it moved slowly and was at every step susceptible to tactics that sought to disrupt or "capture" the democratic process in order to advance narrow interests. Participants in OSI, like citizens in democratic states, were forced to delay, compromise, and even sacrifice their own ambitions in order to secure enough votes to achieve a workable consensus and democratic "progress." Open systems, therefore, were more than a manifestation of the economic and strategic ambitions of critics of centralized control. OSI, by virtue of its organization within international standards committees, democratized the process of technological innovation in a way that had never been attempted, let alone achieved.

Democracy did not work well in this setting. The aspirations of international cooperation collapsed under the pressures of nationalism and the realities of global competition in emerging markets for digital networks. By the mid-1980s, the OSI project began to stall and fail to live up to its promises of

---

[6] Frederick P. Brooks, Jr., *The Mythical Man-Month: Essays on Software Engineering Anniversary Edition* (Boston: Addison-Wesley, 1995), 40–50; William Lehr, "Standardization: Understanding the Process," *Journal of the American Society for Information Science* 43 (1992): 550–555.

[7] See Chapters 2 and 3 in this book, and Craig N. Murphy and JoAnne Yates, *The International Organization for Standardization (ISO): Global Governance through Voluntary Consensus* (New York: Routledge, 2009).

[8] Hubert Zimmermann, "OSI Reference Model – The ISO Model of Architecture for Open Systems Interconnection," *IEEE Transactions on Communications* 28 (1980): 425–432; Charles Bachman, "Domestic and International Standards Activities for Distributed Systems," September 28, 1978, Bachman Papers, Box 18, Folder 12.

a grand future. Its demise was hastened by a second effort to create standards for computer networks during this time period: the Internet. Unlike OSI, the Internet was nurtured in an *autocratic* setting, sponsored lavishly by the American Department of Defense and administered by a "council of elders" who flatly rejected basic features of democracy such as membership and voting rights.[9] Before we consider the development of the Internet's standard-setting process in Chapter 9, however, we need to do something that existing histories of the Internet do not: look closely at the ideological and technological origins of OSI in the late 1970s and the factors that led it to its decline in the 1980s and 1990s.

Open Working, Distributed Systems, Open Systems

We have already seen how the experimental and diffuse character of computer networking technologies in the 1970s hindered international standardization. The most pressing problem was not one of absence, but rather one of abundance. Users wanted to exchange data between different types of computers and networks, but there was no single organization that could settle competing jurisdictional claims and incompatible protocols. The Dutch computer scientist Andrew Tanenbaum famously captured the irony of the situation in his 1981 textbook, *Computer Networks*: "The nice thing about standards is that you have so many to choose from; furthermore, if you do not like any of them, you can just wait for next year's model."[10]

Factors of political economy shaped the realm of the possible, and help explain differences in the varieties of national responses to the convergence between computing and communications. The American response, detailed in Chapters 5 and 6, featured a disorganized and inconsistent mix of policy measures that constrained the monopoly Bell System and created space for entrepreneurship and market competition. The French response, in contrast, was centrally coordinated from government offices in Paris. French policy makers feared the prospect of IBM domination, and pursued a two-pronged strategy – heavy subsidies for its "national champion" computer firm and an expanded jurisdiction for its monopoly PTT – in order to protect its domestic industries from international competition.

Postal and telecommunications service in 1970s Britain was, like France, under the exclusive jurisdiction of the state-owned General Post Office. However, unlike the French (and more like the Americans), British leaders tolerated greater institutional and technological diversity as well as a more competitive computer manufacturing industry that did not feature one national champion. As a result, packet-switching experiments flourished under the leadership of Derek

---

[9] Ed Krol (1993), "FYI on 'What Is the Internet?'" RFC 1462, http://www.ietf.org/rfc/rfc1462.txt (accessed September 25, 2013).

[10] Andrew S. Tanenbaum, *Computer Networks* (Englewood Cliffs, NJ: Prentice Hall, 1981), 168.

Barber and Donald Davies in the prestigious National Physical Laboratory, and a variety of small companies scrambled to meet customer demand for time-sharing and other computer network products and services. The British computer scientist Peter Kirstein recalled that he and his fellow researchers benefited from British institutional pluralism and close contact with American researchers. Within the British research community, Kirstein wrote in 1999, "there was no perceived threat of transatlantic dominance. This avoided much of the political infighting that had dogged the French and German scenes at the time; here, the struggle was seen between European Standards and U.S. dominance. We avoided that dilemma; in fact, we capitalized on it."[11]

The British Standards Institute (BSI), the national standardizing body for the Great Britain, was the institution that was best positioned to be an advocate for the country's fledgling computer industry. BSI, like its American counterpart ANSI, was a voluntary consensus organization that restricted membership to trade associations, engineering societies, government agencies, and other groups that professed to rise above any single proprietary interest. Of course, participants in subcommittees could and did advocate for their own interests – hence the requirement for each committee to maintain a balance of interests. In early 1977, a small group of engineers and computer researchers who were active in BSI – including a few who had been active in INWG – decided that their best bet for creating computer network standards on an international scale was to channel their work through ISO, the international umbrella organization for national standardization bodies.

Within ISO Technical Committee 97 (TC 97), which was responsible for Computers and Information Processing, jurisdiction over standards for digital data transmission was assigned to Subcommittee 6.[12] Engineers within ISO/TC97/SC6 (SC6) had been discussing computer-communication standards since 1973. After years of difficult technical and political negotiation, the group finally was close to reaching an agreement over the High-Level Data Link Control protocol, a connection-oriented point-to-point protocol. As a result, SC6 appeared to be a realistic alternative venue for INWG researchers who hoped to standardize their datagram network protocols. Indeed, when INWG researchers sent protocol proposals to CCITT in 1976, they sent

---

[11] Martin Campbell-Kelly, "Data Communications at the National Physical Laboratory (1965–1975)," *IEEE Annals of the History of Computing* 9 (1988): 221–247; Peter T. Kirstein, "Early Experiences with the Arpanet and Internet in the United Kingdom," *IEEE Annals of the History of Computing* 21 (1999): 38–44; Eli Noam, *Telecommunications in Europe* (New York: Oxford University Press, 1992); Mark Thatcher, *The Politics of Telecommunications: National Institutions, Convergences, and Change in Britain and France* (New York: Oxford University Press, 2000).

[12] Todd Shaiman, *The Political Economy of Anticipatory Standards: The Case of the Open System Interconnection Reference Model* (MA thesis, Oxford University, 1995): 44–45; Business Equipment Manufacturers' Association, *Committee X3 Organization Manual* (New York: American National Standards Institute, 1970).

identical proposals to ISO's SC6 (as we saw in Chapter 6). Therefore when the telecommunications professionals in CCITT rejected the INWG proposal, many INWG veterans resolutely turned to SC6.[13]

It soon became apparent, however, that the members of the SC6 Committee – including two influential leaders in the American delegation who were employed by IBM – were cool toward the ambitious new proposal for connectionless datagram standards. As a result, the leaders of the British computer standards committee proposed to bypass SC6 and organize a new international subcommittee that would begin work on "network standards needed for open working." Their use of the neologism "open working" was conspicuous, for it prioritized "the ability of the user or program of any computer to communicate with the user or the program of any other." The primary motivation of the British, in other words, was strategic. The British computer engineer Jack Houldsworth argued that "open working" was needed as an alternative to "the traditional computing scene" that was "dominated by installations that were planned and implemented as self-contained, 'closed' systems with little regard for the possibility of their interworking with each other." The conclusion, in Houldsworth's eyes, was obvious: open network standards "will be of significant economic value to computer manufacturers and users," the very parties that were prepared to represent themselves most effectively through ISO. All that was needed, the British engineers argued, was to create a new subcommittee that would not have to reckon with the entrenched political powers in SC6.[14]

At its March 1977 meeting in Sydney, Australia, ISO/TC97 accepted the British request and created Subcommittee 16 (SC16), Open System Interconnection. SC16's jurisdiction was unusually vague. Its scope, to "investigate the need for standardization in the area of open systems," required the group to define relevant terms, identify existing standards work, and proceed on the basis of a thorough "study of the structure and the nature of the standards required." Curiously, the framers of the committee left their central

---

[13] Louis Pouzin, "TC 6 Contributions to ISO," June 15, 1976, INWG 122, Alexander McKenzie Collection of Computer Networking Development Records, Charles Babbage Institute, University of Minnesota, Minneapolis [hereafter, McKenzie Collection]; Louis Pouzin, "Virtual Circuits vs. Datagrams: Technical and Political Issues," January 1976, INWG 106, McKenzie Collection; "List of ISO/TC 97/SC 6 Reference Documents," August 15, 1977, Bachman Papers, Box 19, Folder 1.

[14] Jack Houldsworth, "Standards for Open Network Operation," *Computer Communications* 1 (1978): 5–12; T. J. McNamara, "UK Proposals to New Open System Networking Standards (TC 97/SC 16)," June 20, 1977, Bachman Papers, Box 19 Folder 3; Bachman, "Domestic and International Standards Activities for Distributed Systems"; James Pelkey, "Entrepreneurial Capitalism and Innovation: A History of Computer Communications, 1968–1988" (2007), chapter 6, http://www.historyofcomputercommunications.info/Book/6/6.0-Overview.html (accessed September 25, 2013); Charles Bachman, interview by Andrew L. Russell, April 9, 2011, IEEE Computer Society History Committee, available from http://www.computer.org/comphistory/pubs/2012-03-russell.pdf (accessed September 25, 2013).

term – "open systems" – undefined. They did, however, acknowledge that the committee would be entering a contested organizational field and should "take into due account the activities of TC97/SC6 and CCITT."[15]

The creation of SC16 triggered a scramble among leaders of the French, British, and American representatives to secure key positions within it. Authority within ISO committees was a function of bureaucratic maneuvering and administrative delegation: ISO/TC97 chose one nation to serve as the committee secretariat, responsible for administrative tasks such as mailings and meeting arrangements. The nation chosen as the secretariat also selected the committee chairman, who had the power to set the agenda, select the leaders of subcommittees, and establish the pace and direction of work. The American, British, and French delegations to ISO, deeply aware of the strategic implications, each volunteered to take on the secretariat duties for SC16. ISO's leaders decided to accept the American proposal, in part because the Americans held a small number of ISO secretariats compared to the other countries.[16]

By the time the United States became the SC16 secretariat in September 1977, the Americans had already met to discuss their goals for the committee and their strategy for controlling its work.[17] It will be useful to consider the formation of the American delegation before returning to SC16 itself and the fate of its ambitious project for Open Systems Interconnection.

Charles Bachman and the American SPARC

In the creation of international standards, administrative details matter every bit as much as technical details. ANSI, the official American member body in ISO, delegated all responsibility for computer and information processing standards to its members who were active in ANSI's X3 Committee – consistent with the procedures described in Chapter 3. The X3 Committee, in turn, hosted dozens of Technical Committees with jurisdiction over specific areas such as programming languages, terminology and symbols, and data transmission. In addition to its Technical Committees, X3 also maintained three standing committees: the International Advisory Committee, which guided X3's position in the international arena; the Standards Steering Committee, which ensured that all X3 subgroup activities adhered to ANSI procedures and expectations; and the Standards Planning and Requirements Committee (SPARC), which

---

[15] "Resolution 11. Establishing Subcommittee 16."
[16] Pelkey, "Entrepreneurial Capitalism and Innovation," chapter 6. AFNOR, the French national standard-setting body, chose French national delegates to ISO. Therefore it was politically possible for computer researchers who worked on the Cyclades project – minus Pouzin – to contribute to discussions in ISO/TC97/SC16, even when their work may have run counter to the interests of the French PTT.
[17] Robert M. Brown, "Draft Minutes: Study Group on Distributed Systems, Meeting #1," August 16, 1977, Bachman Papers, Box 19, Folder 1.

supervised existing and proposed future standardization projects conducted under the auspices of the X3 Committee.[18]

Of these three standing committees, SPARC had the broadest and most important mission. Six times each year, the twenty members of SPARC met for three days to chart the strategic direction of X3 activities. Their duties included reviewing all project proposals, ensuring that X3 projects had a coherent scope and valid purpose, and checking to see that proposed standards would be technically and economically viable.

SPARC, created in 1969, occasionally formed study groups to investigate and make recommendations on specific emerging technologies. These recommendations, even though they were the product of teams of technical experts, did not always find acceptance as industrywide standards. For example, in 1972, SPARC created a Study Group on Database Systems to determine whether database standards would be feasible or necessary. Warren Simmons, a computer programmer at U.S. Steel and an influential member of SPARC, recruited Charles W. Bachman, a prominent database expert, to lead the effort. Despite his best efforts, Bachman later recalled, the SPARC recommendations were "never standardized, because IBM pushed back against it and IBM was the biggest force because they had both manufacturing people and also customers. And they're always aware of the IBM customers that are laid in there, so they had a predominance of the voting power in the committee."[19] Bachman's experience taught him that IBM's overwhelming power continued to pose problems for voluntary, industrywide computer standards in the 1970s, just as it had in the 1960s (see Chapter 5). IBM's strength was a function of its deep relationship with customers who appreciated that it could deliver some level of standardization to them. For the most part, it did not matter to these customers if standards were de facto IBM designs or if they were the product of an international committee. Consistency was most important because standards made things simpler and allowed IBM's customers to focus on other problems. Nevertheless, SPARC persisted in its efforts to bring together leading American computer engineers and focus attention on pressing problems in computing,

[18] Gary S. Robinson, "Accredited Standards Committee for Information Processing Systems, X3," *Computer Standards and Interfaces* 5 (1986): 189–193; Business Equipment Manufacturers' Association, *Committee X3 Organization Manual*, 1–7.

[19] One success of the SPARC Committee was its abstract "three-schema" design (known as the ANSI-SPARC architecture) that became foundational for the design of most modern database management systems – despite the fact that it was not a formal American standard. Bachman interview, IEEE Computer Society History Committee; Charles W. Bachman interview by Thomas Haigh, September 24–25, 2004, ACM Digital Library; Charles W. Bachman, "Summary of Current Work, ANSI/X3/SPARC/Study Group – Database Systems," *Bulletin of ACM SIGMOD* (1974): 16–39; Charles W. Bachman, "The Origin of the Integrated Data Store (IDS): The First Direct-Access DBMS," *IEEE Annals of the History of Computing* 31 (2009): 42–54; "Facts on the Reorganization of ANSI Standards Committee X3," December 1969, Box 1, Folder 2, Charles A. Phillips Collection, CBI 39, Charles Babbage Institute, University of Minnesota, Minneapolis.

even though it lacked the authority to force IBM (or anyone else, for that matter) to adhere to its recommendations.

In 1977, when the leaders of SPARC learned about the British proposal to organize a new international subcommittee to create "network standards for open working," they decided to prepare a cohesive American position. In July, SPARC created a Study Group on Distributed Systems with a three part mission: to develop an "analytic model for distributed systems" and identify protocols and interfaces that should be standardized within ANSI, to develop proposals for making such standards, and to represent the United States at the international meetings of committee SC16.[20]

The SPARC Study Group on Distributed Systems, known as DISY (for DIstributed SYstems), fulfilled its mission with remarkable speed. An initial meeting on August 15, 1977, was uneventful: the participants introduced themselves, shared information about related work in other standards committees, pondered the meaning of the title of their group (which, according to meeting minutes, was "deemed uninformative"), and began to "investigate the interest, feasibility, and need for a project to develop proposed American National Standards on this subject." The pace of work then picked up considerably, fueled in part by ISO's announcement in mid-September that it had selected the United States to be the secretariat for SC16. DISY met three times before the end of the year to prepare the administrative details of the first SC16 meeting to be held in Washington, D.C., from February 28 to March 2, 1978. DISY also moved quickly to clarify its liaison relationships with existing American standards committees, define its strategy for contributing to SC16's technical work, and look for additional experts who were interested in work on distributed systems and open systems or who wished to become members of the DISY Committee. In short, it provided a model of how the ANSI voluntary consensus process was designed to work.[21]

As it gained momentum, engineers working in DISY grappled with a broader problem of terminology. The foundational term of the international effort – "open systems" – was ambiguous, unstable, and contested. As we have seen, many leading researchers did not conceive of their work in computer-communication systems in those terms. British researchers hoped to enable "open working"; the American military-sponsored researchers Vinton Cerf and Robert Kahn (who chose not to participate in DISY or OSI) referred to their experiments as "internetworking"; French researchers, led by Louis Pouzin, spoke of a concatenation of networks (or Catenet); and Bachman referred to his work in this area as "distributed systems." Although SPARC chose Bachman

[20] "Charter: Study Project on Distributed Systems," July 20, 1977, Bachman Papers, Box 19, Folder 1.

[21] Robert M. Brown to X3 Members, Observers, and Other Interested Persons, "First Meeting to Investigate Possible Standards on 'Distributed Systems,'" August 5, 1977, Bachman Papers, Box 19, Folder 1; Robert M. Brown, "Draft Minutes, Meeting #1."

as its chair and Distributed Systems as the name of his committee, records from the committee's first meeting reflect a state of semantic confusion. Because the group "agreed that the descriptive terms, Distributed Systems, Open System Interconnection and related terms, are used with various meanings," it assigned a member of the committee "to collect commonly-used descriptions and propose definitions for uniform usage." As in the past, work to standardize language, terminology, and concepts proceeded alongside work to create technical standards.[22]

Discussions within DISY in these early meetings produced a fundamental point of agreement: distributed systems would facilitate the interconnection of existing computer networks. The committee, therefore, prepared an American position that would standardize gateways and interfaces – the points where networks meet. A frank summary of the second DISY meeting in late September 1977, prepared by Honeywell engineer and standards veteran Thomas McNamara, named three "principal movers" who "do not intend to wait for U.K. proposals to react to." Interestingly, the three principal movers all worked for major *users* of computer systems from different branches of the American federal government: Richard des Jardins from NASA, Steven Kimbleton from the National Bureau of Standards, and Ralph Hayward from the Navy. Here, as we have seen so many times in earlier standardization projects, officials working for the federal government played key roles. Representatives from the major computer manufacturers such as IBM, Digital, Honeywell, Sperry Univac, and Burroughs were all in attendance, but the consensus of the committee was to work on a system that could accommodate a range of "proprietary, unique protocols."[23]

McNamara's report from the second DISY meeting provides a rare glimpse of the ways that corporate engineers used national standards committees to advance their proprietary goals. Beginning in 1975, a team of Honeywell engineers had been designing the Honeywell Distributed Systems Architecture (HDSA) as a rival to the proprietary architectures introduced by IBM, Digital, and other companies. McNamara saw that Honeywell could use the DISY process to create distinct advantages in the marketplace:

If [Honeywell] is to have any influence on minimizing divergence of the industry standards work from our HDSA/HDNA product offerings, we must participate actively. [Charles] Bachman has been invited to be Chairman of this U.S. SPARC/Distributed Systems Study Committee.... Since the United States has been assigned the Secretariat of this new ISO committee, this would mean Charlie Bachman would be the logical Chairman of that ISO committee. If we decline, we must realize that a different energetic chairman may spur the work along at a pace which does not allow for adequate

---

[22] Robert M. Brown to X3 Members, Observers, and Other Interested Persons, "First Meeting to Investigate Possible Standards on 'Distributed Systems, Bachman Papers'"; Robert M. Brown, "Draft Minutes, Meeting #1."

[23] T. J. McNamara to I. M. Wyman, September 29, 1977, Bachman Papers, Box 19, Folder 2.

weighing of technical alternatives and require a costly gateway from HDSA/HDNA to the standard(s). Alternatively, a passive chairman of the U.S. Committee will most certainly cause the Europeans to push for an active European (UK probably) to chair the ISO committee.[24]

McNamara concluded with two clear recommendations for Honeywell to use the committee to advance its proprietary aims: "Bachman and [Honeywell engineer Gerald] Yon [should] be given freedom to influence protocol discussions toward HDNA protocols"; and "The wider concepts of HDSA [should] not be offered to the standards community at this time."[25]

Bachman, a veteran of earlier SPARC study groups on databases, had been a leader in the Honeywell group developing Honeywell Distributed Network Architecture (HDNA). In the process, he learned about IBM's System Network Architecture (SNA) and came to admire its design: "It looked like it was well-architected. It had a principle at each layer that only talked to the layer below it. And so the layer below it could be modified as long as it maintained that interface. And so it would essentially swap out a layer with an improved implementation that would support different things." Bachman became convinced that DISY should mimic SNA's basic design principle – a layered architecture – but that DISY (and Honeywell) should make some changes in order to make their design incompatible with IBM's.[26]

Bachman's fundamental alteration was to insert a new layer, the "presentation layer," to the five-layer SNA model. The presentation layer, Bachman recalled in a 2011 interview, "could be empty if the two applications were speaking the same language [or using] the same formats" but would be essential for translating data for applications that used incompatible formats. In October and November, Bachman and his colleagues at Honeywell worked on the details of a layered architectural reference model that they presented to DISY members, who then debated whether it should have six or seven layers.[27]

The prominence of an architectural reference model was Bachman's seminal contribution to computer networking. Whereas earlier efforts such as INWG 96 and CCITT's X.25 sought to define specific protocols for point-to-point connections, Bachman's reference model – which would soon become the basis of the ISO reference model – defined an architecture into which a variety of protocols and interfaces could fit. In other words, the reference model set the ground rules for network interconnection and left the specific

[24] McNamara to Wyman.
[25] McNamara to Wyman.
[26] Bachman interview, IEEE Computer Society History Committee.
[27] Bachman interview, IEEE Computer Society History Committee; Charles Bachman to DISY, November 28, 1977, Bachman Papers, Box 19, Folder 3; Charles Bachman, "Meeting ANSI/X3/SPARC/Study Group on Distributed Systems," February 27, 1978, Bachman Papers, Box 19, Folder 3.

terms of interconnection to standards that fit within the predefined network architecture.[28]

Bachman's leadership style as chairman of the DISY and SC16 committees built on his experience with distributed computer databases in corporate settings. He was unusually comfortable and skillful in his use of abstract concepts that promised to bring order to an inherently disorderly and unstable technical practice. At the same time, his decades of work with computers and control systems at Dow Chemical, General Electric, and Honeywell gave Bachman a pragmatic streak that many of his contemporaries in networking did not share. Because he spent his entire career in industry, Bachman never had the luxury to pursue fundamental research or perform academic experiments. Instead, his projects always responded to practical applications of existing computer technology rather than the experimental style of individualistic "interactive computing" that J. C. R. Licklider and his colleagues persuaded the Pentagon to sponsor. "In some sense," Bachman recalled, "OSI was more oriented towards business transactions and putting files back and forth.... My background came from a business operating mode, so it was conceived to me as fundamentally a business operating thing."[29]

Two additional values that influenced the early design choices in OSI were the product of Bachman's background: the project should promote conceptual clarity – rather than a cut-and-try approach – and it should create advantages for IBM's competitors. "I had an architectural vision that I was trying to move with," Bachman recalled, which motivated him to become an unusually active chairman. "In fact, some criticism I had from some of the people was that the chairman should just be a chairman, just kind of orchestrate things, and I was wearing an architect's hat more seriously than I was wearing the chairman's hat." Further, Bachman was blissfully unaware of the struggles that researchers in INWG had endured to create international standards. He was unconcerned with the technical and political legacies of communication networks and completely unaffected by the politics that stifled Louis Pouzin and other European researchers who opposed the dominance of European telephone monopolies and resented the bias toward virtual circuits in the X.25 standard.[30]

Bachman discussed his proposal for an architectural reference model in October and November 1977 with other leading members of DISY, including des Jardins from NASA, Kimbleton from the National Bureau of Standards, and William F. "Bud" Emmons from IBM. There was some sense of urgency in their task: since the first SC16 meeting was to begin on February 28, and ISO

---

[28] Zimmermann, "OSI Reference Model"; Richard des Jardins, "Overview and Status of the ISO Reference Model of Open Systems Interconnection," *Computer Networks* 5 (1981): 77–80.
[29] Bachman interview, IEEE Computer Society History Committee.
[30] Bachman interview, IEEE Computer Society History Committee. For a statement in support of conceptual clarity from the same era, see Brooks, *Mythical Man-Month*, 40–50.

rules dictated that proposals should be circulated sixty days before the meeting, any American proposal needed to be complete by mid-December. DISY members supported Bachman's architectural reference model, which was by far the most detailed, comprehensive, and ambitious that the group considered. When Bachman met with British computer engineer Hugh McGregor Ross in mid-December to report on DISY's work, Ross reported to the British standards committee that "the US presentation [of its reference model] is more vivid, thorough, wide-ranging and detailed than ours."[31] Ross also was blunt with Bachman about British concerns over American leadership and participation in ISO SC16 and suspicions that IBM would once again dominate the American delegation and use the process to protect their market position:

I told Mr Bachman that most of those seriously involved in this work within Europe believe that the US are intent to slow down, or at least dominate, this work. And that, if there has been a change of heart, he will have to do a hard sell to overcome this deeply held opinion. I mentioned that something we find difficult in any standardization work is the American tendency to work out their own scheme to near-completion, and then slap it on the table for international consideration.[32]

Ross's concerns were only partly justified. Bachman and his American colleagues in DISY had no intention of moving slowly and had already set a goal to complete an ISO reference model within one year. Subsequent ISO standardization work on all interfaces and protocols should, Bachman hoped, be completed by the end of 1980. SPARC voted in their meeting on January 18, 1978, to support the rapid schedule that Bachman had established, but Bachman's notes from the meeting suggest the Americans were open to international review and input. "The American reference model," Bachman wrote in his report to DISY, "is being presented as a 'straw man' to start the development of an ISO reference model." He was further "pleased to see a great similarity" in documents that were distributed by the British Standards Institute and the European Computer Manufacturers Association (ECMA) and was optimistic about the prospects of achieving a rapid international consensus within ISO SC16.[33]

At the same time, Bachman continued to use his position as chairman of DISY and ISO SC16 to bolster the competitive position of his employer, Honeywell. He intended to build Honeywell's networking product to "near-completion"

[31] Charles W. Bachman to ANSI/X3/SPARC/Study Group – Distributed Systems, November 28, 1977, Bachman Papers, Box 19, Folder 2; "Draft Minutes, SPARC/DISY Meeting #4, 30 Nov. – 1 Dec 1977," December 27, 1977, Bachman Papers, Box 19, Folder 1.

[32] Hugh McGregor Ross, "Report on Meeting with Mr Charles W Bachman, 16.12.77," December, 16, 1977, Bachman Papers, Box 19, Folder 6.

[33] Charles W. Bachman, "ANSI Reference Model for Distributed Systems," December 30, 1977, Bachman Papers, Box 19, Folder 6; Charles W. Bachman, "Minutes of 1978 January 18–20 Meeting," February 10, 1978, Bachman Papers, Box 18, Folder 10; Charles W. Bachman to ANSI Study Group – Distributed Systems, "Meeting Notes – BSI DPS/20," February 1978, Bachman Papers, Box 18, Folder 10.

in order to, as the British feared, "slap it on the table for international consideration." In a February 6 memo to his colleagues at Honeywell, Bachman noted the close resemblance of the new ANSI reference model to the existing HDNA. He declared himself "pleased that the ANSI model is such a good representation of HDNA" – the details of which, following the strategic position recommended by McNamara in late 1977, Honeywell engineers decided not to offer to the standards community. The ANSI model, Bachman continued, "has been a super test of HDNA and its authors should be pleased at how well it has stood the test. More tests will come at the ISO level.... It will take a lot of selling and explaining – here and in Europe," he warned. Bachman concluded his report by urging his fellow Honeywell engineers to brace themselves for the challenge: "Let's be sure we understand our own story."[34]

Bachman's tactical instincts were exquisite. His "straw man" reference model – which he had nurtured from a slight permutation of IBM's SNA to a secret Honeywell networking product to a design that gained the support of ANSI's committees – was the central topic of technical discussions at the first meeting of SC16 from February 28 to March 2, 1978, that the Americans hosted in Washington, D.C. Delegates from ten countries (Canada, France, Germany, Italy, Japan, The Netherlands, Sweden, Switzerland, the UK, and the US) participated, as well as observers from four international organizations (ECMA, CCITT Study Group VII, IFIP, and ISO TC97/SC6). As usual, technical, administrative, and strategic discussions were inextricably linked. First, the committee agreed unanimously to accept the premise that their first order of work would be to create a Provisional Model for Open System Architecture. Bachman later reported with understated delight that the American reference model for distributed systems received the bulk of the group's attention – three hours of presentation and discussion time out of the eight hours allotted – "because of its depth of study and detailed documentation.... As a result, it heavily influenced the provisional model approved for review."[35]

SC16 also resolved – again with unanimous votes – to create three working groups in which interested delegates could get into the details of the protocol and interface requirements for open systems. Working Group 1, Architecture, assumed responsibility for a seven-layer ISO reference model for Open System Interconnection, to be based on the proposal originally prepared by Bachman. Hubert Zimmermann, the Frenchman who had worked as Louis Pouzin's lieutenant on the Cyclades project and a protagonist in INWG, was approved unanimously as chairman. Working Group 2, Users of Transport Service,

[34] Charles W. Bachman to Honeywell Distribution List, "ANSI Reference Model – Distributed Systems," February 6, 1978, Bachman Papers, Box 19, Folder 6.
[35] Charles W. Bachman, "Minutes of February 27 and March 3 Meetings," March 7, 1978, Bachman Papers, Box 18, Folder 10; "Report of Meeting ISO/TC 97/SC 16," March 1978, Bachman Papers, Box 18, Folder 10; "Resolutions (Inaugural Meeting ISO/TC 97/SC 16)," March 10, 1978, Bachman Papers, Box 18, Folder 10; ISO/TC97/SC16, "Provisional Model of Open Systems Architecture," February 1978, Bachman Papers, Box 21, Folder 8.

assumed responsibility "to examine the functionality, interfaces and protocols found within the upper three layers of the provisional reference model." The British computer scientist Alwyn Langsford was approved, also unanimously, as chairman. Working Group 3, Transport Service, was responsible for the lower four levels – a task that would necessarily lead them to work closely with the liaisons to SC6 and CCITT, which had both defined lower-level transport protocols. George White, an American computer scientist at the National Communications System[36] who was active in American standardization committees, was approved as chairman by yet another unanimous vote.[37]

The Americans were exceptionally well organized. In addition to DISY meetings in the fall of 1977 and early 1978, eighteen members of DISY met the day before and after the first SC16 meeting in Washington D.C. DISY set up its own task forces that mirrored the organization of the three working groups in SC16. In his summary of their meetings, Bachman noted that "strong, well prepared position papers by ANSI are important if the US interests are to be protected in the ISO negotiations." The resolve of Bachman and his colleagues was so strong, he continued, that the DISY task forces would also be able to function "if an ANSI standard is to be prepared on a go-it-alone basis because of a breakdown of the ISO thrust." Bachman also urged his colleagues to recruit members from their own organizations and from other ANSI groups in order to "represent a cross section of US interests." Bachman's first goal, in other words, was to build an international consensus for his system architecture. If that failed, DISY would be prepared to push forward with a consensus for American architecture and leave the Europeans to squabble among themselves. "Surprisingly," Zimmermann recalled in a 1988 interview, "the ARPANet gang did not join ISO, except just a few, like John Day. Most of them just stayed out of the process." He continued, noting the ARPA program managers Vint Cerf and Bob Kahn "had passed contracts to various people here and there to implement [TCP/IP], and probably they were reluctant to see people going to another place and discuss things which might well be different from their stuff."[38]

Fortunately, the delegates at the first SC16 meeting readily agreed on the basic point of Bachman's plan: an architectural reference model should be defined first, before any specific standards that might fit within the model. Bachman's "Provisional Model of Open System Architecture" had the technical virtue of conceptual clarity because it established the parameters of their

---

36 President John F. Kennedy created the National Communications System in 1963 to oversee the communications capabilities of agencies in the American federal government and to ensure their survivability.

37 Bachman, "Minutes of February 27 and March 3 Meetings"; Zimmermann, "OSI Reference Model."

38 Bachman, "Minutes of February 27 and March 3 Meetings"; Charles W. Bachman, "The Structure of the ANSI/X3/SPARC/Study Group – Distributed Systems," April 20, 1978, Bachman Papers, Box 19, Folder 6; Hubert Zimmermann, interview by James Pelkey, May 25, 1988, courtesy of James Pelkey.

project instead of haphazardly defining various pieces. It also had an important organizational virtue, as Bachman wrote in his notes from the February 27 DISY meeting: "It permits independent groups to develop specifications for standards, which are complementary." Such independence and organizational modularity was crucial for OSI's diplomatic challenge of bringing diverse constituencies together. The advantages of technical and organizational modularity had been well understood in the computer industry for more than a decade, most visibly in the design of the IBM System/360 in the mid-1960s and in the work of the Dutch theorist Edsger Dijkstra. Modular architectures partitioned the numerous and varied tasks of network building into a scalable structure of "black boxes," which, by virtue of their small and specialized character, could become the endless number of small building blocks needed to create a functional network architecture. In the formative discussions of the reference model in 1977 and 1978, there was an overwhelming consensus that the group should begin with the definition of a layered and modular system architecture (see Figure 7.1 for an early drawing of the seven layer model).[39]

The OSI reference model progressed steadily, if not rapidly, along the standards track. In July 1979, TC97 approved the reference model as a working draft, and authorized SC16 to make revisions and incorporate suggestions. Zimmermann, as chairman of the SC16 Architecture Working Group, led the charge. In November 1980, Zimmermann was able to shepherd the revised model to the status of draft proposal. The reference model then advanced to a draft international standard in January 1982, before finally being formally approved by ISO in May 1983 as an international standard, "ISO 7498: The Basic Reference Model for Open System Interconnection." Bachman, in a 2004 interview, recalled how well his vision held up as the reference model proceeded along the path to international standardization: "I would say that 75% was based on my work at Honeywell. It was all well documented. We worked, as you sometimes have to do, to cover the traces."[40]

As the OSI reference model progressed from a proposal in 1977 to an international standard in 1983, hundreds of engineers, computer scientists, and government officials from more than two dozen countries joined the SC16 working groups and national delegations in SC16's parent body, ISO TC97. In these negotiations, the clean theoretical lines of Bachman's layered architecture gave way to the messy reality of technological uncertainty and strategic posturing.

---

[39] Bachman, "Minutes of February 27 and March 3 Meetings"; David Beech, "Modularity in the Design and Standardisation of Open Systems," *Computer Networks* 8 (1984): 49–55; Bruno Latour, *Science in Action: How to Follow Scientists and Engineers through Society* (Boston: Harvard University Press, 1988); Richard N. Langlois, "Modularity in Technology and Organizations," *Journal of Economic Behavior & Organization* 49 (2002): 19–37; Andrew L. Russell, "Modularity: An Interdisciplinary History of an Ordering Concept," *Information & Culture: A Journal of History* 47 (2012): 257–287.

[40] Pelkey, "Entrepreneurial Capitalism and Innovation," chapters 6 and 9; Zimmermann, "OSI Reference Model"; Bachman interview, ACM Digital Library.

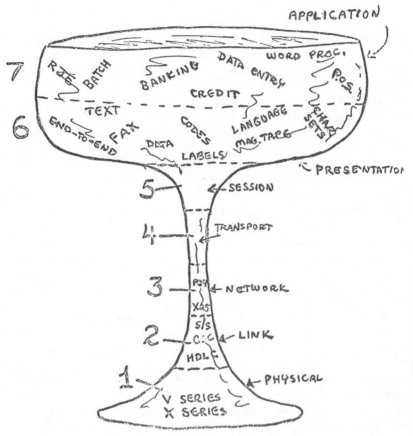

FIGURE 7.1 Seven-level reference model as Margarita Glass, 1979.
Source: John Aschenbrenner.
Courtesy of the Charles Babbage Institute, University of Minnesota, Minneapolis.
"Report from July 18, 1979, second meeting of TC 97/SC 16, June 11–15, 1979,"
Bachman Papers, Box 18, Folder 18/19.

Bachman noted in the autumn of 1978 that the bulk of the group's attention
would soon be shifting from the reference model itself to dozens of standardization projects for functionality within the seven layers of the reference model
and interfaces across layers. Moreover, the momentum that Bachman was able
to generate by working within a small and collegial group would decelerate as
SC16 would be forced to confront powerful institutions such as CCITT and
IBM that wanted to exercise control over international data communication
standards. "The task in establishing such standards is enormous," Bachman
warned.

The organizational problem alone is incredible. The technical problem is bigger than any one previously faced in information systems. And the political problems will challenge the most astute statesmen. Can you imagine trying to get the representatives from ten major and competing computer corporations, and ten telephone companies and PTTs, and the technical experts from ten different nations to come to any agreement within the foreseeable future?[41]

In the same essay, which Bachman circulated among the American standards community and published later in 1978, he warned of specific problems that faced the committee. He was particularly concerned with the organizational tasks of "marshalling the people and gaining the release of currently proprietary information" that the committee would need to be successful. Characteristically, Bachman presented forceful and well-supported suggestions. First, he urged the committee to move quickly, given the threat of "too many ad hoc or one company solutions" that were being prepared outside of SC16. Second, he justified the rapid schedule in SC16 by arguing that members of standards committees were most effective when they limited their involvement to three years. "If the project schedule is longer than that," he continued, "it becomes a career, to be savored but not to be finished, or an impossible task which doesn't deserve much attention." Finally, Bachman relayed the thoughts of Maurice Wilkes, the distinguished computer scientist at Cambridge University, who had recently learned of the OSI project: "He said: 'You people are not doing classical standards work, you are doing an international, industry-wide, design effort under the guise of standardization!' That just about says it all."[42]

Conflicting Strategies, Tenuous Alliances, Stalled Momentum

OSI's organizing principle, as Wilkes noted perceptively, turned the traditional standardization process upside down. Groups such as ANSI and ISO were designed only as umbrella organizations that could coordinate and codify existing practices; they did not promulgate new standards. By undertaking a new design project within ANSI and ISO, which had rigorous requirements for inclusivity and international review, OSI started big and grew bigger. OSI called for collaboration and cooperation on a scale never before accomplished in the international computer industry. Because it invited every conceivable stakeholder into its design process, ISO's SC16 Committee immediately became a home for conflicting strategies, technical disputes, and bureaucratic delays.[43]

[41] Charles W. Bachman to SPARC Chairmen and Members, "Report on SC16 and DISY Status," November 20, 1978, Bachman Papers, Box 19, Folder 6; Bachman, "Domestic and International Standards."

[42] Bachman, "Domestic and International Standards."

[43] For a discussion of the trade-offs involved with anticipatory standards, see Carl F. Cargill, *Information Technology Standardization: Theory, Process, and Organizations* (Bedford, MA: Digital Press, 1989).

SC16's size alone generated friction. Almost immediately, it grew to include well over 200 people from more than 20 countries, with working groups of 20 or more members. This was far bigger than a typical subcommittee in ISO, which might include twenty people and Working Groups with only eight to ten members.[44] Strains became evident between Bachman's call for urgent action and the complex discussions taking place in the working groups. George White, the American chairman of SC16 Working Group 3 on the lower layers, resented the stress and complained that it "puts unbearable pressure on a group that wants to get the work right, and in the shortest amount of time – and without taking any short cuts." That task, White continued, "means debate, and understanding, and the forging of technically acceptable compromises – without foot-dragging and at an expeditious pace, of course. But not under the gun – we don't work that way."[45]

White was not the only one put off by Bachman's aggressive approach. Members of the British delegation were skeptical of the motives of their American counterparts and continued to worry that Bachman was pushing the group to move too quickly. At the beginning of 1979, Alwyn Langsford wrote to Bachman to "set down my points for concern," which included a list of projects that "have had no serious, international discussion," a fundamental disagreement about the meaning of "open systems," the lack of a "mechanism for ruthless editing" of draft standards, and the need for an "alternative model" to the seven-layered model that Bachman had been advocating.[46]

No record of Bachman's reply to Langsford survives, but there is evidence that Bachman began to view such concerns as needless foot-dragging. An April 1981 article in *Computerworld U.K.* quoted Bachman as saying "the British have a reputation for complaining about trivia.... They like the substance of the document but they don't like its format." The *Computerworld* article went on to explain that British objections were linguistic in nature, and the dispute "appears to be a question of 'you will comply' versus 'you should.'" Whereas Bachman perceived a trivial editorial detail, Bryan Wood, chairman of the UK delegation, perceived a symptom of a broader malaise. "In fact," he complained to Bachman, "I am feeling increasingly uneasy.... It seems to me that a quite unreal sense of crisis is being generated and this is leading to an atmosphere of confrontation rather than conciliation and consensus." Rather than a small or isolated incident that could be brushed aside, Wood argued that "the situation, in fact, reflects problems in the operation of SC16 itself and highlights the need to review its operation." The bulk of technical discussions were taking

44 John Day, interview by James Pelkey, July 11, 1988, courtesy of James Pelkey.
45 George White, "Report to ISO/TC 97/SC 16 from Convenor 97/16/3," June 18, 1979, Bachman Papers, Box 18, Folder 18.
46 Hugh McGregor Ross, "Report on Meeting with Mr Charles W Bachman, 16.12.77," December 16, 1977, Bachman Papers, Box 19, Folder 6; A. Langsford to C. W. Bachman, January 3, 1979, Bachman Papers, Box 18, Folder 3.

place within the "too large and unwieldy" working groups, Wood continued, and the main committee of SC16 "meets too infrequently and its procedures are too cumbersome for it to be able to control and coordinate their activities effectively." Modularity, in other words, was harming the cohesiveness of the overall effort.[47]

Such seemingly petty disputes cloaked a more fundamental and severe problem that dogged OSI from the start: the motives of the major powers in SC16 and its working groups were not well aligned. At a certain level, most participants shared the common goal of altering the power dynamics of the industry and preventing IBM from continuing to impose its proprietary designs on users, suppliers, and competitors. The very notion of an open system, however defined, was an implicit rejection of IBM's dominance. OSI thus was conceived as a critique of IBM's proprietary and centralized control. Open systems provided alternatives: a new method of governance, a new means for producing standards, and a new and more decentralized industry structure for the data communications industry – a "level playing field" in the era's resilient cliché. In the moments when OSI's architects were reflecting on the historical precedents of their project, they might have noticed some conceptual similarities between OSI and the American Open Door policy toward China from the turn of the twentieth century. In both cases, policy makers tried to prevent a promising market from dividing into exclusive spheres of influence between competing powers; the alternative was to share jurisdiction over the open market, based on rules, procedures, and protocols that all parties agreed to follow.

The most striking feature of OSI's production – the open committee structure of ANSI and ISO – meant that it issued a standing invitation for foxes to visit the henhouse. The movement for open systems had cast IBM and the telecommunications monopolies as enemies, but the core principles of open systems demanded that all parties deserved a seat at the negotiating table. Strategists working for IBM, as well as their counterparts working for the European telecommunications monopolies, recognized that they could not afford to miss the opportunity to shape the design of such a large and ambitious standardization project.

Joseph De Blasi, IBM's director of standards from 1976 to 1988, recalled in a 2009 interview that his job was to "represent IBM in the external standards community" in order to ensure that "the standards that were being developed would not adversely impact IBM." According to Marc Levilion, an engineer working for IBM France at the time, IBM was "very concerned about OSI," which IBM saw as "much more dangerous than X.25." Levilion recalled that De Blasi built and managed "an extremely vast structure" that coordinated IBM's standardization strategies across each country in which it had a presence. Levilion's position within this vast structure, as he recalled in a 2012 interview, was to "make sure that the IBM companies of Europe would understand the

[47] Bryan Wood to Charles Bachman, April 28, 1981, Bachman Papers, Box 21, Folder 6.

significance of OSI in general and to IBM in particular. I would make sure that they would send representatives to the local ISO committees in Italy, in Germany, in Sweden, in Denmark, in the U.K., in France, and so on." He also was responsible for ensuring that "they would know the strategies and the principles and make sure that we had united views on how to discuss and to defend IBM positions."[48]

De Blasi's defining challenge with OSI, therefore, was to defend against the well-coordinated attacks of virtually every other protagonist in the international standards arena. This was no small task, as De Blasi himself explained:

You have to remember at that time IBM was considered the Goliath of the industry and the Europeans and competitors all wanted to make inroads in areas where our technology was pretty dominant. One way they could do that would be if they could get a standard accepted that were [*sic*] contrary to IBM's technology. Like the I/O interface standard, like our SNA networking standard, they wanted the Open Systems Interconnection standards. So when you represented IBM, there were two things that really came into play. The first was to protect IBM's business interest; to make sure that things were not done in a way that would adversely impact the investment we had in our major technologies. The second was to manage that process so there was a good feeling on both sides and we were not being obstructionists.[49]

De Blasi thus was committed to a delicate task: obstruct competitors without giving the appearance of obstruction. Or rather, as Levilion remembered, "If we can't stop it, we help it. We help it in such a way that it can work and in such a way that it works the same way all over the world ... let it be, but let it be well."[50]

IBM's powerful, skillful, and well-coordinated interventions into the standardization process for OSI eventually attracted grudging admiration from a vocal critic of IBM, the American computer scientist John Day. Day was a veteran of the Arpanet and INWG communities who was one of the self-described "young Turks" that rose to powerful positions in OSI subcommittees and working groups. As a key member of the American delegation to OSI from 1980 to 1992, Day began to perceive large gaps between IBM's earnest rhetoric and self-interested actions: "IBM knew that the [OSI] peer network architecture was an anathema to SNA, so their primary purpose was to stonewall and delay." IBM's subtle interventions, in Day's recollection, often sought to heighten conflict and delay consensus. Michel Gien, a member of the French Cyclades team, remembered that "when [we] started to get some alignment, then IBM comes with totally new contributions ... and then everyone has to work on it to kill it. It's because they were trying to slow down the process." Gien continued, articulating IBM's position of strength: "The power of IBM

48 Marc E. Levilion, oral history interview by Andrew L. Russell, April 2, 2012, Paris, France. Charles Babbage Institute, University of Minnesota, Minneapolis.
49 Joseph De Blasi, interview with Arthur Norberg, June 4, 2009, ACM Digital Library.
50 Levilion interview, Charles Babbage Institute.

was that they could push the same idea within the French AFNOR, within the BSI in Britain, in the U.S.... So if they lose in one ground, then they come in the backdoor with the other guys." Day remembered that IBM representatives expertly and regularly intervened amid disputes among various delegates who were "fighting over who would get a piece of the pie.... IBM played them like a violin," he continued. "It was truly magical to watch." Day later wrote a networking textbook but imagined that he "could write a 'Discourses on Livy' for standards committees."[51]

IBM's approach, which appeared Machiavellian to IBM rivals such as Day, worried the architects of OSI. These architects, including Bachman and Zimmermann, conceived of the project as a way to prevent IBM from extending its control into the future. Far more worrying for OSI's leaders, however, were the delicate political and organizational relationships they were forced to build with the community of telecommunications experts representing the interests of the incumbent monopolies. If one of the founding ideals of OSI was to create a viable architectural alternative to IBM's SNA, an equally important ideal was to ensure that future standards for computer networks be under the jurisdiction of *computer* professionals, not *telecommunications* professionals. Nevertheless, the convergence of computer and telecommunication technologies and services once again forced both groups to come together and try to reconcile the incompatibilities between their respective technical practices, work cultures, political economies, and motivations for creating international standards.

Proponents of open and distributed systems were successful in creating their own forum for deliberations – SC16 – but they always operated under the explicit requirement to take the activities of telecom monopolies "into due account."[52] At its first meeting in February and March 1978, OSI leaders negotiated an agreement with members of CCITT Study Group 7 (the committee responsible for X.25) to share jurisdiction over proposed standards that would "have an impact on the basic design and features of Data Processing Systems." The two groups also named official liaisons in 1978 and 1979 that would maintain a healthy flow of information and policy of consultation. Subsequent personal correspondence and official resolutions established additional points of agreement: both groups supported the urgent development of a layered architectural reference model, and both groups possessed the organizational capabilities to sustain a modular approach to the technical and social challenges of open system standardization.[53]

---

[51] Michel Gien, oral history interview by Andrew L. Russell, April 3, 2012, Paris, France. Charles Babbage Institute, University of Minnesota, Minneapolis. John Day, *Patterns in Network Architecture: A Return to Fundamentals* (Upper Saddle River, NJ: Prentice-Hall PTR, 2007), 358; Day interview, 1988.

[52] "Resolution 11. Establishing Subcommittee 16."

[53] V. C. MacDonald, "ISO News," *Computer Networks* 2 (1978): 416–417; Charles W. Bachman to T. deHaas, September 18, 1978, Bachman Papers, Box 18, Folder 12.

The collaboration between ISO and CCITT created an opportunity to revisit an old and bitter debate between the advocates of connectionless datagram protocols and the advocates of connection-oriented virtual circuits. Pouzin's close colleague Hubert Zimmermann, a veteran of INWG and chairman of the OSI Architecture Working Group, took personal responsibility for soothing the rift. Zimmermann, like Pouzin, had been a leading proponent of datagram service. Unlike Pouzin, Zimmermann was able to learn and master the art of standards diplomacy – a realm in which Pouzin's bombastic, confrontational style had not served him well. In 1977 and 1978, Zimmermann surprised many of his colleagues by supporting the virtual circuit service already defined by X.25 as a key component of transport in the OSI model. Some suspected that Zimmermann had simply given up on the prospects for connectionless standards; in fact, Zimmermann was one step ahead. As Vint Cerf later recalled, "He was much more politically astute than I was at that point." Zimmermann recognized that if he could foster a consensus for the layered model, it would not be difficult to add connectionless datagram service sometime down the road.[54]

Moreover, Zimmermann's career took a turn that crossed the boundaries of the computer and telecommunications sectors. In 1980, he left his position at IRIA, the French national computing research institute, and began a new position at the research organization of the French PTT, named the Centre National d'Etudes des Telecommunications (CNET). Five years earlier it would have been unimaginable for Zimmermann to go to work for the French telecom monopoly. By 1980, however, French bureaucrats began to rethink their protectionist approach to the convergence of computing and communications. As a result, Zimmermann sensed that the French PTT resolved to "get more of the data processing culture" into its operations. By hiring Zimmermann and allowing him to stay involved with standards work in OSI, the French telecom monopoly sent a clear signal that it would be more accommodating to the computer research community than it had been in the late 1970s. Zimmermann thus became a living emblem of the new French national strategy. He recalled the significance of the change: he was "still in charge of the OSI reference model group, and people knew that I had moved to the PTT. They could see that it did not change my way of managing and pushing things, and it was clear that it was supported by the French PTT."[55]

Two other veterans of the Cyclades project, Pouzin and Gien, also moved to CNET with Zimmermann. Gien recalled that although the "computer network war was over," decisively won by the French telecom monopoly, the trio still had some ability to pursue their own interests. While Zimmermann continued his work in standards committees, Gien recalled that he had complete freedom in his new job, which allowed him to hire people and to build a French

---

[54] Pelkey, "Entrepreneurial Capitalism and Innovation," chapters 6 and 9.
[55] Zimmermann interview, 1988; Pelkey, chapters 6 and 9.

community around the Unix operating system and microcomputers. Pouzin, on the other hand, was more constrained. Gien recalled that Pouzin "had a lot of people that didn't like him for political reasons" because he was "always saying in public things that you don't say in public, and people were very embarrassed." Pouzin agreed: "I was not in a position to get that [ability to participate in OSI committees] from the CNET management. Perhaps I could have tried, but I knew that if I tried it probably would have backfired onto me. I preferred to keep a low profile." Gien summarized, "Hubert was more efficient because he tried to change the things more slowly, where Louis was," Gien chuckled, "kind of, you know, a mess."[56]

Once he was secure in his new position, Zimmermann was able to champion a proposal, put forth by ECMA, to accommodate connectionless datagrams within OSI's transport layer. Despite securing tentative agreements in August 1980 for CCITT and SC16 to support an option for connectionless service, the version of the reference model that was approved as an ISO standard in May 1983 did not include a specification of a connectionless protocol. Such a protocol would need to be approved as an addendum, which would, in turn, need to proceed through the multiple stages of ISO's process for moving from a working draft to a draft proposal to a draft international standard and finally to an international standard. The OSI reference model was not as lean as the Cyclades datagram design, but it was doing a better job of bringing more stakeholders into a widespread agreement. And as digital convergence continued with the force of a runaway train, there was scarcely time to pause and plan for the future: competitive pressures and consumer demand for network services continued to grow, and the allure of first-mover advantages were too attractive to resist.[57]

Despite the increasing duration of the project and seemingly endless levels of bureaucracy and diplomacy needed to create consensus international standards, the leaders and advocates of OSI still had cause for optimism. By 1984, ISO and CCITT formally merged their separate committees working on OSI standards. The combined effort promised to streamline the standardization and user adoption process by developing common text for the reference model and for affiliated standards. Cooperation, in theory, would reduce confusion and expedite international approval across the computing and communications communities.[58]

In reality, the discourses of open systems spread much more quickly through the industry than the actual protocols that manufacturers and users would

---

[56] Gien interview, Charles Babbage Institute; Louis Pouzin, oral history interview by Andrew L. Russell, April 2, 2012, Paris, France. Charles Babbage Institute, University of Minnesota, Minneapolis.

[57] Pelkey, "Entrepreneurial Capitalism and Innovation," chapters 6 and 9; Zimmermann interview, 1988.

[58] John Larmouth, *Understanding OSI* (Boston: International Thomson Computer Press, 1996), chapter 1; David M. Piscitello and A. Lyman Chapin, *Open Systems Networking: TCP/IP and OSI* (Reading, MA: Addison-Wesley, 1993); Pelkey, "Entrepreneurial Capitalism and Innovation," chapter 9.

need to implement OSI as a fully functional networking architecture. In the United States, ANSI's DISY Committee reconstituted itself in the autumn of 1979 as the Open Systems Interconnection Committee (OSIC).[59] Large firms in the computer networking industry, including Sun Microsystems and Intel, also made the discourse of openness prominent in their marketing and public relations campaigns in the mid-1980s.[60] IBM's chief scientist Lewis Branscomb embraced the rhetoric of openness in a 1980 keynote address at the International Conference on Computing Communication, even as he defended the utility of IBM's proprietary network architecture: "As we enter the arena of mixed data and text, and unformatted documents," Branscomb reasoned, "SNA [IBM's System Network Architecture] continues to be expanded, and to include more and more peer capability, and more and more openness."[61] Marc Levilion, the IBM France engineer who was head of the French delegation to ISO, saw the widespread embrace of the language of open systems as an indication of Zimmermann's diplomatic skills. "The definition of an open system as a set of computers and equipments of all sorts, which was in fact coined by Zimmermann," Levilion recalled, "is very interesting because it allows all possible interpretations.... So everybody was happy. And that's the genius of Zimmermann."[62]

Another indication that Open Systems Interconnection was on the brink of international acceptance came at the end of 1983, when a special issue of the *Proceedings of the IEEE* assessed the current state of OSI. The issue featured an impressive range of papers on topics such as the OSI architecture, protocol standards that had been developed for individual layers, business implementations of OSI, government policies to adopt OSI, and several symposia and public demonstrations scheduled for 1984 to showcase OSI implementations. Despite the many signs of progress, however, a few of the articles had a distinctly defensive tone to them. Richard des Jardins, who had succeeded Charles Bachman as SC16 chairman in 1982, began his article with the statement "OSI Is Very Real." Des Jardins seemed to protest too much. It would have been obvious to his readers that OSI's critics had struck a nerve by claiming that OSI was an overengineered paper fantasy rather than a set of operational protocols.[63]

59 J. R. Aschenbrenner to Members of OSIC, "Minutes of First Meeting," November 8, 1979, Bachman Papers, Box 18, Folder 20; Richard des Jardins to OSIC Members and Observers, "Plan of Action for Initiation of X3T5," Bachman Papers, Box 20, Folder 20; "USA National Activity Report, ANSI X3T5," Bachman Papers, Box 19, Folder 13.

60 Christopher M. Kelty, *Two Bits: The Cultural Significance of Free Software* (Durham, NC: Duke University Press, 2008), 143–178.

61 Lewis M. Branscomb, "Computer Communications in the Eighties – Time to Put It All Together," *Computer Networks* 5 (1981): 3–8; Richard N. Langlois, "External Economies and Economic Progress: The Case of the Microcomputer Industry," *Business History Review* 66 (1992): 1–50.

62 Levilion interview, Charles Babbage Institute.

63 Harold C. Folts, "Scanning the Issue," *Proceedings of the IEEE* 71 (1983): 1331–1333; Richard des Jardins, "Afterword: Evolving Towards OSI," *Proceedings of the IEEE* 71 (1983): 1446–1448.

Even as OSI was gathering momentum along the path to international standardization, it was proving to be very difficult to implement. To be successful, voluntary standards need to be clearly specified and then widely adopted by users. Unfortunately, OSI's size and modular development inadvertently institutionalized the divide between designers and users of its standards. In the mid-1980s, for example, a variety of user groups in the United States, Europe, and Asia held workshops and specified implementation guides for OSI.[64] The "formation of implementor's workshops distinct from the standards organizations," according to John Day, tended to splinter ideas about OSI. "As it usually turns out," Day continued, "the designers had a much better idea of how to implement than the implementors did ... [the workshops] merely served, as predicted, as a forum to re-hash the same arguments in different groups, adding to the confusion."[65]

In an attempt to overcome the confusion and guide OSI procurement and implementation, OSI "profiles" emerged in the mid-1980s. Large users of networks – including industrial firms and government agencies in the United States and Europe –knew they had an important role to play. They knew from experience that the inclusion of OSI profiles in procurement requirements would stimulate the market in OSI products and services. A vibrant market would, in turn, resolve some of the confusion caused by OSI's complexity and tendency to publish standards that could be misinterpreted. In the U. S. National Bureau of Standards, John Heafner organized workshops and demonstrations to make it simpler for the federal government to make purchasing decisions for network equipment. Large industrial firms such as General Motors and Boeing faced similar problems with a lack of industry standards, which prompted their leadership in two implementations of OSI networking standards: the Manufacturing Automation Protocol (MAP) profile was for factory operations, and the Technical Office Protocol (TOP) profile was for engineering and office systems. All that was needed was for networking companies to build products that integrated MAP and TOP, but markets for OSI equipment remained sluggish in the mid-1980s.[66]

European conceptions of an active and well-coordinated government role sharpened after 1985, when officials in the European Commission rejected the protectionist national champion strategies of the recent past in favor of a New Approach to Technical Harmonization and Standards. By making commitments to adopt common technical standards, Europeans believed they could create a larger common market that would, in turn, allow European computer

---

[64] Examples included General Motors' Manufacturing Automation Protocol, Boeing's Technical and Office Protocol, and workshops sponsored by the Corporation for Open Systems.

[65] Day, *Patterns in Network Architecture*, 358.

[66] Lee Mantelman, "MAP: GM and Boeing Promise a Real Four-Bus Circus," *Data Communications* (October 1985): 78–79; Ray Walker, "Standardization in IT – Current Developments and Users' Response," *Computer Standards & Interfaces* 5 (1986): 277–279; Pelkey, "Entrepreneurial Capitalism and Innovation," chapter 12.

companies to develop the economies of scale and scope that would help them compete against IBM. Such logic led to strong European support for OSI and related networking projects.[67]

Nevertheless, there were signs that these and other collective efforts by users and governments could not overcome the complex institutional, technical, and strategic obstacles between OSI's design and its ultimate implementation by users. OSI's unwieldy scope and unmanageable complexity were compounded by another source: competitive pressure from the unexpected growth of the TCP/IP Internet. The Internet protocols had been designed for American defense networks, but in the late 1980s and early 1990s, as we will see in Chapter 8, Internet protocols were gaining popularity among academic researchers and equipment manufacturers. Louis Pouzin, who had been reassigned to be the dean of technology at a business school near the south coast of France, wrote an article in 1991 to assess OSI's progress after ten years of existence. Could the American-sponsored Internet protocols, he wondered, serve as a viable alternative to OSI for public and commercial computer networks? "TCP has never been considered as a potential candidate for becoming an international standard," Pouzin correctly observed. Nevertheless, he continued, TCP/IP had already been bundled into the popular UNIX operating system and adopted by manufacturers looking to enable ad hoc connections between different types of computer networks. "Therefore," Pouzin predicted, "TCP is likely to grow faster than OSI till about 1995." After that, Pouzin predicted, "one should expect a sharp drop in TCP sales, and a sharp increase in OSI."[68]

Pouzin, like so many other networking experts in the early 1990s, misjudged the future of OSI. Levilion, who worked very hard to align OSI with IBM products, led an effort to "make specific products inside IBM France for some of the levels of OSI.... That new offer was cumbersome, expensive, and not very efficient," Levilion recalled. "And at the same time came Internet. Free! So on one side you have something that's free, available, you just have to load it. And on the other side, you have something which is much more architectured, much more complete, much more elaborate, but it is expensive. If you are any director of computation in a company, what do you choose?"[69]

The fate of the American Government Open Systems Interconnection Profile (GOSIP), the American government's standard implementation of OSI, indicates that Levilion's experience was typical. Large network users in industry

---

[67] Michel Audoux, "Recent Developments in European Community Policy to Promote a Harmonized Implementation of OSI Standards," *Computer Networks & Interfaces* 5 (1986): 225–227; Alfred D. Chandler, Jr., *Inventing the Electronic Century: The Epic Story of the Consumer Electronics and Computer Industries* (New York: Free Press, 2001), 177–189; Wayne Sandholtz, *High-Tech Europe: The Politics of International Cooperation* (Berkeley: University of California Press, 1992).

[68] Louis Pouzin, "Ten Years of OSI – Maturity or Infancy?" *Computer Networks and ISDN Systems* 23 (1991): 14.

[69] Levilion interview, Charles Babbage Institute.

and government slowly but surely realized that their support of OSI was unlikely to pay off. The National Bureau of Standards created GOSIP during the late 1980s, and American federal agencies were required to procure GOSIP-compliant products beginning in August 1990.[70] Through this procurement requirement, government officials intended to stimulate the market for OSI products and enforce some uniformity across the diverse computing departments of American federal agencies. However, because products designed around TCP/IP were more familiar and readily available in the marketplace – and OSI products were not – the government's mandate failed to overcome the momentum that was building for the TCP/IP Internet. As we will see in Chapter 8, the early 1990s was a crucial phase in the Internet's history, where it moved decisively and finally into the world of global commerce. In 1994, the National Institute of Standards and Technology (as the National Bureau of Standards was renamed in 1988) recognized the Internet's momentum and utility and abandoned its GOSIP program. By that point, it had become clear that the market for network protocols had tipped in favor of the Internet, and the grand future planned for OSI had vanished.[71]

Conclusions

To understand why Pouzin and so many other networking experts were wrong about OSI's fate, we need to examine the development of the TCP/IP Internet and the sources behind its widespread adoption by computer researchers, manufacturers, and users. Before doing so, however, we should note the divergence between the open systems that existed in industry rhetoric and the project to create open international networking standards within OSI. By the early 1990s, open systems existed more in rhetoric than they did in technical practice. As the OSI project itself encountered difficulties and delays in the mid-1980s, the rhetoric and ideology of open systems continued to be mobilized as a justification for a critique of the existing order through the production of technical

---

[70] For numerous articles on GOSIP (Government Open Systems Interconnection Profile) and the progress of OSI more generally, see the collection of *Connexions: The Interoperability Report* that spans 1987–1996, hosted by the Charles Babbage Institute and available from http://www.cbi.umn.edu/hostedpublications/Connexions.

[71] "U.S. Government Open Systems Interconnection Profile," U.S. Federal Information Processing Standards Publication 146, Version 1, August 1988, cited in Vinton Cerf and Kevin Mills (1990), "Explaining the Role of GOSIP," RFC 1169 http://tools.ietf.org/rfc/rfc1169 (accessed September 25, 2013). See also Phillipe Janson, Refik Molva, and Stefano Zatti, "Architectural Directions for Opening IBM Networks: The Case of OSI," *IBM Systems Journal* 31 (1992): 313–335; David A. Mills, oral history interview by Andrew L. Russell, February 26, 2004, Newark, Delaware. Charles Babbage Institute, University of Minnesota, Minneapolis; Piscitello and Chapin, *Open Systems Networking*, 3–29; John S. Quarterman, "The Demise of GOSIP," *Matrix News* 4 (October 1994): 6; David C. Wood, "Federal Networking Standards: Policy Issues," *StandardView* 2 (1994): 218–223; and Libicki, *Information Technology Standards*, 108–119.

and ideological alternatives. In 1994, Sun Microsystems standards strategist Carl Cargill summarized the movement's prevailing myths and ideals: "The concept of 'open systems' has become a convenient icon to express all that is good about computing and the promise that computing can hold. It, too, has undergone significant shifts in its meaning, but it has always been held up as the ideal to which all computing should subscribe."[72] By the early 1990s, the OSI seven-layer reference model was featured in a new generation of computer networking textbooks and enshrined in computer networking curricula worldwide. But whereas the reference model was standardized and easily understood, its implementations were not. On closer inspection, the apparent success of the model was in fact symptomatic of OSI's fatal flaw: inadequate mechanisms for managing the political and technical complexity of competing visions for the future. The paradox was all too familiar – and all too clear. Standardization could overcome excess diversity, but there was no standard process for creating standards in a fair and efficient manner. Ideas about standardization were equally diverse and difficult to reconcile. One textbook, T. A. Critchley and K. C. Batty's 1993 *Open Systems: The Reality*, captured the state of confusion by collecting nearly 100 definitions of "open systems." In textbooks, as in dozens of articles in industry magazines and professional journals, authors were able to articulate a variety of theoretical notions of open systems that they identified with the OSI reference model. By the early 1990s, they increasingly turned their attention to other open systems, especially the UNIX operating system.[73]

The movement for open systems proclaimed a new era of democratic inclusivity, but the project to create standards for Open Systems Interconnection was, in the end, unable to manage the rivalries and incompatibilities that usually emerge in heterogeneous communities. Rather than becoming a model of international cooperation, contemporaries believed that OSI demonstrated the pitfalls of "anticipatory standardization," a term they coined to describe projects that try to shape new technologies instead of codifying existing practice. They were right: ISO was not designed for projects that attempted to standardize practices that were not yet common. The Cyclades veteran Gérard Le Lann put the point succinctly in a 2012 interview: "We need [standards bodies] for

---

[72] The quote first appeared in Carl Cargill, "Evolution and Revolution in Open Systems," *StandardView* 2 (1994): 3; and later in modified form in Carl Cargill, *Open Systems Standardization: A Business Approach* (Upper Saddle River, NJ: Prentice Hall PTR, 1997), 70.

[73] T. A. Critchley and K. C. Batty, *Open Systems: The Reality* (New York: Prentice Hall, 1993). See also Marshall Rose, *The Open Book: A Practical Perspective on OSI* (Englewood Cliffs, NJ: Prentice Hall, 1990); Pamela Gray, *Open Systems: A Business Strategy for the 1990s* (New York: McGraw-Hill, 1991); R. J. Cypser, *Communicating for Cooperating Systems: OSI, SNA, and TCP/IP* (Reading, MA: Addison-Wesley, 1991); John Quarterman, *UNIX, POSIX, and Open Systems: The Open Standards Puzzle* (Reading, MA: Addison-Wesley, 1992); and Piscitello and Chapin, *Open Systems Networking*. Cypser's book acknowledges sponsorship by IBM, and both Critchley and Batty were employed with IBM in the United Kingdom when they wrote their book.

well-established technology. But when you are pushing for new stuff, this is not the most efficient way of doing things."[74]

Throughout the 1970s and 1980s, OSI's critics and advocates alike questioned the organizational competence of ISO committees to host such a large and ambitious project. Despite the diplomatic efforts of leaders such as Bachman and Zimmermann, there were basic incompatibilities between technical ideas and organizational strategies. American computer scientist John Day neatly summarized the source of OSI's problem as a series of jurisdictional conflicts: "The strife within OSI was not limited to connection versus connectionless, but also Japan versus Europe versus the United States, and the computer industry versus the phone companies and everyone against IBM."[75] Industry participants and observers who were frustrated with OSI used the occasion to raise deeper questions about the international regime for setting standards. They complained that its formal procedures, which were designed to foster an unshakable international consensus, were far too slow and bureaucratic to permit effective collaboration in such a dynamic area of technology. IBM engineer John Aschenbrenner recognized the crux of the problem as early as 1979:

The combination of the complexity and urgency of the SC16 effort also brings into focus the fact that the standards process itself needs a thorough review. It was agonizing to sit through a straight fourteen hours of a line by line editing of N227 [a revision of the OSI reference model] in full plenary (which only completed about half of the document).... All of the above are but signs of a big problem.... The standards process needs to be reexamined.[76]

For the most staunch advocates of the connectionless approach to datagrams – the approach that was most heavily resisted by the telecommunications companies and by IBM engineers – the chief problem of the ISO process was that the entrenched powers embraced their opportunity to oppose a disruptive form of innovation. As OSI's first chairman Charles Bachman anticipated, OSI's open process turned out to be its fatal flaw. An open committee structure guaranteed that the organizations with the most to lose from the success of OSI – the computer manufacturer IBM and the national telecom monopolies – were able to place their engineers in influential committee positions, slow down the committee's work, and, by some accounts, sabotage it altogether. John Day, who tried to outmaneuver the IBM and telecom strategists in several different committees in the 1980s, summarized in 2007 the "main lesson" that he took away from the OSI experience: "Never include the legacy in a new effort....

---

[74] Gérard Le Lann, oral history interview by Andrew L. Russell, April 3, 2012, Paris, France. Charles Babbage Institute, University of Minnesota.
[75] Day, *Patterns in Network Architecture*, 358.
[76] John Aschenbrenner, "Report of Second Meeting of ISO/TC97/SC16," July 18, 1979, Bachman Papers, Box 18, Folder 19.

No matter how reasonable it may appear, collaborating with the legacy will lead to failure."[77]

In the European context it was impossible to avoid the legacy powers, but in the American context the situation was different. It may be the case that the leaders of the Internet community, Vint Cerf and Robert Kahn, anticipated this insight when they stepped back from international negotiations in the late 1970s. In the 1970s and 1980s, they devised their own structures and procedures for creating and implementing standards. Internet advocates, as we will see, assumed their approach was new, when it was, in effect, a return to a traditional mode of industrial standardization – abandoned by OSI – where implementation preceded standardization. By the 1990s, their efforts had led to the explosive adoption of Internet standards by manufacturers and users worldwide. And by the early twenty-first century, outsiders began to interpret the Internet's idiosyncratic standards process not merely as an exception but rather as the first instance of a new set of rules for standardization in the digital age.

[77] Day, *Patterns in Network Architecture*, 123, 359.

8

The Internet and the Advantages of Autocratic Design, 1970s–1990s

The standards elephant of yesterday – OSI.
The standards elephant of today – it's right here.
As the Internet and its community grows, how do we manage the process of change and growth?

- Open process – let all voices be heard.
- Closed process – make progress.
- Quick process – keep up with reality.
- Slow process – leave time to think.
- Market driven process – the future is commercial.
- Scaling driven process – the future is the Internet.

We reject: kings, presidents and voting.
We believe in: rough consensus and running code.

– David D. Clark, "A Cloudy Crystal Ball," 1992[1]

By all accounts, David Clark's plenary address at the July 1992 meeting of the Internet Engineering Task Force (IETF) was a transformative moment in the history of networking. There was a palpable tension in the sessions leading up to Clark's talk, which occurred at the end of the final session on the fourth day of a contentious five-day meeting. Clark, a senior research scientist in MIT's Laboratory for Computer Science, had been a leading member of the Internet technical community since 1976. Clark's title, "A Cloudy Crystal Ball," suggested uncertainty, but his "alternate title" – "Apocalypse Now" – better encapsulated the tone of his remarks. He used the occasion to cajole Internet engineers into paying more attention to technical problems that threatened the

---

[1] David D. Clark, "A Cloudy Crystal Ball: Visions of the Future (Alternate Title: Apocalypse Now)," July 16, 1992, Proceedings of the Twenty-Fourth Internet Engineering Task Force, Cambridge, MA, July 13–17, 1992, available from http://www.ietf.org/proceedings/24.pdf (accessed November 17, 2011).

Internet's continued growth and success. Like many people in the audience, Clark worried about the strains that the commercialization of the early 1990s was placing on the technical capabilities of the Internet. However, he was more deeply concerned by fundamental problems with the security of the Internet, as one of his slides made plain: "Security is a CRITICAL problem. Lack of security means the END OF LIFE AS WE KNOW IT."[2]

Clark's dire appraisal of the Internet's technical problems was provocative, but his plenary address was transformative for other reasons. In the stirring conclusion to his talk, Clark turned away from technical subjects in order to articulate his thoughts on the political philosophy of Internet standards. Some context is helpful for understanding why Clark's political turn was both highly appropriate for the occasion and historically significant. In the summer of 1992, the Internet was at a mature yet unstable phase in its development. The Internet community had grown from a few dozen ARPA-sponsored researchers in the 1970s to hundreds of engineers from industry, academia, and governments around the world. The new composition of the community reflected the widespread deployment of Internet protocols, which had been put into use in a wide variety of experimental computer networks in the late 1980s and early 1990s. The National Science Foundation (NSF) assumed control over the Internet backbone after the Arpanet was decommissioned in 1990, and the rapid adoption of Internet protocols convinced scientists and policy makers at the highest levels of the American government that a commercialized Internet could become an "information superhighway" and the new "information infrastructure" for American innovation in the 1990s and beyond.[3]

The combined force of these pressures brought the Internet community to an unprecedented moment of self-reflection in the summer of 1992. Clark's plenary address was transformative precisely because he was able to articulate the spirit of the Internet community like no one had done before. He captured the imagination of the community with a phrase that appeared on one of his final slides: "We reject: kings, presidents and voting. We believe in: rough consensus and running code."[4]

It didn't take long for "rough consensus and running code" to take on a life of its own. The slogan resonated deeply with the audience of Internet engineers and soon became a central ingredient of the IETF's *imaginaire* – that is, the social vision that Internet experts used to promote the Internet as a social and technological ideal. "Rough consensus and running code" was so popular that Marshall Rose, a vocal IETF participant, created the ultimate

---

[2] Clark, "A Cloudy Crystal Ball."

[3] Janet Abbate, "Privatizing the Internet: Competing Visions and Chaotic Events, 1987–1995," *IEEE Annals of the History of Computing* 32 (2010): 10–22; Craig Lyle Simon, *Launching the DNS War: Dot-Com Privatization and the Rise of Global Internet Governance* (PhD dissertation, University of Miami, 2006).

[4] Clark, "A Cloudy Crystal Ball."

form of computer geek approval: T-shirts with the phrase emblazoned across them. Lawrence Lessig, in his influential 1999 book *Code and Other Laws of Cyberspace*, declared that "rough consensus and running code" was "a manifesto that will define our generation."[5]

"Rough consensus and running code" generated and sustained this level of enthusiasm because it helped Internet engineers articulate their shared values and shared sense of identity. As I will explain in this chapter, the context of Clark's talk – as well as his claim that Internet engineers "reject: kings, presidents and voting" – provide clues that can help explain why his slogan resonated so deeply as a political philosophy of Internet standards. It is impossible to grasp the significance of "rough consensus and running code" unless we understand Clark's slogan as an act of *critique*. The target of Clark's critique was the Internet's rival, Open Systems Interconnection (OSI), which was itself the product of a formal and democratic process for setting international standards through intricate networks of "industrial legislatures" in national and international standard-setting bodies. Clark's critique suggested that OSI had strayed from the real purpose of standardization: to establish agreements around technologies that had already been tried and tested.

The final chapter of my story – the technical, cultural, and political origins of the Internet – is the starting point of many accounts that identify the emergence of the Internet as an inflection point, a driver of a new world order, the harbinger of the rise of the Network Society, and the infrastructure and symbol of openness in the digital age. In this chapter, I set Clark's slogan into a historical context that will shed light on the peculiar mode of governance that underpins the Internet standards process. I do not follow the many scholars who suggest that the Internet standards process is an idealized form of distributed control and participatory democracy that emerged organically from the interactions of Internet engineers.[6] Nor do I believe that the discourse of "consensus" in Internet governance is new, exceptional, or in any way specific to the Internet. Instead, I emphasize that many key features of the Internet standards process, including

---

[5] Lawrence Lessig, *Code and Other Laws of Cyberspace* (New York: Basic Books, 1999), 4; Patrice Flichy, *The Internet Imaginaire* (Cambridge, MA: The MIT Press, 2007); Erik Huizer (1996), "IETF-ISOC Relationship," RFC 2031, http://tools.ietf.org/html/rfc2031 (accessed September 25, 2013); Scott Bradner, "The Internet Engineering Task Force," in Chris DiBona, Sam Ockman, and Mark Stone, eds., *Open Sources: Voices from the Open Source Revolution* (Sebastopol, CA: O'Reilly Media, Inc., 1999), 50; and Harald Alvestrand (2004), "A Mission Statement for the IETF," RFC 3935, http://tools.ietf.org/html/rfc3935 (accessed September 25, 2013).

[6] Brian Carpenter, ed. (1996), "Architectural Principles of the Internet," RFC 1958, http://tools.ietf.org/html/rfc1958 (accessed September 25, 2013); Anthony Rutkowski, "Today's Cooperative Competitive Standards Environment and the Internet Standards-Making Model," in Brian Kahin and Janet Abbate, eds., *Standards Policy for Information Infrastructure* (Cambridge, MA: The MIT Press, 1995); A. Michael Froomkin, "Habermas@Discourse.Net: Toward a Critical Theory of Cyberspace," *Harvard Law Review* 116 (2003): 749–873; Simon, *Launching the DNS War*; Tim O'Reilly, "The Architecture of Participation," (June 2004), http://oreilly.com/pub/a/oreilly/tim/articles/architecture_of_participation.html (accessed January 3, 2012).

its motto of "rough consensus and running code," can be best understood as acts of critique – that is, as an accumulation of ad hoc competitive responses to their chief antagonist, OSI. In this light, we can see clear continuities – rather than disruptions or radical breaks – between the Internet standards process and the history of industrial standardization that I have described throughout the various chapters in *Open Standards and the Digital Age.*

Unlike OSI's slow, unwieldy, and democratic bricolage of institutions and technical proposals, the Internet's growth was grounded in a fluid set of institutions that resembled a form of *autocratic* and centralized control over system architecture and standardization.[7] In its early phases, Internet engineers enjoyed the generous and persistent patronage of the American military, thus enabling Internet leaders such as Vinton Cerf and Robert Kahn to force the community to focus all of their attention on the development and implementation of Internet standards such as the Transmission Control Protocol (TCP) and Internet Protocol (IP). In the process, the Internet community developed an esprit de corps that cast the Internet as a superior alternative to the OSI "standards elephant" (to steal a phrase from Clark's plenary address).

By the mid-1990s, the turmoil of the Internet-OSI standards war had subsided, with surprising results: OSI's international alliance stalled because of fundamental incompatibilities among its constituents, while Internet protocols became the global de facto standards for computer network interconnection. Commercial firms – such as the router company Cisco – that forged early ties with the Internet technical community reaped the rewards of their strategic investments in Internet-based equipment and devices. In a strange plot twist, the autocratic Internet emerged as a symbol of open systems and an exemplar of a new style of open-systems standardization, as the project titled Open Systems Interconnection crumbled under the weight of its democratic structure and process. The Internet soon became known as an unqualified success story, and its advocates quietly recast Internet history as a story of open and democratized innovation – ironic, as its sponsor, the American Department of Defense, was the quintessential Cold War closed world institution, and the leaders of the Internet community repeatedly rejected basic formalities of democracy such as membership and voting rights.

Because these events have occurred so close to the present day, it is difficult to use the historian's most valuable tool – long-term perspective – to assess their significance. The historian's task is complicated further by the fact that the Internet-OSI standards war provides a very recent example of history literally having been written by the winners. Ever since the mid-1990s, Internet advocates and observers have claimed that the Internet standards process represents something more than a deviation from the formal standardization process. The Internet, in this view, is a new norm, an emblem of a broader "regime change"

---

[7] Frederick P. Brooks, Jr., *The Mythical Man-Month: Essays on Software Engineering Anniversary Edition* (Boston: Addison-Wesley, 1995), 40–50.

that has revealed the inadequacies of tradition-bound standards organizations such as ISO. Nevertheless, uncomfortable contradictions persist in Internet governance, including profound security flaws, international opposition to American dominance, and the strong affinity of online communities for benevolent technocratic dictators.[8] Internet advocates and users can make better sense of these contradictions only when they understand how the prevailing *imaginaire* of the open Internet obscures its autocratic and centralized origins.

Internet Protocols and Institutional Evolution, 1975–1992

By the mid-1970s, Vint Cerf was deeply enmeshed in the international community of packet-switching researchers. He was the energetic chairman of the International Network Working Group (INWG) and a regular correspondent and collaborator with European networking experts such as Peter Kirstein, Hubert Zimmermann, Louis Pouzin, and Gerard Le Lann (who spent a year as a visitor in Cerf's lab at Stanford). As I described in Chapter 6, Cerf tried unsuccessfully to persuade his colleagues in INWG to endorse the Transmission Control Program (TCP) that he and Robert Kahn wrote and eventually deployed within the Arpanet. When it became clear in late 1975 that INWG overwhelmingly supported INWG 96, an end-to-end protocol that departed from some aspects of TCP's design, Cerf resigned as chairman and accepted Kahn's offer to move to Washington and direct ARPA's new Internet program.

After his frustrating education in the realpolitik of international standards setting, and some angst over the pressure that would come with his new job in Washington, Cerf ultimately must have been relieved to be able to concentrate on the development of TCP and related "Internet" standards within the small community of like-minded ARPA contractors and researchers. In this environment – which was shaped decisively by J. C. R. Licklider, as discussed in Chapter 6 – administrators and researchers could skip the time-consuming process of peer review. Instead of making its researchers and their work accountable to outsiders, the ARPA culture and procedures required frequent reporting to program managers. Democratic oversight existed only through the hierarchical apparatus of the military-industrial-academic complex.[9]

---

[8] Jack Goldsmith and Tim Wu, *Who Controls the Internet? Illusions of a Borderless World* (New York: Oxford University Press, 2006); Julian Dibbell, "A Rape in Cyberspace: How an Evil Clown, a Haitian Trickster Spirit, Two Wizards, and a Cast of Dozens Turned a Database into a Society," *The Village Voice* (December 23, 1993); Andrew L. Russell, "Constructing Legitimacy: The W3C's Patent Policy," in Laura DeNardis, ed., *Opening Standards: The Global Politics of Interoperability* (Cambridge, MA: The MIT Press, 2011); Gary Wolf, "Why Craigslist Is Such a Mess," *Wired* 17 (2009), http://www.wired.com/entertainment/theweb/magazine/17-09/ff_craigslist (accessed January 3, 2012).

[9] On the structure and culture of computing research at ARPA, see Arthur L. Norberg and Judy E. O'Neill, *Transforming Computer Technology: Information Processing for the Pentagon, 1962–1986* (Baltimore: The Johns Hopkins University Press, 1996).

To build on their initial networking experiments with the Arpanet, Cerf and Kahn led an ARPA research effort to explore and design networking concepts for satellite and radio technologies. One of their most important objectives was to make the network invisible to users, including their military patrons who would be more interested in accessing information over the network than in the operation of the network itself. In July 1977, they ran a successful "inter-networking" demonstration that sent packets across ARPA's satellite network (SATnet), a radio network (PRnet), and a land-based network (the Arpanet). In 1977, Cerf and two other ARPA-funded researchers, Danny Cohen and Jon Postel, decided to split the functions of the Transmission Control Program into two protocols, the Transmission Control Protocol (TCP) and the Internet Protocol (IP), that worked together and became known as TCP/IP[10]

In addition to their technical experiments, Cerf and Kahn presided over organizational experiments and ad hoc mechanisms to facilitate the testing, review, and specification of Internet protocols. By the late 1970s, Kahn and Cerf had come to recognize that regular meetings of researchers in the community would be desirable. Kahn recalled,

When we started the Internet program in the mid 1970s, originally it was just me in the office running the program. And after Vint was hired, then it was just Vint running the program with me to kibitz. And he was so good at what he did that he basically had everything in his head. What I worried about was what would happen if he got hit by a truck? Number two, what would happen if he would ever have to leave? And number three, how was anybody else in the community ever going to be part of the thinking process. So he set up, after some discussions, a kind of kitchen cabinet, if you will, of knowledgeable people that he would convene periodically. These were mostly the workers in the field, the key people who were implementing protocols.... When [Cerf] left, that group stayed intact.[11]

Kahn's characterization of Cerf's advisory group as a "kitchen cabinet" is suggestive of the broader outlines of the Internet's governance model in the 1970s and 1980s. The term "kitchen cabinet" dates from the administration of President Andrew Jackson, who preferred to consult with an informal group of advisors – in the White House kitchen, according to legend – instead of his official "Parlor" cabinet. Jackson's managerial style privileged personal ties and practical experience at the expense of formality and bureaucratic protocol.

[10] Janet Abbate, *Inventing the Internet* (Cambridge, MA: The MIT Press, 1999), 129–130; Vinton G. Cerf, "Protocols for Interconnected Packet Networks," *Computer Communication Review* 18 (October 1980): 10–11; Jon Postel, ed. (1980), "DOD Standard Internet Protocol," RFC 760, http://tools.ietf.org/rfc/rfc760 (accessed September 25, 2013); RFC Editor, "Internet Engineering Notes," ftp://ftp.rfc-editor.org/in-notes/ien/ (accessed September 25, 2013). For a critical appraisal of the TCP/IP split, see Fred Goldstein and John Day, "Moving Beyond TCP/IP," April 2010, available from http://pouzin.pnanetworks.com/images/PSOC-MovingBeyondTCP.pdf (accessed November 22, 2012).

[11] Robert Kahn, oral history interview by Judy E. O'Neill, April 24, 1990, Reston, Virginia. Charles Babbage Institute, University of Minnesota, Minneapolis.

Kahn recognized that the institutional innovation of "King Andrew" (as his political opponents called him) was thus especially well suited to ARPA's preference to rely on handpicked experts rather than democratically elected representatives.[12]

Cerf's "kitchen cabinet" had an official-sounding, if nondescript name: the Internet Configuration Control Board (ICCB). Created in 1979, the ICCB expanded control over Internet development by bringing more of the network users – technical experts distributed throughout universities, firms, and government agencies – into Cerf's inner circle in a more structured manner. Cerf's close associates remained in charge: Clark, as chairman, earned the title "Internet Architect"; Jon Postel, the assistant chairman, became the "Deputy Internet Architect." Despite a modest diffusion of control, the ICCB was merely an advisory group that, in the end, was in no way accountable to the types of commercial and political interests that complicated standards development within the democratic processes of ISO and the American National Standards Institute (ANSI).[13]

The explanations given by Cerf and Kahn cast the ICCB's creation as a logical response to the internal dynamics of the Internet's growth. Their accounts, however, neglect the extent to which external developments may have prompted their institution building. In late 1975, researchers who had been active within INWG (see Chapter 6) split into roughly two factions: a small group of ARPA-funded followers of Cerf and Kahn who worked on the Internet, and a larger group of researchers who were closer to Hubert Zimmermann and moved with him into ISO to pursue international standards for Open Systems Interconnection. Thanks to the leadership of Zimmermann and Charles Bachman, the OSI project moved quickly from a brainstorming phase to a phase of institutionalization and formalization. OSI's first meeting was in early 1978; Cerf and Kahn created the ICCB in 1979. It certainly is possible that the creation of the ICCB represented something more than "an informal committee to guide the technical evolution of the protocol suite," as Cerf explained; in practice, it appears that the ICCB also might function as an institutional counterweight to the new committees organized to advance OSI.[14]

[12] Robert V. Remini, *Andrew Jackson: The Course of American Freedom, 1822–1832* (Baltimore: The Johns Hopkins University Press, 1998), 315–330.

[13] Robert E. Kahn, "The Role of the Government in the Evolution of the Internet," *Communications of the ACM* 37 (August 1994): 16; Jon Postel and Joyce Reynolds (1984), "ARPA-Internet Protocol Policy," RFC 902, http://tools.ietf.org/html/rfc902 (accessed September 25, 2013). In a 2004 interview, Internet engineer David Mills commented on the ICCB's name: "That name is revealing. Because it says we're not controlling anything. We didn't control anything. We couldn't call it something that the government could construe as managing. We didn't manage the Internet. We were just a committee down here that solves configuration problems." David A. Mills, oral history interview by Andrew L. Russell, February 26, 2004, Newark, Delaware. Charles Babbage Institute, University of Minnesota, Minneapolis.

[14] Vint Cerf (1989), "The Internet Activities Board," RFC 1120, http://tools.ietf.org/html/rfc1120 (accessed September 25, 2013).

To be sure, the chief advantage that Cerf and Kahn had over their rivals in OSI was the generous sponsorship of ARPA. One advantage of having a single benefactor – as opposed to the heterogeneity of motivations and sponsors represented in OSI – was that it could, when necessary, force its users to follow orders. Arpanet history is rich with anecdotes in which ARPA managers, cognizant that their contractors depended heavily on military funding, learned to use this leverage to get their way. For example, Larry Roberts demurred when he was first recruited to leave his position at MIT Lincoln Lab to join the ARPA staff in the autumn of 1966. His stance changed when his recruiter, the ARPA program manager Robert Taylor, convinced ARPA director Charles Herzfeld to call the Lincoln Lab's director, remind him that 51 percent of the Lincoln Lab's budget came from the Department of Defense, and "tell him it's in Lincoln Lab's and ARPA's best interests for him to tell Larry Roberts to come down and do this." [15] Although he was "not particularly eager for the job," Roberts found himself working in the Pentagon for ARPA by January 1967. [16]

Roberts quickly mastered the tactic that had forced his move to Washington. When Roberts and Taylor wanted to build a network – the Arpanet – to connect all of the computers at IPTO-sponsored sites, they encountered "tremendous resistance" from their academic contractors who were unwilling to devote the effort required. Rather than bending to their demands, Roberts responded from a position of overwhelming strength. He recalled:

Well, the universities were being funded by us, and we said, "We are going to build a network and you are going to participate in it. And you are going to connect it to your machines. By virtue of that we are going to reduce our computing demands on the office. So that you understand, we are not going to buy you new computers until you have used up all of the resources of the network." So over time we started forcing them to be involved, because the universities in general did not want to share their computers with anybody. [17]

By the early 1980s, successive ARPA managers had learned to exploit the dependencies created by the power of their patronage. The event that is widely acknowledged as the "birth of the Internet" on January 1, 1983, provides another example of the Internet's foundational dependence on a coercive and autocratic style of management. Since its implementation in 1970, NCP had proved to be an adequate mechanism for transmitting packets across the

[15] Robert Taylor, oral history interview by William Aspray, February 28, 1989, San Francisco, California. Charles Babbage Institute, University of Minnesota, Minneapolis.

[16] L. Roberts, "Expanding AI Research and Founding ARPANET," in Thomas C. Bartee, ed., *Expert Systems and Artificial Intelligence: Applications and Management* (Indianapolis, IN: Howard W. Sams & Company, 1988), 229.

[17] Lawrence G. Roberts, oral history interview by Arthur L. Norberg, April 4, 1989, San Francisco, California. Charles Babbage Institute, University of Minnesota, Minneapolis; Vint Cerf, quoted in Stephen Segaller, *Nerds 2.0.1: A Brief History of the Internet* (New York: TV Books, 1998), 59.

Arpanet. Since 1974, however, Cerf, Kahn, and other ARPA researchers had been promoting TCP as an alternative method that could, as discussed earlier, transmit packets between different types of networks (a practice Cerf and Kahn called "internetworking"). TCP had been tested and demonstrated in different environments, but many Arpanet users had been slow to abandon NCP and switch to TCP. Once again, Arpanet users objected because of the exorbitant amount of labor they would need to devote to the implementation and testing of the new protocols; once again, their resentment had no bearing on the orders they had to carry out.[18]

When the Department of Defense adopted TCP and IP as military standards in 1980, Cerf and Kahn were anxious to make sure they delivered the flexible and functional data communications network that their military patron had sponsored. In November 1981, Jon Postel – the editor of the RFC Series and a key member of the Internet's leadership team – published RFC 801, "NCP/TCP Transition Plan," to inform Arpanet users that they were responsible for the implementation of TCP/IP on their own host computers by January 1, 1983. The centrally planned transition had a clarity of purpose that Bancroft Gherardi, chief engineer of the Bell System in the 1920s and 1930s, would have admired. Arpanet users, conscious of the months of labor that were being forced on them, continued to resist but ultimately obeyed. The transition was famously traumatic for programmers such as Dan Lynch, who memorialized the event by distributing buttons that bragged "I survived the TCP transition."[19]

Amid the drama that arose *within* the Arpanet community during the TCP transition, we should not lose sight of the broader contextual factors – particularly competition from the rapid development of OSI – that pressured Arpanet managers to demonstrate the viability of TCP and IP. As we saw in Chapter 7, Bachman and Zimmermann were steering very capably the OSI reference model through the various stages of consensus building required by the formal rules of international standardization. In November 1980, Zimmermann was able to shepherd the revised model to the status of draft proposal, before it advanced to a draft international standard in January 1982 and finally was approved formally by ISO in May 1983 as an international standard, "ISO 7498: The Basic Reference Model for Open System Interconnection." During the same few years, ISO standards for transport protocols that would fit within the OSI

---

[18] Amy Slaton and Janet Abbate, "The Hidden Lives of Standards: Technical Prescriptions and the Transformation of Work in America," in Michael Thad Allen and Gabrielle Hecht, eds., *Technologies of Power: Essays in Honor of Thomas Parke Hughes and Agatha Chipley Hughes* (Cambridge, MA: The MIT Press, 2001), 95–144; Stephen J. Lukasik, "Why the Arpanet Was Built," *IEEE Annals of the History of Computing* 33 (2011): 4–21.

[19] Slaton and Abbate, "The Hidden Lives of Standards"; Jon Postel (1981), "NCP/TCP Transition Plan," RFC 801, http://tools.ietf.org/rfc/rfc801 (accessed December 13, 2011). The "TCP/IP Digests," an electronic mailing list that Arpanet engineers used to trade stories and advice, capture the "trauma of the birth of the Internet proper on January 1, 1983." See ftp://ftp.rfc-editor.org/in-notes/museum/tcp-ip-digest/ (accessed September 25, 2013).

reference model – alternatives and direct competitors to TCP/IP – appeared as draft proposals in June 1982.[20] Most Arpanet engineers who were struggling to implement TCP during 1982 did not pay much attention to the technical or procedural nuances of the work on OSI taking place within ISO; they certainly were not encouraged to do so by the Arpanet project managers such as Cerf and Kahn. It is inconceivable, however, that Cerf and Kahn were themselves unaware of the standardization of the OSI reference model and transport protocols. Therefore, it seems likely that their ambition to force Arpanet users to convert to TCP/IP was, at least in part, a competitive response to the accomplishments of OSI.[21]

The perception that the Internet was falling behind in its competition with OSI was made explicit by the results of a study by the National Research Council that had been commissioned by the U.S. Defense Department in 1983. The study, requested to help the Defense Department prepare a procurement strategy for computer network equipment, compared TCP and its functional counterpart in OSI, a protocol called TP-4. The National Research Council's final report, published in 1985, presented three options: keep TCP and TP-4 as costandards, adopt TP-4 as soon as it was shown to be ready for military networks, or keep TCP and defer indefinitely a decision on TP-4.[22] The Department of Defense – despite the millions of dollars it had spent subsidizing the Arpanet and the Internet – supported option two as part of a plan to "move ultimately toward exclusive use of TP-4." In other words, the report reflected the conventional wisdom among most computer experts in the early 1980s: the Internet was an interesting experiment, but technological momentum and a distinct sense of inevitability was amassing around OSI.[23]

Internet engineers grew to relish their role as the underdog in a fight with the official international standards body. They had good reasons to be optimistic: Internet users were able to take advantage of the network effects generated by the Arpanet's transition to TCP in 1983 as well as its support for common protocols to support file transfer (FTP), mail (SMTP), and virtual terminal access (Telnet). In addition to these technical accomplishments, ARPA managers revised the fragile, ad hoc institutions they had been using to develop protocols

[20] Alex McKenzie, ed. (1984) "ISO Transport Protocol Specification ISO DP 8073," RFC 905, http://tools.ietf.org/html/rfc905 (accessed September 25, 2013).

[21] Hubert Zimmermann, "OSI Reference Model – The ISO Model of Architecture for Open Systems Interconnection," *IEEE Transactions on Communications* 28 (1980): 425–432.

[22] Board on Telecommunications and Computer Applications, Commission on Engineering and Technical Systems, National Research Council, *Transport Protocols for Department of Defense Data Networks: Report to the Department of Defense and the National Bureau of Standards Committee on Computer-Computer Communication Protocols* (Washington, DC: National Academy Press, 1985).

[23] Jon Postel (1985), "A DoD Statement on the NRC Report," RFC 945, http://tools.ietf.org/rfc/rfc945 (accessed September 25, 2013); Michael Witt, "Moving from DoD to OSI Protocols: A First Step," *Computer Communication Review* 16 (April/May, 1986): 2–7; Mills interview, Charles Babbage Institute.

and make decisions. As the Internet (and the number of people connected to it) continued to grow, Kahn realized he needed to reconsider "the process that ARPA was using to manage the evolution of the network." The resulting effort created more sophisticated and durable institutions – clear indications of the Internet's growing complexity and maturity.[24]

Barry Leiner, who replaced Vint Cerf as head of the ARPA Internet program in 1983, assisted Kahn in this rethinking of the Internet's management. In September 1984, Leiner disbanded the ICCB and, in its place, created the Internet Advisory Board (IAB).[25] David Clark continued his close involvement as the first chair of the IAB – a position that continued to confer the title of "Internet Architect." Within the IAB, Leiner created a number of small groups (called task forces) focused on various aspects of Internet technologies, such as gateway algorithms, end-to-end protocols, privacy, and security. As chairman, Clark was responsible for selecting the chairs of each task force and for deciding whether new people could join the IAB. The broad jurisdiction of the IAB chairman, and the group's insular process for making decisions and selecting members, prompted the author and computer scientist Ed Krol to describe it as a "council of elders."[26]

During the mid-1980s, the Internet's leaders continued to nurture a strong esprit de corps and a shared recognition that the community's efforts were dedicated to the continued growth of the Internet rather than any particular financial ambition. This esprit de corps – and the noncommercial orientation of the Arpanet community of the early 1970s – became strained as the Internet grew and more people sought to have a voice in its design and standardization process. A major source of new interest came from researchers who had been working on the NSFNET, a packet-switched network created in 1984 by the National Science Foundation and open to university researchers. By 1986, the convergence of the Arpanet and NSFNET meant that the number of networks connected via the Internet grew dramatically, with several thousand host computers connected to more than 400 networks via 120 gateways located within the network.

[24] Kahn, "Role of the Government," 16.

[25] The acronym "IAB" has remained consistent since 1984, but the "A" – and the meanings behind it – have changed. From 1984 to 1986 the IAB was the Internet Advisory Board; in 1986, its name changed to the Internet *Activities* Board; in 1992 it changed once again, this time to the Internet *Architecture* Board. See Internet Architecture Board, "A Brief History of the Internet Advisory/Activities/Architecture Board," http://www.iab.org/about/history/ (accessed January 3, 2012).

[26] Internet Architecture Board, "A Brief History"; Barry M. Leiner, Vinton G. Cerf, David D. Clark, Robert E. Kahn, Leonard Kleinrock, Daniel C. Lynch, Jon Postel, Larry Roberts, and Stephen Wolff, "A Brief History of the Internet," http://www.isoc.org/internet/history/brief.shtml (accessed September 25, 2013); Kahn interview, Charles Babbage Institute; Vinton Cerf (1990), "The Internet Activities Board," RFC 1160, http://tools.ietf.org/rfc/rfc1160 (accessed September 25, 2013); Ed Krol (1993), "FYI on 'What Is the Internet?'" RFC 1462, http://tools.ietf.org/rfc/rfc1462 (accessed September 25, 2013).

This expansion of the network and its user community – particularly users in universities – prompted the Internet's leaders to experiment with governance reforms that were, to continue a theme developed throughout this book, an institutional mix of command hierarchies and uncoordinated markets. User interest grew especially around the gateways that linked local networks to the Internet. Therefore, interest in the activities of the IAB's Gateway Algorithms and Data Structures Task Force grew rapidly. David Mills, the task force's chairman, decided to split into two groups: the Internet Architecture Task Force (INARC) and the Internet Engineering Task Force (IETF). INARC maintained an interest in general topics related to network research, but the IETF proved to be far more popular for people who wanted to hammer out the messy details of day-to-day implementation and propose specific technical standards for immediate adoption.[27]

Because of its broad mandate and importance for the engineering of the present, the IETF became the main focus of attention for newer members of the Internet community. After its first meeting in January 1986, the IETF met four times in 1986 and 1987 and three times annually in every subsequent year. Twenty-one people attended the January 1986 meeting; thirty-five attended the meeting in October. Attendance increased dramatically in 1987, from thirty-five at the February meeting to 101 at the July meeting. At first, only invited participants could attend the IETF meetings; by the end of 1987, anyone interested in the Internet – including representatives from commercial firms not under government contracts – were welcome to join and participate.[28]

The makers of Internet standards – mostly university-based Arpanet veterans – were joined by engineers working in companies selling equipment that implemented Internet standards. In 1988, Arpanet veteran Dan Lynch organized the first TCP/IP Interoperability Conference, or Interop, where equipment makers could show off their wares. The rapid growth of Interop strengthened ties between Internet engineers and equipment vendors and thus contributed to the growing size and power of the community that was making, using, standardizing, and selling Internet technologies.[29] It also signaled that the Internet community had, by and large, shifted its focus away from its research orientation of the 1970s and toward a different set of problems – scaling and commercialization – that required technical fundamentals to remain constant. As newcomers joined the Internet community, they would have been forgiven for assuming that the Internet's technical fundamentals were sound and exemplified the most advanced research in networking. They were, for the most part, oblivious to the insights generated by Louis Pouzin's Cyclades projects, the

[27] Mills interview, Charles Babbage Institute.

[28] IETF Proceedings are available from http://ietf.org/meetings/past.meetings.html.

[29] See James Pelkey, "Entrepreneurial Capitalism and Innovation: A History of Computer Communications, 1968–1988" (2007), available from http://www.historyofcomputercommunications.info/ (accessed September 25, 2013).

debates that raged within INWG, and the protracted "datagram versus virtual circuit" controversy of the 1970s.

As the Internet and its user community grew and matured during the late 1980s, its standard-setting institutions came to embody two different modes of governance. The first mode, the structure led by Cerf that coordinated the development of the Internet at ARPA, was a self-selected, experienced group – a so-called council of elders. Frequently described as a meritocracy, this close-knit network of people had worked together since the early Arpanet days (many as graduate students at MIT or UCLA or as engineers at Bolt, Beranek and Newman, the consulting firm that had designed aspects of the Arpanet) and provided the bulk of the technical and bureaucratic leadership in the ICCB and IAB. "Relatively few, competent, highly motivated people were involved," recalled Larry Press, "and they had considerable autonomy."[30]

In contrast to the top-down mode of governance practiced in the IAB, the IETF embraced a decentralized, bottom-up mode of governance. After 1986, the IETF became an effective mechanism that could engage and direct the energies and interests of an ever-increasing constituency of Internet engineers. As a result, the IETF, as a task force for "Internet engineering," asserted jurisdiction over the distributed, hands-on tasks involved in the engineering and implementation of protocols. The IAB maintained its character as a council of elders, but the growing ranks of Internet engineers began to question the IAB's motives and decisions. A long-term pattern was becoming clear: the size and energy of the IETF was pushing Internet governance away from the autocratic style pioneered by Arpanet managers in the 1970s and early 1980s, and closer to a messier style that Internet advocates imagined to be participatory and consensus oriented. For a while, any tensions that arose between the two groups of Internet engineers – and their two different visions of control over the Internet's technical development – were obscured by their increasing disdain for a shared nemesis: OSI.

The Internet-OSI Standards War

Scholars tend to frame "standards wars" – competitions between two or more technical standards and their respective advocates – as conflicts that are essentially economic or technical in character. By framing their topic in this way, they miss the significance of sociological and cultural rifts that grow within and between different communities of engineers. Consequently, they also fail to account for the ways that conflict against an outside enemy (or "other") can facilitate the creation of an "imagined community" that coheres around shared values, practices, and traditions. The Internet-OSI standards war – like most

[30] Norberg and O'Neill, *Transforming Computer Technology*; Thomas P. Hughes, *Rescuing Prometheus* (New York: Pantheon Books, 1998); Larry Press, "Seeding Networks: The Federal Role," *Communications of the ACM* 39 (October 1996): 11–18.

aggressive encounters – was fueled by ignorance, tribalism, belligerence, and intransigence.[31]

The first generation of Internet engineers in the 1970s were led by Cerf and Kahn who, as we have seen, had extensive and productive interactions with network engineers in Europe. Their transatlantic working relationships weakened after the mid-1970s, but available evidence suggests that personal relations between one-time collaborators, such as Cerf and Louis Pouzin or Cerf and Hubert Zimmermann, remained cordial if not close. By the mid-1980s, however, a generational shift in the Internet technical community was underway. Cerf left ARPA in 1982 to take a lucrative offer to join the telecommunications company MCI; Kahn left in 1986 to create the Corporation for National Research Initiatives (CNRI), a nonprofit organization dedicated to supporting research and development for information networks. One of CNRI's first activities was to function as the secretariat for the IETF, thus providing administrative and financial support for the Internet standards process. Also in 1986, Cerf accepted an offer to join Kahn as the vice president of CNRI.[32]

As Cerf and Kahn stepped away from their roles in the daily technical and managerial tasks of Internet engineering, a new generation of Internet advocates inherited Cerf and Kahn's single-minded ambition to see the Internet grow, prosper, and defeat all rivals – including the international community of engineers working on OSI. Tribalism grew as familiarity faded and personal ties weakened. The combination of competition and ignorance, often expressed in unconventional ways, nurtured conflict. For example, in a 1983 article that explored similarities between the ARPA and ISO protocol architectures, the Internet engineers Danny Cohen and Jon Postel criticized the ISO model as a rigid abstraction that was inappropriate for use "as a model for all seasons."[33] In their article, Cohen and Postel – both of whom were instrumental in the early history of TCP/IP – took their comments on OSI far beyond the conventional limits of scholarly or scientific discourse. For example, they speculated that "mystical" traditions such as early Zoroastrianism, New Testament celestial beings, and the Christian seven deadly sins might have "shaped the

---

[31] Carl Shapiro and Hal Varian, "The Art of Standards Wars," *California Management Review* 41 (1999): 8–32; Victor Stango, "The Economics of Standards Wars," *Review of Network Economics* 3 (2004): 1–19; Angelique Augereau, Shane Greenstein, and Marc Rysman, "Coordination versus Differentiation in a Standards War: 56K Modems," *RAND Journal of Economics* 37 (2006): 887–909; Benedict Anderson, *Imagined Communities: Reflections on the Origins and Spread of Nationalism* (New York: Verso, 1983); Emmanuel Levinas, *Totality and Infinity: An Essay on Exteriority* (Pittsburgh: Duquesne University Press, 1969).

[32] "Robert E. Kahn," Corporation for National Research Initiatives, http://www.cnri.reston.va.us/bios/kahn.html (accessed December 16, 2011); "Vinton G. Cerf," Internet Corporation for Assigned Names and Numbers, http://www.icann.org/en/biog/cerf.htm (accessed December 16, 2011).

[33] Daniel Cohen and Jonathan Postel, "The ISO Reference Model and Other Protocol Architectures," in R. E. A. Mason, ed., *Information Processing 83: Proceedings of the IFIP 9th World Computer Congress* (New York: North-Holland, 1983), 34.

choice of Seven" for the number of layers in the OSI reference model. Evidently, Cohen and Postel were so detached from the technical discussions surrounding OSI that they either didn't know or didn't care that Charles Bachman proposed a seven-layer model as a deliberate strategic choice.[34]

The most articulate and ruthless attacks on OSI came from Michael A. Padlipsky, an MIT graduate and self-described "Network Old Boy" who was a veteran of the Arpanet and Internet communities. Padlipsky's 1985 book, *The Elements of Networking Style*, is a series of hilarious and technically astute critiques of OSI and its proponents. It was very important for Padlipsky – as it was for other Internet advocates – to argue that Internet protocols were simple and OSI standards were not. In a typical display of witty contempt, Padlipsky recommended that the ISO reference model, or ISORM, be pronounced "Eyesore-mmm." His objection to the X.25 protocol – which predated OSI by several years but that Padlipsky tarred with the same brush – also was typical of his elaborate and uncompromising style:

When the title of my actually quite serious critique of one of the overly-known communications protocol X.25 appeared on one of the pubs sheets [circulated by his employers] as "Low Standards: A Critique of X.25," a few of the old-line engineers (a.k.a. Old Poops) objected so strongly that I decided, in the interests of getting the thing cleared for public release so that eventually you could get to see it, to drop the lovely pun before the colon (and, being moderately scrupulous, the colon as well).[35]

Padlipsky's objections to OSI transcended technical and aesthetic considerations. He was genuinely offended by ISO's open and democratic character, and lamented that "ISO, as a Voluntary Standards Organization, has to take *whomever* it gets sent and let them decide on the merits and demerits of various proposals – irrespective of the depth, breadth, or duration of their technical expertise and/or experience." ISO's organizational culture and its presumption of authority deeply offended Padlipsky and his fellow Internet engineers. In *The Elements of Networking Style*, Padlipsky opined that the Internet was based on experience and therefore "Descriptive," but OSI was a confused "Prescriptive" plan for a future that was yet to be built. "Another way of putting it," he continued, "is that whereas the Descriptive approach is suitable for technology, the Prescriptive approach is suitable for theology." The Internet engineer David Mills agreed in a 2004 interview: "Internet standards tended to be those written for implementers. International [OSI] standards

[34] Cohen and Postel, "The ISO Reference Model," 30; Charles Bachman, oral history interview by Andrew L. Russell, April 9, 2011, IEEE Computer Society History Committee, available from http://www.computer.org/comphistory/pubs/2012-03-russell.pdf (accessed September 25, 2013).

[35] M. A. Padlipsky, *The Elements of Networking Style, And Other Animadversions on the Art of Intercomputer Networking*, 2nd ed. (Lincoln, NE: iUniverse, 2000; 1st ed. Prentice-Hall, 1985), quoted on pp. 3 and 8; M. A. Padlipsky (1982), "A Critique of X.25," RFC 874, http://tools.ietf.org/html/rfc874 (accessed September 25, 2013).

were written as documents to be *obeyed*." Although the technical functions of the OSI international standards and an IETF Request for Comments (RFCs) were functionally equivalent from a technical perspective – since they both defined protocols to allow computer users to exchange data across different types of networks – they nevertheless reflected deeply divergent views of how power should operate within the standards-setting process.[36]

The American computer scientist Richard des Jardins, an early contributor to the American team that proposed a reference model in 1978, captured the intensity of this rivalry in 1992. Des Jardins compared the "OSI Bigots" and the "IP Bigots" to people who objected to "the convergence of cultures and races in the world at large." His proposed resolution to the conflict started with a sensible, practical suggestion, but ended with a flash of violence: "Let's continue to get the people of good will from both communities to work together to find the best solutions, whether they are two-letter words or three-letter words, and let's just line up the bigots against a wall and shoot them."[37] In the Internet-OSI standards war, as in so many other cases throughout human history, ignorance sustained and emboldened intolerance and bigotry. Nevertheless, a small number of "people of good will from both communities" tried to reach across the stark battle lines and the deeply incompatible political cultures. The IETF established working groups in 1987 to integrate Internet and OSI technologies for electronic mail, directory services, and fundamental network protocols. Cerf, the most widely respected member of the Internet community and chairman of the IAB from July 1989 to July 1991, threw his considerable authority behind the integration effort: for instance, he cowrote an RFC in August 1990 to justify the goal of interoperability between OSI and the Internet.[38]

Despite Cerf's leadership and the earnest participation of respected network engineers such as Ross Callon, Lyman Chapin, Erik Huizer, and David Piscitello, the work to integrate the Internet with OSI was far from universally admired among IETF participants. A rift of sorts began to emerge between the council of elders in the IAB, who were open to compromise with OSI, and growing ranks of new Internet engineers, who were not. Most participants in the IETF had no personal experience with OSI or its advocates and therefore no meaningful understanding of the internal problems that shaped its political and technical direction.

For example, Marshall T. Rose, who published his first RFC in 1983 when he was a graduate student at the University of California, Irvine, consciously

---

[36] Padlipsky, *Elements of Networking Style*, xv and 11; Mills interview, Charles Babbage Institute.

[37] Richard des Jardins, "OSI is (Still) a Good Idea," *ConneXions* 6 (1992): 33.

[38] Vint Cerf and Kevin Mills (1990), "Explaining the Role of GOSIP," RFC 1169, http://tools.ietf.org/html/rfc1169 (accessed September 25, 2013). See also Erik Huizer, "The IETF Integrates OSI Related Work," *ConneXions* 7 (1993): 26–28; David Piscitello and Erik Huizer, "OSI Integration Area," March 1992, ftp://isi.edu/ietf/92mar/area.osi.92mar.txt (accessed September 25, 2013).

positioned himself in the late 1980s as an interpreter of OSI for Internet engineers. His efforts were met with howls of derision from Internet partisans and engineers, who were in daily contact by way of the tcp-ip electronic mailing list. The unrestrained "flame wars" that took place on that email list would be a worthy subject for historians who wish to dig deeper into the emotional depths of engineering discourse during a standards war. For my purposes here, it is enough to say that most Internet engineers were deeply skeptical of the technical merits of OSI, deeply resentful of mandates to implement it, and openly hostile toward people (such as Rose) who positioned themselves between the two communities.[39]

In the meantime, Padlipsky was delighted to sling one rhetorical barb after another. In a typical contribution to the tcp-ip mailing list, Padlipsky reminded his colleagues of "a msg Jon [Postel] sent some time back containing a quotation to the effect that the trouble with 5,000-member standards committees is that they spend all their time debating whether to have croissants or doughnuts for breakfast." Padlipsky characteristically couldn't resist the opportunity to up the ante: "I'm sure you'll agree that the quotation couldn't have been about ISO or CCITT, since they'd have been debating croissants vs. brioches – doughnuts would give the American vendors too big a headstart."[40] Even Rose, the "Pied Piper of OSI" who devoted tremendous effort to reading and interpreting OSI documents for the benefit of Internet partisans, never brought himself to attend an OSI meeting and see how things looked from their perspective. "Quite frankly," Rose admitted in 1987, "I'm not even tied in that well to the OSI bunch." Instead, he was content to complain about the absurdity of the development of electronic mail standards in OSI to the friendly crowd of Internet engineers on the tcp-ip mailing list:

As has been pointed out in the past by Jon [Postel], these guys don't even use computers. Seriously, it's the stupidest catch-22 I've ever heard of: you can't start an electronic mail interest group for use by OSI zealots because either:

- they don't have working X.400 [electronic mail] systems
- their systems don't talk to each other yet
- they won't agree to use one organization's system for political reasons
- they won't use the ARPAnet mailsystem (even though they probably all could get legitimate access) because even though it works it's not OSI and they can't use it 'cause that would be admitting that the ARPAnet stuff works.[41]

[39] D. E. Cass and M. T. Rose (1986), "ISO Transport Services on Top of TCP," RFC 983, http://tools.ietf.org/rfc/rfc983 (accessed September 25, 2013); Michael A. Padlipsky, "Re: GOSIP," March 2, 1987, http://www-mice.cs.ucl.ac.uk/multimedia/misc/tcp_ip/8702.mm.www/0076.html (accessed December 21, 2011).

[40] Michael Padlipsky, "Re: GOSIP vs TCP/IP," March 10, 1987, http://www-mice.cs.ucl.ac.uk/multimedia/misc/tcp_ip/8702.mm.www/0127.html (accessed September 13, 2013)

[41] Marshall Rose, "Re: A Defense of GOSIP," March 15, 1987, http://www-mice.cs.ucl.ac.uk/multimedia/misc/tcp_ip/8702.mm.www/0146.html (accessed September 25, 2013).

Such rhetoric certainly accomplished its goal to foster a sense of self-righteousness among Internet engineers at the time. But its exaggerations ("these guys don't even use computers") also reveal how little Rose (and many other Internet engineers) actually knew about the computer experts who worked within OSI committees, including the Turing Award winner Bachman and French computer scientists such as Zimmermann, Michel Gien, and Najah Naffah, who moved on to successful and lucrative careers after leaving the Cyclades project at the end of the 1970s.[42]

Nevertheless, Rose continued his 1990 *The Open Book: A Practical Perspective on OSI* along the same lines: "The Internet community tries its very best to ignore the OSI community. By and large, OSI technology is *ugly* in comparison to Internet technology." Rose was by no means alone in his ritualistic disparagement of OSI. For example, Carl Malamud, in his 1992 "technical travelogue" of networking in twenty-one countries across the world, suggested that trying to implement OSI over slow, low-quality lines was "akin to looking for a hippopotamus capable of doing the limbo."[43]

In summary, Internet engineers in the 1980s and 1990s developed a well-worn critique of OSI – a critique that operated simultaneously on technical, institutional, and aesthetic levels. That these critiques were constructed on a series of unfair and underinformed caricatures of OSI and its protagonists was immaterial; their point was to advance the Internet's esprit de corps, to draw clear battle lines across the increasingly blurry edges of the Internet and OSI projects, and to advance their long-standing and single-minded ambition for the Internet to win.

Lyman Chapin, an engineer who participated in both OSI and Internet standards committees before becoming IAB chairman in 1991, summarized the difference between the two approaches in 1990: "It didn't take long to recognize the basic irony of OSI standards development: there we were, solemnly anointing international standards for networking, and every time we needed to send electronic mail or exchange files, we were using the TCP/IP-based Internet!" Another Internet pioneer, Einar Stefferud, captured the irony more concisely: "OSI is a beautiful dream, and TCP/IP is living it." Chapin was among a small group of people who were nevertheless determined to build bridges between the Internet and OSI. He continued, "I've been looking for ways to overcome

---

42 Najah Naffah, oral history interview by Andrew L. Russell, April 2, 2012, Paris, France. Charles Babbage Institute, University of Minnesota, Minneapolis; Michel Gien, oral history interview by Andrew L. Russell, April 3, 2012, Paris, France. Charles Babbage Institute, University of Minnesota, Minneapolis. For a richer and more balanced view of X.400 message handling standards, see Susanne K. Schmidt and Raymund Werle, *Coordinating Technology: Studies in the International Standardization of Telecommunications* (Cambridge, MA: The MIT Press, 1998), 229–262.

43 Marshall Rose, *The Open Book: A Practical Perspective on OSI* (Englewood Cliffs, NJ: Prentice-Hall, 1990), 591–592; Carl Malamud, *Exploring the Internet: A Technical Travelogue* (Englewood Cliffs, NJ: Prentice- Hall PTR, 1992), 191.

this anomaly ever since; to inject as much of the proven TCP/IP technology into OSI as possible, and to introduce OSI into an ever more pervasive and world-wide Internet. It is, to say the least, a challenge!"[44]

"We Reject: Kings, Presidents and Voting"

By custom, the overall architectural direction of the Internet was set by the IAB. As the locus of power within the Internet community shifted in the mid-1980s, the IETF grew to include several hundred participants who communicated regularly via email and met three times a year at weeklong meetings. They were, taken together, an articulate, opinionated, and at times rowdy bunch. In some ways, the two groups mimicked the familiar tensions between the upper and lower houses in a bicameral legislature: the IAB was a body of technical elites, the select and presumably disinterested elders; the IETF, on the other hand, was a ragtag lower house in which dedicated members of the rank and file could raise their voices and make important contributions – but who were not expected to have an expansive or systematic vision about the future of their shared enterprise.

During the Internet's phase of immense growth and commercialization in the early 1990s, criticisms of the "elders" in the IAB became increasingly common. Some critics complained that the IAB was providing weak leadership and alleged that the IAB "failed at times to provide a solid agenda and time-tables of engineering problems" for the IETF to address. Other critics took a different tack, arguing that the IAB was too overbearing and too eager to advance the commercial interests of companies (such as Digital Equipment) who were developing new networking products. Malamud, the author of the 1992 Internet technical travelogue, explained astutely that "the issue was not technical, it was procedural. How to run an informal standards process like the IETF, yet preserve due process safeguards, has been a continuing problem as the body grew in size from the original 13 participants."[45]

Internet insiders tried (and continue to try) to present the Internet's governing institutions in a more positive light. For example, "A Brief History of the Internet," coauthored by a number of Internet pioneers (including Cerf, Clark, and Kahn), characterized the growth of the Internet standards process benignly as a "steady evolution of organizational structures designed to support and facilitate an ever-increasing community working collaboratively on Internet issues." Such statements overstate the grace and ease with which the autocratic style of Internet governance gave way to – or rather was attacked by – the

---

[44] Lyman Chapin, quoted in Gary Malkin (1990), "Who's Who in the Internet: Biographies of IAB, IESG, and IRSG Members," RFC 1336, http://tools.ietf.org/rfc/rfc1336, (accessed September 25, 2013); Einar Stefferud, quoted in Marshall T. Rose, "Comments on '*Opinion*: OSI Is (Still) a Good Idea,'" *ConneXions* 6 (1992): 20–21.

[45] Malamud, *Exploring the Internet*, 1992, 196, 151.

growing numbers of active and opinionated engineers who were implementing
Internet protocols, mocking OSI on mailing lists, and making their voices heard
through IETF committees and working groups.[46]

The tensions within the Internet community came to a head in 1992, when
the IAB was struggling with a key technical shortcoming of the Internet's protocol design. The problem stemmed from the number of network addresses
that could be handled by the Internet Protocol (IP version 4, or IPv4). IPv4,
designed in the late 1970s, did not take into account the vast numbers of computers and networks that would join the Internet. By the early 1990s, Internet
engineers realized that the address space in IPv4 would soon be exhausted.
Moreover, the routing tables that listed the large number of Internet addresses
were becoming too large and bulky to handle easily. Since the exhaustion of the
address space and the explosion of the routing tables threatened to throttle the
Internet's growth, the council of elders assembled in the IAB began to consider
their options with a certain sense of urgency. Lyman Chapin, who served as
IAB chairman from July 1991 to July 1993, stated simply in an interview that
the shortage of Internet addresses was "definitely the most significant engineering problem on the Internet now."[47]

The IAB responded by convening three workshops in January and June
1991 and January 1992 where more than thirty leaders of the Internet community held extensive discussions that were "spirited, provocative, and at
times controversial, with a lot of soul-searching." The Internet elite eventually
reached a consensus around four "basic assumptions regarding the networking world of the next 5–10 years": the TCP/IP Internet and OSI would continue to coexist, the Internet needed to support multiple protocols and include
diverse networking technologies, traffic would be carried both by public and
private networks, and the Internet architecture needed to be able to scale far
beyond its current capacity. As a result of these meetings, the IETF created a
Routing and Addressing (ROAD) working group to explore new approaches
for networking addressing and to propose solutions. The Internet Engineering
Steering Group (IESG), which coordinated the disparate activities taking place

---

[46] Leiner et al., "A Brief History of the Internet." An additional complicating factor at this time –
one that I do not emphasize due to space constraints – concerned the creation of the Internet
Society, designed by Cerf, Kahn, and other Internet veterans to provide financial support and an
institutional home for the IETF and the Internet standards process. See Simon, *Launching the
DNS War*, 101–123; Milton L. Mueller, *Ruling the Root: Internet Governance and the Taming
of Cyberspace* (Cambridge, MA: The MIT Press, 2002); and Vint Cerf, "IETF and the Internet
Society," July 18, 1995, Internet Society, http://www.internetsociety.org/internet/internet-51/
history-internet/ietf-and-internet-society (accessed January 4, 2012).

[47] Dave Clark was "Internet Architect" from 1983 to 1989; Cerf served from 1989 to 1991 and
was followed by Lyman Chapin (through March 1993). Ellen Messmer, "Internet Architect
Gives Long-Term View," *Network World* 9 (May 18, 1992): 37. My analysis of these events
depends on similar documents, and comes to similar conclusions, as the accounts in Simon,
Launching the DNS War, 123–129; and Laura DeNardis, *Protocol Politics: The Globalization
of Internet Governance* (Cambridge, MA: The MIT Press, 2009), 25–70.

in dozens of IETF working groups, summarized the various recommendations from the rank and file of the IETF and sent them along to the IAB. The IAB – pointedly renamed the Internet *Architecture* Board in June 1992 – considered the IESG recommendations at its meeting on June 18 and 19 in Kobe, Japan.[48]

Discussions of routing and addressing dominated the Kobe meeting. The IESG's report on routing recommended the adoption of Classless Inter-Domain Routing (CIDR), a mechanism developed within the ROAD working group. At first, four of the eight IAB members at the meeting objected on the grounds that there was no operational evidence to prove that CIDR would do what its designers hoped. Moreover, CIDR would only provide a short-term solution to the address space problem and thereby postpone rather than solve the more fundamental architectural problems vexing the Internet community. Despite these grave reservations, the IAB eventually agreed that the "IETF should aggressively pursue the work to engineer CIDR" as a short-term solution for the immediate routing and addressing problems that threatened the future of the Internet.[49]

The debate in Kobe over appropriate long-term solutions to Internet addressing problems was more complicated and raised fundamental technical and political issues for the global networking community. The IAB considered a second proposal, written by Ross Callon and developed within the ROAD working group, that proposed a "long-term solution to Internet addressing, routing, and scaling." At the heart of Callon's proposal was the ConnectionLess Network Protocol (CLNP), a descendent of the Cyclades transport protocol that had been modified to serve as the OSI functional counterpart to IPv4. Callon had worked closely with OSI technologies at Bolt, Beranek and Newman and at Digital Equipment Corporation, and was one of a growing number of people who believed that CLNP had the potential to overcome the limitations of the Internet Protocol. Chapin, the IAB chairman, agreed in a May 1992 interview: "The most comprehensive solution is to replace the Internet Protocol in the Internet with the Open Systems Interconnection Connectionless Network Protocol. That idea is already almost universally accepted." There certainly was broad agreement within the small circle of Internet architects that CLNP provided a useful *technical* option, but subsequent events would show that Chapin

---

[48] David Clark, Lyman Chapin, Vinton Cerf, Robert Braden, and Russ Hobby (1991), "Towards the Future Internet Architecture," RFC 1287, http://tools.ietf.org/rfc/rfc1287 (accessed September 25, 2013); Phillip Gross and Philip Almquist (1992), "IESG Deliberations on Routing and Addressing," RFC 1380, http://tools.ietf.org/html/rfc1380 (accessed September 25, 2013).

[49] V. Fuller, T. Li, J. Yu, K. Varadhan (1992), "Supernetting: An Address Assignment and Aggregation Strategy," RFC 1338, http://tools.ietf.org/html/rfc1338 (accessed September 25, 2013); Gross and Almquist, RFC 1380; Internet Activities Board, "Meeting Minutes – June 18–19, 1992," http://www.iab.org/documents/minutes/minutes-1992/iab-minutes-1992-06-18 (accessed January 3, 2012).

badly misjudged the *strategic* appeal of CLNP among Internet engineers who detested all things OSI.[50]

IAB members raised several technical objections to the adoption of CLNP. It was clear that the protocol would need to be reengineered to suit the Internet, but Vint Cerf and others raised a related procedural question: Would the IETF be able to maintain control over future revisions of CLNP? Perhaps worried about the reaction of the Internet loyalists, Cerf proposed that the new protocol be called "IPv7" rather than CLNP. The change in nomenclature, Cerf argued, would clearly indicate that the IETF "owns" the specification. Chapin, pressing the case for adoption, countered that it would be advantageous to keep "CLNP" prominent in the new specification because it would demonstrate stability and consistency to commercial firms that manufactured and sold networking equipment. Here, as in previous generations of industrial standardization, links between designers and users were paramount for the creation of successful standards.

Despite Chapin's efforts, Cerf's experience had taught him to remain skeptical of ISO's willingness to cooperate on equal terms. According to the extensive minutes of the June Kobe meeting preserved on the IAB's Web site, Cerf "observed that when [David] Piscitello took IP to ISO and tried to convince them it represented a piece missing from their architecture, ISO felt comfortable with changing IP to make CLNP. All we are asking for now is 'reciprocity!'" Steven Kent, a security expert and chief scientist for BBN Communications, declared flatly, "We should say, 'We have decided to adopt CLNP, but we are not happy with the details of the protocol. We are currently planning to use the existing protocol, but in the future we expect to explore (at least?) different address formats.'" Chapin agreed that the IAB should insist that "any changes we make to CLNP for IPv7 be incorporated into CLNP in OSI." For Chapin, the overarching strategic advantage was clear: there was "an opportunity to do away with protocol wars" and to achieve a major step toward global interoperability of computer networks.[51]

Chapin, as IAB chairman, drafted a "Proposal on Routing and Addressing" at the end of the Kobe meeting. "Based upon an analysis of the architectural requirements," Chapin summarized, "the IAB furthermore proposes using CLNP (the OSI internetwork protocol defined by the ISO 8473 standard, which uses variable-length addresses that may be up to 20 bytes long) as the starting point for developing IPv7." Obviously anticipating a visceral response from Internet partisans, Chapin hastily added:

It is important to understand that this does NOT mean "adopting OSI" or "migrating to OSI" or "converting the Internet to use GOSIP protocols." The IAB recommends

---

[50] Ross Callon (1992), "TCP and UDP with Bigger Addresses (TUBA), A Simple Proposal for Internet Addressing and Routing," RFC 1347, http://tools.ietf.org/rfc/rfc1347 (accessed September 25, 2013); Messmer, "Internet Architect," 46; Internet Activities Board, "Meeting Minutes – June 18–19, 1992."

[51] Internet Activities Board, "Meeting Minutes – June 18–19, 1992."

only that a new version of IP (IPv7), with much wider addresses and a more extensible header, be based on the existing CLNP.... The IAB does not take this step lightly, nor without regard for the Internet traditions that are unavoidably offended by it. We look forward to a lively discussion of these conclusions during the upcoming IETF meeting in Boston.[52]

Despite Chapin's defensive language and cautious argument, the IAB statement ignited the passions of the masses of Internet engineers. They saw red. These engineers flooded their email lists with howls of protest against what they perceived as a unilateral decree.[53] Christian Huitema, a French computer scientist who was invited to join the IAB in 1991 to provide a richer international perspective, later summarized how quickly the situation deteriorated:

The IAB discussed [the draft proposal to incorporate CLNP] extensively. In less than two weeks, it went through eight successive revisions. We thought that our wording was very careful, and we were prepared to discuss it and try to convince the Internet community. Then, everything accelerated. Some journalists got the news, an announcement was hastily written, and many members of the community felt betrayed. They perceived that we were selling the Internet to the ISO and that headquarters was simply giving the field to an enemy that they had fought for many years and eventually vanquished. The IAB had no right to make such a decision alone.[54]

The fact that this proposal provoked such outrage from hundreds of engineers and computer scientists reflects the passion and commitment of engineers in the pitched battle of a standards war. Internet engineer Carl Cargill preferred a religious metaphor to Huitema's military metaphor: "For the general membership of the IETF," he commented, "this was rank heresy." Both metaphors were apt: where Huitema's military metaphor captured the strategic and organizational tensions between the competing standards bodies, Cargill's religious metaphor captured the emotional commitment that Internet engineers had invested in their underdog network architecture.[55]

Internet partisans vented their frustrations on email lists, raising concerns about the technical capabilities of CLNP as well as the "undemocratic and closed nature of the decision making process" by which the IAB had arrived at such a contentious statement of architectural intent. Many agreed that the Internet community had a severe political problem on its hands – that, as Stanford's Bill Simpson put it, "the IAB is too separated from the organizational

---

[52] Internet Activities Board, "Meeting Minutes – June 18–19, 1992."

[53] Two of the most active email lists were the IETF discussion list, ftp://ftp.ietf.org/ietf-mail-archive/ietf (accessed January 4, 2012); and the Big-Internet list, ftp://munnari.oz.au/big-internet/list-archive (accessed January 4, 2012).

[54] Christian Huitema, *IPv6: The New Internet Protocol* (Upper Saddle River, NJ: Prentice-Hall PTR, 1998), 2.

[55] Carl Cargill, *Open Systems Standardization: A Business Approach* (Upper Saddle River, NJ: Prentice-Hall PTR, 1997), 257; Scott Bradner, "The Internet Engineering Task Force," *OnTheInternet* 7 (2001): 24; Cerf, "IETF and the Internet Society." For an illuminating discussion of religious metaphors in computer engineering discourse, see Christopher M. Kelty, *Two Bits: The Cultural Significance of Free Software* (Durham, NC: Duke University Press, 2008).

and implementation aspects of the IETF to be effective." Simpson continued, "This is hubris of the highest order. We need a vigorous debate at the Boston meeting about our organizational structure." Cerf, Chapin, and others tried to reiterate that CLNP was only to be a starting point for discussion (a point that had been missed by engineers who insisted it was a mandate). They also denied that the proposal meant the IETF would have to "buy into the ISO process," but their response was too late, too disorganized, and insufficient to soothe the outrage.[56]

The revolt of the Internet engineers came to a climax during the July 1992 meeting of the IETF in Boston, a five-day event that attracted nearly 700 participants. Although many IETF participants had technical reservations about CLNP, their resistance was motivated primarily by the fear that the IAB was violating established procedures and turning the Internet into the thing they most despised – an arrogant, centralized bureaucracy. In the view of the Internet rank and file, the demographics of Internet standards had changed so much that the old, autocratic model was inappropriate.[57]

Cerf, who had voiced his deep reservations over the IAB's decision during the Kobe debate in June, took it upon himself to reassure the anxious and angry masses of the IETF in July. He intervened in sensational style. During one session, Cerf stepped up to a microphone to make some informal comments. He then proceeded to remove the layers of his signature three-piece suit, eventually stripping down to reveal a homemade T-shirt that proclaimed: "IP on Everything." The audience was thrilled and the tension was partially eased. The striptease, Cerf later recalled, was designed to reiterate a goal of the IAB: to defend and promote the existence of the Internet Protocol on every underlying transmission medium. The *New York Times* later reported that the audience "went nuts" and that a member of the audience even "rushed to the podium and placed a $5 bill in Dr. Cerf's waistband."[58]

Cerf's outrageous performance, unmatched in the admittedly reserved cultural history of standard-setting committees, sent an unmistakable signal that he would continue to fight to maintain the Internet's distinctive identity and traditions. The Internet Protocol might change, Cerf seemed to be saying, but it would never be abandoned or surrendered. Cerf thus continued his remarkable trajectory in the world of computer network standards, from the loser of the INWG protocol debate in the mid-1970s to the Internet's elder statesman and most talented diplomat in the early 1990s. He would soon become the closest thing the Internet engineering community world has to a celebrity – eventually

---

[56] Steve Deering, "Re: IPv7 (CLNP) Is a Mistake," July 2, 1992; Vint Cerf, "Re: IPv7 == CLNP or CLNP+?," July 6, 1992; William Allen Simpson, "Lessons – Political," July 6, 1992; Lyman Chapin, "Re: IPv7 (CLNP a Mistake)," July 6, 1992; all from ftp://ftp.ietf.org/ietf-mail-archive/ietf/ (accessed September 25, 2013).

[57] Anonymous, "The 2nd Boston Tea Party," *Matrix News* 10 (2000).

[58] Vinton G. Cerf, personal communication, January 27, 2002; Katie Hafner, "For 'Father of the Internet,' New Goals, Same Energy," *New York Times*, September 25, 1994.

taking a position at Google with the revealing title that formalized his role for so many years, "Chief Internet Evangelist."

The other leader of the Internet community who emerged from the July 1992 meeting with his reputation enhanced was David Clark. Clark had been an Internet "elder" since the late 1970s but was no longer an active member of the IAB and therefore untainted by its Kobe proposal. At the final session on the final day of the July 1992 IETF meeting, Clark took to the podium for a technical plenary address. Sensing that the community had a deep need for something more than simple technical advice, he saw his presentation as an opportunity to defend the need for the community to have strong leadership, to "rally the troops," and to reaffirm the shared values of the Internet community, values that had been sullied by the CLNP debacle.[59]

As we saw at the opening of this chapter, Clark framed his talk, titled "A Cloudy Crystal Ball: Visions of the Future (Alternate Title: Apocalypse Now)," in the familiar terms of the drastic architectural choices that faced the Internet community. After outlining the major technical challenges that worried him, Clark turned to the delicate question of how the IETF should "manage the process of change and growth." Clark demonstrated his purity to the Internet engineers by referring to ISO as the "standards elephant of yesterday," but he also warned pointedly that the "standards of today" was "right here." In 2006, Clark recalled that Internet engineers in 1992 "had thought we were running from the OSI" and its "alternative standards process. And part of what we had to defend was our standards process."[60]

Clark proceeded to weigh a sequence of values that the IETF should consider as "the Internet and its community grows." An open process would allow all voices to be heard, but it also needed to limit participation in order to "make progress." A quick process would help the IETF keep up with external forces such as commercialization, but the virtue of a slow process, Clark noted, was that it could "leave time to think." As the Internet community tried to find a balance between these issues, Clark added, they also needed to make sure that they kept up with market forces and, at the same time, be certain that the Internet could scale more gracefully. He did not try to reconcile some of the troubling contradictions in his list, such as his implication that a thoughtful, democratic process would fail to "keep up with reality" and fail to "make progress." Instead, he pressed on to define what he saw as the community's shared values:

We reject: kings, presidents and voting.

We believe in: rough consensus and running code.[61]

[59] Bradner, "The Internet Engineering Task Force," *OnTheInternet*, 24.

[60] Clark, "A Cloudy Crystal Ball." On the occasion of the IETF's twentieth anniversary in 2006, Clark delivered an encore of this presentation. The video of his talk is in possession of the author.

[61] Clark, "A Cloudy Crystal Ball."

Clark's rejection of "kings, presidents and voting" was, on the surface, a direct attack at the formal style of bureaucracy embraced by regimes of centralized control as well as by the Internet's rival and "standards elephant," OSI. As such, it resonated deeply with the Internet engineers who were united by their opposition to OSI. As Clark later emphasized, however, he did not intend for the IETF to interpret his slogan as an "unalloyed compliment. It is not always good to reflexively kill your leaders." The slides of his 1992 talk show that Clark continued to praise the Internet community for its engineering prowess, especially its ability to respond to "short term reality" and "building stuff that works." But he also reminded the audience that they struggled to grow processes to match the size of the community and, more important, to set "long-term direction" for their collective efforts. In other words, Internet engineers still needed the leadership of architects with a cohesive long-term view of the Internet and its development path.

Internet engineers did not latch on to Clark's subtle dig at their leveling tendencies. Instead, they simply agreed that Clark's "rough consensus and running code" slogan perfectly captured their distinctive political culture. "Rough consensus" drew a clear distinction between the Internet and OSI: whereas OSI standardization depended on the formal and arcane voting procedures required by ISO and CCITT rules, Internet standards could be developed through the modest mechanisms followed by the IETF. The Internet community encouraged newcomers to contribute their expertise and approved proposals that enjoyed broad support within the group. IETF veterans placed an acceptable level of agreement at around 80 to 90 percent – a level high enough to demonstrate strong support, but flexible enough to work in the absence of unanimity. Internet engineers did not invent such voting procedures, even if they rarely acknowledged their direct precedents in "voluntary consensus" standards organizations such as ISO. In short, "rough consensus" was an apt description of an informal process in which a proposal must answer to criticisms but need not be held up if supported by a vast majority of the group. The Internet engineers saw features of democratic governance such as presidents and voting as irrelevant, and, echoing Paul Gough Agnew from the 1920s, argued that their mode of "rough consensus" was superior to all alternatives.[62]

As a complement to rough consensus, "running code" meant "multiple actual and interoperable implementations of a proposed standard must exist and be demonstrated before the proposal can be advanced along the standards track."[63] Because most standards began with a proposal from an individual or group within a working group – and not from the IAB or IETF leadership – the party behind the proposal needed to provide multiple working versions of it. This burden of proof on the proposed standard facilitated the adoption of new

[62] Bradner, "The Internet Engineering Task Force," in *Open Sources*, 47–53. See also Dave Crocker, "Making Standards the IETF Way," *StandardView* 1 (1993): 48–54.

[63] Bradner, "The Internet Engineering Task Force," *OnTheInternet*, 26.

IETF standards across the diverse computing platforms on the Internet. The Internet community, in the summary of Peter Salus, believed in specification before implementation.[64]

"Running code" also evoked a major difference between the Internet standards process and the OSI process: whereas Internet protocols represented "the result of intense implementation discussion and testing," the goal of OSI was to standardize a framework (reference model) for protocols that had not yet been created. As a result, OSI committees subverted the traditional mode of standard setting that had been developed in the industrial standards committees described in Chapters 2 and 3, in which standardization came *after* experimentation and market acceptance. In many ways, the differences between the two approaches to engineering practice echo the debates of the late nineteenth century between American mechanical engineers: whereas "shop culture" engineers favored a pragmatic approach to standardization that privileged existing practices (Clark's "running code"), "school culture" engineers believed they should use their own judgment to decide which existing practices were superior.[65]

Clark was perhaps too subtle in expressing his concerns about the Internet community's instinct to reject authority and therefore fail to set a coherent and informed "long-term direction." Nevertheless, "rough consensus and running code" achieved instant fame as a way for Internet engineers to identify their community's unique approach to making standards. It was more than a simple criticism of OSI; it was an example of "constructive snottiness" (to use Padlipsky's delightful term), a declaration of the ways in which Internet engineers believed their standards process was *superior* to OSI. In other words, "rough consensus and running code" generated and sustained enthusiasm within the Internet community because it framed their work as a *critique* of OSI. Moreover, the slogan rejected the unilateral and autocratic tendencies of the style of governance that had been the norm within the Arpanet and Internet communities since the 1960s. Clark's talk thus marks the beginning of a new era for the Internet – an era that sacrificed architectural clarity and diplomatic leadership in order to assert the power of the Internet rank and file and reject with greater vehemence than ever before the technical work of the elephants in OSI.

By the end of the July meeting, the IAB acknowledged that it had failed to properly consult and engage the Internet community. IAB members acknowledged their jurisdictional error, stepped back from the proposal they had painstakingly crafted at their meeting in Kobe, and reassessed the future. Although the jurisdictional rift within the Internet standards had been soothed, the underlying technical problems stubbornly remained. During the next two years, the Internet community debated various proposed designs for the next generation

---

[64] Peter H. Salus, *Casting the Net: From ARPANET to INTERNET and Beyond* (New York: Addison-Wesley Publishing Company, 1995), 123.
[65] Padlipsky, *Elements of Networking Style*, 104.

of the Internet Protocol. By January 1995, the community arrived at a consensus over the broad outlines of the new protocol, known as Internet Protocol version 6 (IPv6), that would be developed within the IETF over the next several years. In the meantime, the IETF developed protocols including the Dynamic Host Control Protocol (DHCP) and Network Address Translator (NAT) as short-term solutions to the Internet address space problem. The technical crisis was averted, but the procedural drama of the 1992 meetings precipitated what Scott Bradner, one of the new generation of IETF leaders who stepped forward after the summer of 1992, aptly characterized as the Internet's "constitutional crisis." Gone were the days that a small council of elders could exercise autocratic authority over Internet architecture and engineering. The mob had become too large and too powerful and succeeded in their palace revolt.[66]

Indeed, many IAB members who supported the CLNP in Kobe left the IAB soon after the Cambridge IETF meeting. Chapin, who in 1991 seemed uniquely qualified for the IAB chairmanship because of his vast experience with both the Internet and OSI communities, left in March 1993. Several other individuals who had tried to engineer an end to the "protocol wars" also faded from prominent positions in the Internet community. The IAB was itself radically expanded and reformed: the defenders of CLNP were replaced ("purged" may be too strong a word), and at least eleven new members joined the IAB in 1993 and 1994. Some had experience with OSI, but that experience was no longer considered to be an asset or virtue.[67]

Between 1992 and 1996, members of the Internet community revised and clarified the Internet standards process so that future constitutional crises might be avoided. These reforms also sought to strengthen and maintain the integrity of the Internet standards process in the face of increasing commercial interest.[68] In 1993, the IETF published an RFC titled "The Tao of the IETF"

---

[66] Tim Dixon (1993), "Comparison of Proposals for Next Version of IP," RFC 1454, http://tools.ietf.org/rfc/rfc1454 (accessed September 25, 2013); R. Droms (1993), "Dynamic Host Configuration Protocol," RFC 1541, http://tools.ietf.org/rfc/rfc1541 (accessed September 25, 2013); Scott Bradner and Allison Mankin (1993), "IP: Next Generation (IPng) Working Paper Solicitation," RFC 1550, http://tools.ietf.org/rfc/rfc1550 (accessed September 25, 2013); K. Egevang and P. Francis (1994), "The IP Network Address Translator (NAT)," RFC 1631, http://tools.ietf.org/rfc/rfc1631 (accessed September 25, 2013); S. Crocker (1994), "The Process for Organization of Internet Standards Working Group (POISED)," RFC 1640, http://tools.ietf.org/html/rfc1640 (accessed September 25, 2013); Scott Bradner and Allison Mankin (1995), "The Recommendation for the IP Next Generation Protocol," RFC 1752, http://tools.ietf.org/rfc/rfc1752 (accessed September 25, 2013); Huitema, *IPv6*; DeNardis, *Protocol Politics*; Simon, *Launching the DNS War*; Bradner, "The Internet Engineering Task Force," *OnTheInternet*, 24.

[67] Internet Architecture Board, "A Brief History"; Krol, RFC 1462; Simon, *Launching the DNS War*.

[68] Scott Bradner (1996), "The Internet Standards Process – Revision 3," RFC 2026, http://tools.ietf.org/rfc/rfc2026 (accessed September 25, 2013); Richard Hovey and Scott Bradner (1996), "The Organizations Involved in the IETF Standards Process," RFC 2028, http://tools.ietf.org/rfc/rfc2028 (accessed September 25, 2013).

that introduced newcomers to its customs and norms. The days of informal discussions among bearded and sandaled graduate students were in the past; rather, a greater proportion of IETF attendees in the late 1990s worked for private companies and wore suits and ties – a vivid reminder that the world of Internet standards had changed.[69]

Even though the Internet community struggled to accommodate rapid commercialization, it was better to struggle with success than to dwell in commercial irrelevance. Subsidies from the U.S. government – including support to implement TCP/IP for the increasingly popular UNIX operating system – continued to be decisive. As the French computer scientist Gérard Le Lann recalled, "Forgetting about its technical weaknesses, TCP/IP has won because it was well-tested, pushed by a powerful community, and it was public domain." In 1991, the National Science Foundation, which had operated the Internet backbone since 1986, lifted its restriction on commercial activity over the network – thus paving the way for the Internet to serve as a new commercial infrastructure. Another turning point for the popularity of the Internet also occurred in 1991, when Tim Berners-Lee released the World Wide Web, an application that used the Internet to allow users to browse hypertext documents.[70]

In the meantime, OSI failed to live up to the high expectations it had engendered. In 1994, the National Institute of Standards and Technology abandoned its GOSIP program in favor of the TCP/IP Internet. The market for network protocols had tipped in favor of the Internet. The Internet-OSI religious war was over, and the grand future planned for OSI had vanished – despite the technical shortcomings of the Internet. Twenty years later, IPv4 remains widely deployed as the host protocol for billions of users of the global Internet, but the Internet community still has been unable to identify, develop, and deploy a solution that can overcome IPv4's architectural limitations.[71] Internet champions have, however, been remarkably successful in convincing outsiders that the Internet's technical and political development stands as one of the greatest accomplishments of the twentieth century. They also have been able to convince outsiders that the Internet standards process could be a model for future

---

[69] Gary Malkin (1993), "The Tao of the IETF – A Guide for New Attendees of the Internet Engineering Task Force," RFC 1391, http://tools.ietf.org/rfc/rfc1391 (accessed September 25, 2013). "The Tao" was subsequently revised in 1993, 1994, 2001, and 2006.

[70] Gérard Le Lann, oral history interview by Andrew L. Russell, April 3, 2012, Paris, France, Charles Babbage Institute, University of Minnesota, Minneapolis; Mills interview, Charles Babbage Institute; Abbate, *Inventing the Internet*, 181–220; Shane Greenstein, "The Emergence of the Internet: Collective Invention and Wild Ducks," *Industrial and Corporate Change* 19 (2010): 1521–1562.

[71] IPv6, in the absence of an autocratic mandate, has seen lackluster adoption. According to Google, barely two percent of Internet users access Google using IPv6. Google, "IPv6 – Google," http://www.google.com/ipv6/statistics.html (accessed September 30, 2013).

attempts to create a technologically enabled style of open, participatory, and democratic governance.[72]

Conclusions

The Internet would not have succeeded as a technological system if the Internet community had not developed a structure and process to sustain the development and standardization of Internet protocols. Beginning in the late 1970s, Internet leaders such as Vinton Cerf, David Clark, Robert Kahn, and Jon Postel created a set of institutions that effectively circumvented the industrial legislatures responsible for the "official" standards process. They consciously decided against leaving their fate to a democratic international effort whose outcome they might not like and could not control. Instead, they designed new institutions to protect the development of TCP/IP from the commercial and political pressures that stalled OSI. Whereas OSI found the time-honored principles of democratic inclusivity to be both necessary and fatal, the Internet flourished by developing their network architecture within a well-funded and homogenous environment that was, in the late 1970s and early 1980s, insulated from commercial and political pressures. Competition with OSI, together with vibrant market demand and continued patronage from the U.S. government, drove TCP/IP into multiple implementations as a de facto standard – an interim solution to the problems of interconnection that OSI was expected eventually to solve.

The Internet's autocratic mode of architectural control, a key to its success in the 1970s and early 1980s, endured successive rounds of critique, reform, and democratization. Control over Internet standardization that was once centralized in a "kitchen cabinet" and "council of elders" shifted to the IETF's jurisdiction during the late 1980s and early 1990s. A different political philosophy prevailed in the rank and file of the IETF, one that encouraged more direct democracy and more transparent decision making. The IETF marked its embrace of a more mature, if messier, form of participatory democracy with a series of constitution-like RFCs published in 1994 and 1996 that specified the Internet standards process and working group guidelines in new detail.[73] Direct representation became more clearly articulated as a key distinction between the

---

[72] See John S. Quarterman, "The Demise of GOSIP," *Matrix News* 4 (1994): 6; David C. Wood, "Federal Networking Standards: Policy Issues," *StandardView* 2 (1994): 218–223; Libicki, *Information Technology Standards*, 108–119; Kai Jakobs, "Why Then Did the X.400 E-mail Standard Fail? Reasons and Lessons to be Learned," *Journal of Information Technology* 28 (2013): 63–73; John Day, *Patterns in Network Architecture: A Return to Fundamentals* (Upper Saddle River, NJ: Prentice-Hall PTR, 2007).

[73] Lyman Chapin (1992), "The Internet Standards Process," RFC 1310, http://tools.ietf.org/html/rfc1310 (accessed September 25, 2013); IAB and IESG (1994), "The Internet Standards Process – Revision 2," RFC 1602, http://tools.ietf.org/html/rfc1602 (accessed September 25, 2013); Erik Huizer and David Crocker (1994), "IETF Working Group Guidelines and Procedures," RFC 1603, http://tools.ietf.org/html/rfc1603 (accessed September 25, 2013); Bradner, RFC 2026;

Internet standards process and the traditional (read: ISO and ANSI) standards process, which relied on a balance of representatives who promoted the interests of particular industrial, national, and international organizations. Despite the IETF's self-conception as a group of individual volunteers concerned only with the good of the Internet, many individuals participating in it nevertheless tried to shape Internet standards to benefit the proprietary interests of their employers. This development came as no surprise to anyone with even a passing familiarity with the realities of industrial standardization.

A useful framework for reflecting on the perils of democratization in the Internet standards process may be found in a popular 1975 book by the IBM software engineer Fred Brooks. In a chapter titled "Aristocracy, Democracy, and System Design," Brooks observed that conceptual clarity flowed from a singular or small design team, but important adaptations, innovations, and fresh concepts come from implementers and users. Despite the important contributions of users, Brooks defended "aristocracy" by insisting that "conceptual integrity is *the* most important consideration in system design." By the late 1990s, Internet engineers and users regularly endorsed the benefits of the old autocratic and centralized style of control. The sudden death of Jon Postel from heart problems in 1998 served as a poignant reminder that times had changed. In 2001, Cerf commented frankly on the jurisdictional void that Postel's departure created: "There is no substitute for an enlightened despot who is well intentioned and capable. Trying to make an institution that balances everyone's interest is incredibly difficult."[74]

It is no wonder that Internet partisans convey mixed messages about two modes of governance – autocracy and democracy – that are theoretical opposites yet, in the development of the Internet, coexisted in practice. One of the most striking ironies in the history of computer networking is the Internet's transformation from its origins in the "closed world" of American Cold War defense research to become a symbol of the "open world" of late twentieth century global capitalism. As we have seen, this transformation was sustained by institutional reforms during the late 1980s and early 1990s when the Internet standards process (and Internet governance more generally) transitioned away from a traditional autocracy and embraced more decentralized and participatory structures.

In addition to the reforms in the structure and process of Internet standardization, the discourse of Internet engineers also changed subtly to embrace the language of "openness." In the 1980s and early 1990s, Internet engineers,

Scott Bradner (1998), "IETF Working Group Guidelines and Procedures," RFC 2418, http://tools.ietf.org/html/rfc2418 (accessed September 25, 2013).

[74] Brooks, *The Mythical Man-Month*, 42; Cerf quoted in Simon, *Launching the DNS War*, 362. On Postel, see Simon, *Launching the DNS War*, 42–86 and 360–364; Goldsmith and Wu, *Who Controls the Internet?*, 29–48. See also Dibbell, "A Rape in Cyberspace"; Russell, "Constructing Legitimacy"; Wolf, "Why Craigslist Is Such a Mess"; and Joseph Michael Reagle, Jr., "The Benevolent Dictator," in *Good Faith Collaboration: The Culture of Wikipedia* (Cambridge, MA: The MIT Press, 2011).

for obvious reasons, had not habitually used the word to describe themselves, their standards process, or their technologies. The first RFC explicitly to ascribe open qualities to the Internet was RFC 1310, "The Internet Standards Process," published in March 1992. The author of RFC 1310 was none other than Lyman Chapin, the IAB chairman who would be deposed less than a year later in the aftermath of the IETF's 1992 palace revolt. Chapin used the word "open" at least eight times in the document, which was the first attempt to define the Internet standards process formally. No doubt following the customary discourse in OSI, he introduced the term "open" to refer to a variety of the Internet's attributes, including its "open protocols"; the "clear, open, and objective" procedures that his document intended to provide; the requirement to discuss proposed Internet standards in "open meetings" before approval; identifying "openness and fairness" as a principle goal of the Internet standards process; and a warning that proprietary vendor contributions did not qualify as "open standards" in the same way that specifications approved by accredited standards bodies such as ANSI, IEEE, and ISO did. Chapin thus appears to be a conceptual innovator – the first person to align the Internet standards process with the historically and symbolically rich discourse of openness.[75]

Notions of an open Internet became popularized in the late 1990s and early 2000s as outsiders became familiar with the Internet and its peculiar governance structure as well as other open systems, such as the UNIX operating system.[76] The earliest references to the open Internet in media and popular discourse can be found throughout 1994. They differed from Chapin's positive identification of openness in Internet protocols and the Internet standards process. Articles that appeared in *Newsday* and *Fortune* described the "wide-open Internet" cautiously as a target for password-stealing hackers and a global web of networks that "hasn't been easy to navigate, and ... hasn't been secure." In contrast to these articles that emphasized the porous – and therefore dangerous – borders of the "wide-open Internet," a student in a fall 1994 MIT course on Ethics and Law on the Electronic Frontier noted that "commercial providers like America Online and Compuserve are beginning to open gateways from their exclusive services to the open Internet." The implication, as with the articles in *Newsday* and *Fortune*, was to identify the Internet as an unruly, uncensored, nonproprietary, and potentially dangerous and disruptive electronic frontier – just as John Perry Barlow would describe it in his influential 1996 cyberlibertarian manifesto, "A Declaration of Independence of Cyberspace."[77]

[75] Chapin, "The Internet Standards Process."

[76] Paulina Borsook, "How Anarchy Works," *Wired* 3 (October 1995); Kelty, *Two Bits*, 143–178.

[77] Joshua Quittner, "The Password is 'Loopholes,'" *Newsday*, March 1, 1994: 61; Rick Tetzeli et al., "Fortune Checks Out 25 Cool Companies for Products, Ideas, and Investments," *Fortune*, July 11, 1994, available from http://money.cnn.com/magazines/fortune/fortune_archive/1994/07/11/79502/index.htm (accessed September 25, 2013); Rosemary W. McNaughton, "Culture Clash on the Changing Internet," Massachusetts Institute of Technology, http://groups.csail.mit.edu/mac/classes/6.805/student-papers/fall94-papers/mcnaughton-culture.html

Policy makers in the Clinton administration – led by Vice President Al Gore and Commerce Secretary Ron Brown – picked up on discourses of openness to the Internet in the context of their plans for a National Information Infrastructure. Academic policy analysts, including a small team of researchers in the Information Infrastructure Project at Harvard University's John F. Kennedy School of Government, followed suit and explored the implications of openness in greater detail. For example, in their introduction to a 1995 volume *Standards Policy for Information Infrastructure*, Harvard faculty members Brian Kahin and Lewis Branscomb (a former IBM chief scientist) praised the virtues of "open competition," noted that users had come to "generally demand open systems," and celebrated the interoperability that had been enabled by the "experimental and open environment of the Internet." Other academics and policy makers, mostly unaware of the Internet's history but eager to shape its future, borrowed the language of open systems and applied it to the Internet.[78]

By the late 1990s, even officials in the Federal Communications Commission who had studiously avoided making jurisdictional claims over the Internet mobilized the discourse of openness to address regulatory questions surrounding broadband Internet access and the "digital divide." Cries to preserve the "open Internet" from censorship and monopolists – echoes of earlier generations of American expectations of open communication – soon became routine. Twenty-first-century lobbyists and advocates for "network neutrality" repeatedly and anachronistically declared that openness had always been a "fundamental principle of the Internet's design." They did not realize that they were testifying to the overwhelming force of the Internet's technological, cultural, and discursive triumph over their old rivals, the proponents of Open Systems Interconnection.[79]

(accessed January 4, 2012); John December, "Have We a Clue?" *CMC Magazine* 2 (1995), http://www.december.com/cmc/mag/1995/jul/ed.html (accessed January 4, 2012). On Barlow's "Declaration," see Goldsmith and Wu, *Who Controls the Internet?*, 13–28.

[78] Lewis M. Branscomb and Brian Kahin, "Standards Processes and Objectives for the National Information Infrastructure," in Brian Kahin and Janet Abbate, eds., *Standards Policy for Information Infrastructure* (Cambridge, MA: The MIT Press, 1995), 7–10; National Telecommunications and Information Administration, *The National Information Infrastructure: An Agenda for Action* (1993), http://www.ibiblio.org/nii (accessed January 4, 2012); Jonathan Band, "Competing Definitions of 'Openness' in the NII," in Kahin and Abbate, eds., *Standards Policy for Information Infrastructure*; Brian Kahin and Ernest Wilson, *National Information Infrastructure Initiatives: Vision and Policy Design* (Cambridge, MA: The MIT Press, 1997); Ole Hanseth, Eric Monteiro, and Morten Hatling, "Developing Information Infrastructure: The Tensions between Standardization and Flexibility," *Science, Technology, & Human Values* 21 (1996): 407–426; Shane Greenstein, "Markets, Standards, and the Information Infrastructure," *IEEE Micro* (December 1993): 36–51.

[79] Federal Communications Commission, "FCC Bureau Chief Says 'Open' Internet Is Primary Goal, Cites Agreement of Consumers and Industry," November 9, 1999, http://transition.fcc.gov/Bureaus/Cable/News_Releases/1999/nrcb9018.html (accessed January 4, 2012); Open Internet Coalition, "Why an Open Internet: Openness Is a Fundamental Principle of the Internet," http://www.openinternetcoalition.org/index.cfm?objectid=8C7857B0-5C6A-11DF-9E27000C-296BA163 (accessed January 4, 2012); David D. Clark, "Network Neutrality: Words of Power and 800-Pound Gorillas," *International Journal of Communication* 1 (2007): 701–708.

9

Conclusion

Open Standards and an Open World

By the mid-1990s, the Internet's advocates had learned to cast Internet history in a most flattering light. They downplayed its autocratic and closed world origins, belittled the work undertaken by their competitors in Open Systems Interconnection (OSI), and reimagined the Internet as an open system. In their revisionist hands, the Internet standards process became a novel form of distributed control and participatory democracy that emerged organically from the interactions of Internet engineers.[1] Internet users, dazzled and enchanted by their sublime new toy, searched for secrets of the Internet's astonishing success that they could apply in other realms.[2] A chorus of academics and policy makers unwittingly latched on to an origin myth – that the Internet was a meritocracy, the product of "nerds" and "hackers" who collaborated through a decentralized and participatory design process.[3]

---

[1] Roy Rosenzweig, "Wizards, Bureaucrats, Warriors, and Hackers: Writing the History of the Internet," *American Historical Review* 103 (1998): 1530–1552; Anthony Rutkowski, "Today's Cooperative Competitive Standards Environment and the Internet Standards-Making Model," in Brian Kahin and Janet Abbate, eds., *Standards Policy for Information Infrastructure* (Cambridge, MA: The MIT Press, 1995); Craig Lyle Simon, *Launching the DNS War: Dot-Com Privatization and the Rise of Global Internet Governance* (PhD dissertation, University of Miami, 2006), 1–5.

[2] Many observers rightly highlighted the importance of ARPA's generous patronage. See for example Arthur L. Norberg and Judy E. O'Neill, *Transforming Computer Technology: Information Processing for the Pentagon, 1962–1986* (Baltimore: The Johns Hopkins University Press, 1996); and Janet Abbate, *Inventing the Internet* (Cambridge, MA: The MIT Press, 1999).

[3] See for example L. Jean Camp and Charles Vincent, "Looking to the Internet for Models of Governance," *Ethics and Information Technology* 6 (2004): 161–173; Jeffrey V. Nickerson and Michael zur Muhlen, "The Ecology of Standards Processes: Insights from Internet Standard Making," *MIS Quarterly* 30 (2006): 469; Christopher M. Kelty, *Two Bits: The Cultural Significance of Free Software* (Durham, NC: Duke University Press, 2008), 33–36 and 57–63; and Vittorio Bertola, "Power and the Internet," *Journal of Information, Communication and Ethics in Society* 8 (2010): 323–337.

The Internet standards process was an especially rich source of inspiration for legal scholars who offered new interpretations of legal philosophy, the process of innovation, and the role of nongovernmental and transnational regulatory regimes in the twenty-first-century global economy.[4] In many cases, the lessons gleaned from the Internet success story flew in the face of accepted scientific and engineering practice – evidence, it would seem, that the Internet's emergence truly marked a new kind of technology-enabled society. For example, Harvard law professor Jonathan Zittrain generalized from his reading of the Internet's architectural history to propose a "procrastination principle" built on the assumption that "most problems confronting a network can be solved later or by others." This principle, Zittrain argued, "suggests waiting for problems to arise before solving them."[5]

Two important points emerge from this new conventional wisdom that casts the Internet as a radically decentralized open system, capable of facilitating a new style of "organizing without organizations." First, it indicates that Internet advocates, popular writers, and the general public have a poor understanding of the Internet's history. The Internet, like many technological novelties that preceded it, provokes fantasies that its history does not support. Its origins were autocratic, not democratic. Its design and standardization indicates the central importance of organizational boundaries and the alliances that mobilize across them; it does not portend the death of organizational gatekeepers or of organizations altogether. The second point follows from the first. We should take with a grain of salt any policy or normative advice – particularly of the counterintuitive variety – that is based on the conviction that the Internet's creation provides new theoretical foundations for action. For example, it would be especially unwise for engineers in other industries (such as those who build bridges and airplanes) to bring the "procrastination principle" into their own design philosophies. The history of networking provides no reason to ignore the IBM software engineer Fred Brooks's insistence that "conceptual integrity is *the* most important consideration in system design."[6]

Rather than add to the chorus of observers who see the Internet as the harbinger and enabler of a techno-utopian "open world," I have chosen to situate

[4] See for example A. Michael Froomkin, "Habermas@Discourse.Net: Toward a Critical Theory of Cyberspace," *Harvard Law Review* 116 (2003): 749–873; Gralf-Peter Calliess and Peer Zumbansen, *Rough Consensus and Running Code: A Theory of Transnational Private Law* (Portland, OR: Hart Publishing, 2010); Tim Büthe and Walter Mattli, *New Global Rulers: The Privatization of Regulation in the World Economy* (Princeton, NJ: Princeton University Press, 2011).

[5] Clay Shirky, *Here Comes Everybody: The Power of Organizing without Organizations* (New York: Penguin Books, 2008); Patrice Flichy, *The Internet Imaginaire* (Cambridge, MA: The MIT Press, 2007); Jonathan Zittrain, *The Future of The Internet – And How to Stop It* (New Haven, CT: Yale University Press, 2008), 31, 135.

[6] Frederick P. Brooks, Jr., *The Mythical Man-Month: Essays on Software Engineering Anniversary Edition* (Boston: Addison-Wesley, 1995), 42.

the Internet within a deeper and more complex set of technical, political, and organizational contexts. Rather than focus on the lessons that past information networks can teach us in the present, I have chosen to study how the designers of networks responded to their own circumstances and how they have seen these through their own eyes. Inspired by Henry Demarest Lloyd's observation that "history is condensed in the catchwords of the people," I have been especially attentive to the discourses and practices of *standardization* through which network architects, engineers, and users sought to exercise power, impose order, create stability, and pursue *openness*.[7] By focusing on ideas as well as their social and material manifestations, I have argued that the lens of standardization allows us to see the history of information networks as ongoing acts of *critique*. These critiques reflected the ideological convictions of network architects, engineers, and users who were responding to the opportunities and constraints they faced. Critiques were simultaneously intellectual and technological, active equally in network technologies and in the institutions that sustained the standardization process.

My understanding of the long-term historical significance of these acts of critique has been guided by historical studies of technological and organizational change, summarized in the "organizational synthesis" outlined by the historian Louis Galambos.[8] The organizational lens that Galambos polished brought into clear focus the deep significance of the "new and elaborate networks of formal, hierarchical structures of authority" that emerged in the nineteenth, twentieth, and twenty-first centuries. Historians have been attentive to the expansion of government capabilities during this era, but, as we have seen in *Open Standards and the Digital Age*, new structures of authority also emerged outside of government control in engineering societies, trade associations, and standards-setting organizations. Ambitious individuals and powerful organizations – and coalitions of organizations – worked across the public, private, and nonprofit sectors. They did not necessarily need to see or act like a state in order to generate powerful organizational capabilities.[9]

---

[7] Henry Demarest Lloyd, *Wealth against Commonwealth* (New York: Harper & Brothers Publishers, 1894), 498.

[8] Louis Galambos, "The Emerging Organizational Synthesis in Modern American History," *Business History Review* 44 (1970): 279–290; Louis Galambos, "Technology, Political Economy, and Professionalization: Central Themes of the Organizational Synthesis," *Business History Review* 57 (1983): 473–491; Louis Galambos, "Recasting the Organizational Synthesis: Structure and Process in the Twentieth and Twenty-First Centuries," *Business History Review* 79 (2005): 1–37; Louis Galambos, *The Creative Society – And the Price Americans Paid for It* (New York: Cambridge University Press, 2012).

[9] Stephen Skowronek, *Building a New American State: The Expansion of National Administrative Capabilities, 1877–1920* (New York: Cambridge University Press, 1982); Brian Balogh, "Reorganizing the Organizational Synthesis: Federal-Professional Relations in Modern America," *Studies in American Political Development* 5 (1991): 119–172; James C. Scott, *Seeing Like a State: How Certain Schemes to Improve the Human Condition Have Failed* (New Haven, CT: Yale University Press, 1999); Timothy Schoechle, *Standardization and Digital Enclosure: The*

An organizational lens is particularly helpful for bringing into focus two key conceptual questions at the heart of this book: What organizational options did network designers and engineers have at their disposal, and how did they choose among those options? Many historians of capitalism and innovation, most notably Alfred D. Chandler, Jr., have tended to emphasize large, vertically integrated enterprises that were directed by layers of managerial hierarchies. Such hierarchies emerged, in Chandler's view, in situations where managers believed that hierarchies would provide more efficient coordination mechanisms than one-time exchanges guided only by the invisible hand of the market. In response to Chandler, the historians Naomi Lamoreaux, Daniel Raff, and Peter Temin have sought to "move beyond the simple markets versus hierarchies dichotomy that undergirds Chandler's analysis," and bring more attention to coordination mechanisms that exist in a conceptual middle ground between markets and hierarchies. Lamoreaux, Raff, and Temin conceptualized these mechanisms as hybrids of markets (which facilitate singular transactions) and hierarchies (which provide permanent structures for repeated transactions).[10]

As I have demonstrated at various points in the book, standards-setting committees are hybrid organizations – neither market nor hierarchy – that the organizational historians have identified as both deeply significant and as yet understudied. Standard-setting organizations often are studied in terms of their economic function, but they should also be considered in cultural and political terms. I have argued that we should understand these hybrid organizations – and the standards they create – as value-laden expressions of *ideology*, or ideas about how society should be ordered and how power should be exercised. I also have argued that innovation in network standards is a form of *critique*; these innovations do not merely challenge what is, they take productive action and make what could be. In this way, *Open Standards and the Digital Age* is an attempt to bring themes from business, economic, and organizational history into closer conversation with constructivist histories of science and technology. In the process, we can bring a richer historical perspective to the question of how control persists in the midst of decentralization.[11]

Privatization of Standards, Knowledge, and Policy in the Age of Global Information Technology (Hershey, PA: IGI Global, 2009).

[10] Ronald H. Coase, "The Nature of the Firm," *Economica* New Series, 4 (1937): 386–405; Oliver E. Williamson, "Transaction Cost Economics: The Natural Progression," in Karl Grandin, *Les Prix Nobel* (Stockholm: Nobel Foundation, 2010); Naomi R. Lamoreaux, Daniel M. G. Raff, and Peter Temin, "Beyond Markets and Hierarchies: Toward a New Synthesis of American Business History," *American Historical Review* 108 (2003): 404–433; Naomi R. Lamoreaux, Daniel M. G. Raff, and Peter Temin, eds., *Learning by Doing in Markets, Firms, and Countries* (Chicago: University of Chicago Press, 1999); Sally H. Clarke, Naomi R. Lamoreaux, and Steven W. Usselman, eds., *The Challenge of Remaining Innovative: Insights from Twentieth-Century Business History* (Stanford, CA: Stanford University Press, 2009).

[11] Steven Shapin and Simon Schaffer, *Leviathan and the Air-Pump: Hobbes, Boyle, and the Experimental Life* (Chicago: University of Chicago Press, 1989); Wiebe E. Bijker, Thomas P. Hughes, and Trevor Pinch, eds., *The Social Construction of Technological Systems: New*

With these methodological considerations in mind, this book follows two stories anchored in the American experience: the history of a collaborative form of "consensus" standardization and the history of information and communication networks. The combination of these stories, when viewed in a long-term perspective, suggests the outlines of three phases or modes of standardization, particularly with reference to communication network architectures and standards: the colonial era to roughly 1900, 1900 to the 1980s, and the 1980s to the present.[12]

Colonial Era–1900: De Facto Standards

American colonial encounters with British imperial authority – such as the Stamp Act crisis of 1765 – conditioned American politicians and citizens of the early Republic to mistrust concentrated government power. The Post Office Act of 1792 defined standard postage rates, procedures, and penalties for interfering with the mails, but Congress declined to create the organizational capabilities needed to set federal standards for weights, measures, and duties until well into the 1830s. Americans never pursued standardization with the level of coordination employed by governments in Western Europe. Instead, de facto standards for American industrial production emerged from the practices of interchangeable parts manufacturing in a variety of machine tool–based industries.[13]

Rather than depending on the federal government to coordinate economic activity, antebellum Americans were prolific formers of private voluntary organizations. Some of these groups, such as the Chicago Board of Trade and the New England Cotton Manufacturers' Association, facilitated stable relationships among a wide variety of market participants. Others, such as

Directions in the Sociology and History of Technology (Cambridge, MA: The MIT Press, 1989); John F. Kasson, *Civilizing the Machine: Technology and Republican Values in America, 1776–1900* (New York: Hill and Wang, 1999); Alexander R. Galloway, *Protocol: How Control Exists after Decentralization* (Cambridge, MA: The MIT Press, 2006); Richard R. John, *Network Nation: Inventing American Telecommunications* (Boston: Harvard University Press, 2010); Cyrus C. M. Mody, *Instrumental Community: Probe Microscopy and the Path to Nanotechnology* (Cambridge, MA: The MIT Press, 2011); Lee Jared Vinsel, "The Crusade for Credible Energy Information and Analysis in the United States, 1973–1982," *History and Technology* 28 (2012): 149–176; Andrew L. Russell, "Networks, Standards, and Critique," *IEEE Annals of the History of Computing* 34 (2012): 79–80.

[12] The predominant characteristics of these three chronological phases are not mutually exclusive. In many cases, de facto standards, consensus standards, and de jure standards co-exist. See for example Lars Heide, "The Danish Welding Institute and Force Technology, 1940–2005: Technical Standardization and the Shaping of Business," *Enterprises et Histoire* 51 (2008): 57–68.

[13] John Perry, *The Story of Standards* (New York: Funk & Wagnalls Company, 1955), 56–72; David Hounshell, *From the American System to Mass Production, 1800–1932* (Baltimore: The Johns Hopkins University Press, 1984).

Philadelphia's Franklin Institute, conducted investigations and issued technical proposals on topics such as steam boiler explosions and a standard design for screw threads. Although they performed significant economic and technical functions, none of these voluntary groups possessed the legal authority to enforce compliance with their suggestions for standards.[14]

After decades of "wasteful" competition, the telegraph industry consolidated under the monopoly control of Western Union immediately following the American Civil War. As telegraph networks spread across the continent, managers developed standards to rationalize human and technical components of the telegraph system. Whereas some standards – such as rules for telegraph operators and uniforms for messenger boys – emerged from the demands of managers at a private firm, other standards – such as agreements for Morse code and the ohm – were negotiated in international scientific committees. Once formed, standards were neither fixed nor permanent; rather, specifications and practices that arose as local practices spread in an uneven and unorganized way throughout industrial practice. In the process, a new class of economic actors led by engineers, scientists, and corporate managers assumed power over the fundamental tasks of telegraph delivery and innovation.

A variety of trade associations and engineering societies formed standard-setting committees in the 1880s and 1890s as they elaborated the moral, economic, and technological foundations of their professional practice. Committees of engineering elites set standards for pipe threads, electrical units, and steel rails. The first generations of the rules and bylaws that governed engineering standards committees reflected a broader progressive shift away from the chaos and mistrust of the market toward – sometimes stormy – widespread voluntary cooperation and deference to the orderly designs of expert minds. These institutional innovations, so characteristic of their era, were critiques of the industrial order of late-nineteenth-century America and proposals for a newly organized approach to the twentieth century.

1900–1980s: *Consensus Standards, Critique, and Convergence*

Before 1900, de facto standards were the norm: the federal government refused to use its authority to back specific standards, and no organizations existed that could harmonize industrial practice. As the pace and scope of standardization activities increased, jurisdictional overlaps and conflicts became common. Engineers became convinced that technical standards were essential for safe and efficient production, but, ironically, a lack of human coordination

---

[14] Louis Galambos, *Cooperation and Competition: The Emergence of a National Trade Association* (Baltimore: The Johns Hopkins University Press, 1966), 20–30; William Cronon, *Nature's Metropolis: Chicago and the Great West* (New York: W. W. Norton, 1991), 104–142; Bruce Sinclair, *Philadelphia's Philosopher Mechanics: A History of the Franklin Institute, 1824–1865* (Baltimore: The Johns Hopkins University Press, 1974).

threatened to undermine the ordering and rationalizing impulse at the core of industrial standardization.

The permanent establishment of standards committees in American professional engineering societies nevertheless marked a fundamental shift in the American system of standardization – the emergence of collaborative, "consensus" standardization for industrial production. The most important institution for the development of American consensus standards was the American Engineering Standards Committee (AESC), formed in 1918. The AESC was an organizational response to – and critique of – the adversarial and disorderly spirit of American capitalism in the late nineteenth century. Its critique was founded in the moralistic spirit of progressive engineering and inspired by the organizational successes of American mobilization for World War I. The AESC grew steadily in its first ten years before it was reconstituted as the American Standards Association (ASA) in 1928 and aligned itself more closely and explicitly with trade associations and agents of commerce and capital. In the process, it clarified its mission as a group to coordinate agreement around existing practices rather than to create new standards. With the reforms of 1928, the ASA's revised definition of a legitimate standard was one that could demonstrate "a consensus of those substantially concerned with its scope and provisions."[15]

Factors of political economy – particularly the emerging outlines of a corporatist order that fostered cooperation within and across the diverse sectors of the American industrial economy – facilitated the maturing of new organizational capabilities for standardization on an industrial, national, and international scale. The ASA's champions, led by "Mr. Standards," its longtime Secretary Paul Gough Agnew, brilliantly masked internal technical and jurisdictional conflicts by mobilizing discourses of consensus and cooperation to align the practices of consensus standardization with dominant trends in American corporate liberalism. As the ASA's mode of consensus standardization grew and flourished in the 1930s and 1940s, Agnew remained the clearest proponent of the ideology and practices of consensus standardization. He grew fond of arguing that cooperation among engineers within "industrial legislatures" was superior to adversarial conflicts in existing judicial and political institutions. Agnew's critique contained the powerful assertion that industry self-regulation would assure American prosperity and social harmony.

Elite telephone engineers were not among the founders and early advocates of voluntary consensus standardization. Instead, American Telephone and Telegraph engineers and executives led by John Carty, Frank Jewett, and Bancroft Gherardi struggled to create and enforce standards within and across the disparate units of the monopoly Bell System. Although their organizational setting was different from Agnew's ASA (AT&T was a regulated monopoly;

---

[15] American Standards Association, "Procedure," *American Standards Year Book* (New York: American Standards Association, 1929), 84.

the ASA was a coalition of groups from industry and government), their desire to enhance reliability, safety, and efficiency through planning and rationalization were the same. So too was their understanding that standards were simply agreements between producers and consumers – between makers and users. Gherardi's careful, systematic, and centralized style of network engineering and standardization generated a vast collection of Bell System Practices in the 1920s and 1930s that specified standards for an unimaginable variety of tasks and tools. AT&T's critics, suspicious that it was using its dominant position to inflate rates and suppress innovation, grew wary of the Bell System's power over network architecture and standardization. Between the 1930s and the 1970s, American policy makers, entrepreneurs, and intellectuals articulated critiques of centralized control which took the shape of new modes of regulation and new products and services. The persistence of these critiques – and their increasing intensity during the 1960s and 1970s – eventually contributed to the divestiture of the Bell System in the early 1980s.

IBM dominated the postwar computer industry and, like AT&T, projected an all-encompassing aura (which it had cultivated for marketing purposes) that many users resented. A wide variety of interested parties – competitors, antitrust regulators, bureaucrats in European governments, and the growing ranks of computer "hackers" – all articulated critiques of IBM's centralized control in the 1950s, 1960s, and 1970s. As part of their critiques, each of these groups also set into motion plans to topple IBM's dominant position. One of most significant critiques of IBM developed slowly within standard-setting committees organized by the International Federation for Information Processing and the International Organization for Standardization (ISO). These committees provided fertile ground for alliances of government researchers, academic scientists, and industrial engineers to develop new approaches to computer standardization. Because they were organized outside the scope of existing monopolies, they were ideally positioned to bring new ideas to the central technological development of their era: the convergence of computer and communication networks around new, digital transmission technologies. Within these committees, experts investigated designs for new, packet-switched networks. They debated at great length the relative merits of connection-oriented, "virtual circuit" designs – the favorite of engineers in established firms such as IBM and the telecom monopolies – versus a new style of connectionless, "datagram" designs championed by irreverent characters such as the Frenchman Louis Pouzin.

The examples discussed in Chapters 5 through 8 indicate the technological and diplomatic difficulties that faced anyone who used standard-setting committees as a staging ground to attack the established, centralized powers of telecommunications and computing in the 1960s, 70s, and 80s. In each of these chapters we see conditions of *persistent instability* in technology, business, politics, and particularly in the cultures of various standard-setting committees. Engineers in this era did not know what to call these new, digital technologies (compunications? A Catenet? Internetworking? Distributed systems?), but they knew that the new networks would have profound social and economic consequences. The network architectures and protocols developed between the 1960s and the 1970s – including the French Cyclades network, the American Arpanet, and the international X.25 protocol – therefore were not simply technological innovations. They also were organizational and political critiques to advance specific social and economic priorities.

The most significant of these critiques took shape in 1977, under the auspices of an international "industrial legislature" organized within an ISO subcommittee that pursued "standardization in the area of open systems," and "system interconnection."[16] The two major themes in this book – the history of a collaborative form of consensus standardization and the history of information and communication networks – converged in the creation of this subcommittee on Open Systems Interconnection. OSI was the symbol and driver of the movement for open systems that emerged in the 1970s as a critique of control centralized within IBM and the telecom monopolies. OSI's promoters – especially in France, Britain, and the United States – hoped to use its emergence to bring some coherence to the fragmented state of de facto and proprietary designs for computer network architectures. They decided to call their effort "open systems" because the term signified two things at once: their procedural value, stressing that any interested party should have a voice in the preparation of the new standards, and their technical value, emphasizing that the new design should be able to accommodate the widest possible diversity of equipment. OSI was the closest thing to global participatory democracy that the networking world had ever seen. This radical openness, in retrospect, was both its noblest goal and the seed of its demise.

The ideology of open systems sustained the collective endeavor of OSI and lent it some discursive coherence. But the bureaucratic procedures of the international standards process frustrated OSI's architects and ultimately prevented its commercial success. Computer professionals, more accustomed to the entrepreneurial frenzy of the computer business, were uncomfortable within formal standard-setting organizations. Oral history interviews with OSI's protagonists reveal the clash of cultures produced by the convergence of the manic computer industry and the buttoned-up norms of international standards meetings. John Day, recalling an OSI meeting in 1978, observed:

Standards people met from 9:00 to 4:00, took their leisurely time, and talked the issues. It was an old-boys club, but at those times, while the standards were important, they didn't have the economic impact that [OSI] was going to have. So the rules had changed substantially. There was big money involved, and everybody knew it.... Everybody realized that where we drew the lines for the layers, and how we did the technical solutions,

---

[16] "Resolution 11. Establishing Subcommittee 16 – Open System Interconnection," Box 19, Folder 1, Charles W. Bachman Papers, 1951–2007, CBI 125, Charles Babbage Institute, University of Minnesota, Minneapolis.

determined market lines, determined economics, determined money in somebody's pocket, so everybody was out for blood. It was no longer this nice old-boys club.[17]

1980s–Present: Consortia and a Disorganized Open World

The political economy and political culture of international standardization changed significantly in the 1980s and 1990s. The rise of the Internet was a symptom rather than a cause of these changes – although once the Internet user base expanded in the 1990s, it provided the infrastructure for "virtual" meetings of standard-setting organizations and therefore contributed to the pace and extent of change.

During the 1980s, Europeans used the power of national governments – and supranational governments – more explicitly as they forged a "new approach" to standardization that was linked to the European ideological, economic, and political program of harmonization and regionalization. Europeans created new organizations such as the European Telecommunications Standards Institute and the Groupe Spéciale Mobile to fill the chasm between the old style of protective "national champion" authority and the new ambitions for transnational and global standards. These organizations, well aware that the success of their standards was tied to the commercial alliances they could nurture, took the unprecedented step of including representatives from private companies in their standards-making machinery – another example of institutional innovation between markets and hierarchies.[18]

American policy makers at this time echoed their predecessors of the early twentieth century by placing their faith in cooperation among rival firms in the private sector to bolster American economic fortunes. Industry consortia became increasingly prominent, particularly in industries such as telecommunications, computing, and electronics. It was in these industries that innovations in technology and political economy – digital technologies and the global trend toward liberalization and privatization of state monopoly services – prompted professionals to consider fundamental reorganization. They needed new coordination mechanisms to bring order and control to a world whose technological, economic, political, and cultural foundations appeared to be in flux. Their experiments with consortia took them to what we now see as familiar territory, a middle ground between markets and hierarchies.[19] A 1995 summary

[17] John Day, interview by James Pelkey, July 11, 1988, courtesy of James Pelkey.

[18] Jacques Pelkmans, "The GSM Standard: Explaining a Success Story," *Journal of European Public Policy* 8 (2001): 432–453; Jeffrey L. Funk, *Global Competition between and within Standards* (London: Palgrave, 2002).

[19] Stanley M. Besen and Garth Saloner, "The Economics of Telecommunications Standards," in Robert W. Crandall and Kenneth Flamm, eds., *Changing the Rules: Technological Change, International Competition, and Regulation in Communications* (Washington, DC: Brookings Institution, 1989); Eli Noam, *Telecommunications in Europe* (New York: Oxford University Press, 1992); Christopher H. Sterling, "The FCC and Changing Technological Standards," *Journal of Communication* 32 (1982): 137–147; Gail Crotts Arnall and Lawrence M. Mead,

of consortia captured their diversity, as well as their overarching purpose: "Consortia come in many flavors. They may be horizontal (among competitors), vertical (between integrators and suppliers), or comprised of firms providing complementary products and services. They may develop specifications, patentable technology, or tools and platforms. They may be structured as stock companies, exclusive non-profit organizations, open trade associations, or ad hoc interest groups."[20]

Professionals in the computer, electronics, and telecommunications industries – even the standards engineers and executives at IBM – embraced the dynamic organizational capabilities offered by standards consortia. The modular structure of the 1980s computer industry depended on shared technical standards and interfaces in order to build stable information platforms and viable commercial products. But there were limits to this style of international, interindustry cooperation. By the 1990s, technology professionals around the world saw OSI's demise as a cautionary tale of "anticipatory standardization" and an illustration of the perils of a large and democratic design process. OSI's failure suggested that no amount of organization could overcome the profound incompatibilities that continued to exist in an age of technological convergence. Instead, OSI's fate suggested that the route to successful standardization was to use proven technologies, stay nimble, and be wary of the entrenched industrial incumbents that opposed radical change.[21]

OSI's failure was not simply a result of irreconcilable differences among its diverse constituents; it also faltered because it could not produce standards in the speedy manner demanded by manufacturers, vendors, and users of computer equipment. Its chief competitor was the Internet, sustained by a standardization process that was an exception to – and critique of – requirements for consensus, due process, and a balance of interests that were developed by voluntary consensus bodies such as the American National Standards Institute (ANSI) and ISO. Individuals, rather than companies or nation-states, were the only recognized participants. The Internet's work culture – with its affinity for late-night bar meetings, informal hallway conversations, and benevolent elders – inverted the stiff diplomacy and representative democracy embodied in the formal procedures of the venerable national and international bodies.

"The FCC as an Institution," in Leonard Lewin, ed., *Telecommunications: An Interdisciplinary Text* (Dedham, MA: Artech House, 1984).

[20] Michelle K. Lee and Mavis K. Lee, "High Technology Consortia: A Panacea for America's Competitiveness Problems?" *Berkeley Technology Law Journal* 6 (1992): 335–372; Peter Grindley, David C. Mowery, and Brian Silverman, "SEMATECH and Collaborative Research: Lessons in the Design of High-Technology Consortia," *Journal of Policy Analysis and Management* 13 (1994): 723–758; Lewis Branscomb and Brian Kahin, "Standards Processes and Objectives for the National Information Infrastructure," in Brian Kahin and Janet Abbate, eds., *Standards Policy for Information Infrastructure* (Cambridge, MA: The MIT Press, 1995), 11–12.

[21] Andrew S. Grove, *Only the Paranoid Survive* (New York: Doubleday, 1996).

By the mid-1990s, the Internet political philosophy of "rough consensus and running code" – and its circumvention of monopolies, governments, and established channels for setting international standards – was widely admired and increasingly celebrated as a new norm and a new model of standardization for the information technology industries. Some observers even saw the slogan as a panacea for the more general challenges of global "multistakeholder" governance.[22]

OSI was conceived as an open system developed through an open process, but Internet advocates in the 1990s appropriated the discourse of openness as they rejected technical ideas associated with OSI and mocked ISO's adherence to democratic conventions such as presidents and voting. Internet architects and engineers endorsed a new vision of openness, one that was based not on democratic procedures but rather on the technical prescriptions of a community of like-minded experts. "Meritocracy" served effectively as rhetorical cover for what was always designed to be a structure in which old-timers and insiders could maintain control.

The former "Internet Architect" David Clark warned the Internet community in 1992 that it might become a "standards elephant," just like OSI, if it failed to reconcile competing demands of speed and consensus. If Tim Berners-Lee's experience with the Internet standards process was any indication, Clark was right to be concerned. Berners-Lee, the inventor of the World Wide Web, its hypertext markup language (html), and the hypertext transfer protocol (http), submitted his ideas as proposals for Internet standards in 1992 and 1993. He soon grew frustrated with the "endless philosophical rat holes down which technical conversations would disappear." Rather than leave the fate of his creation to the Internet standards process, Berners-Lee decided to create a new standards consortium – the World Wide Web Consortium, or W3C. Rather than mimicking the Internet's rejection of membership and voting, the W3C asked private firms to pay $50,000 annually to become a member and therefore gain voting rights in W3C deliberations. For Berners-Lee, the ability to quickly mobilize a community of self-interested companies was more important than a ponderous and open-ended standards process. He was not only the Web's inventor; he was also a dot-org entrepreneur.[23]

---

[22] William Lehr, "Compatibility Standards and Interoperability: Lessons from the Internet," in Kahin and Abbate, eds., *Standards Policy for Information Infrastructure*; Tim Berners-Lee, *Weaving the Web: The Original Design and Ultimate Destiny of the World Wide Web by its Inventor* (New York: HarperOne, 1999), 98; and Calliess and Zumbansen, *Rough Consensus and Running Code*.

[23] Andrew L. Russell, "Dot-Org Entrepreneurship: Weaving a Web of Trust," *Enterprise et Histoire* 51 (2008): 44–56; Andrew L. Russell, "Constructing Legitimacy: The W3C's Patent Policy," in Laura DeNardis, ed., *Opening Standards: The Global Politics of Interoperability* (Cambridge, MA: The MIT Press, 2011); Raghu Garud, Sanjay Jain, and Arun Kumaraswamy, "Institutional Entrepreneurship in the Sponsoring of Common Technological Standards: The Case of Sun Microsystems and Java," *Academy of Management Journal* 45 (2002): 196–214.

New standards consortia picked up the slack where older modes of hierarchical and consensus-driven standardization failed, and anxious engineers searched for alternatives. They shared the same spirit that animated the Internet critique of the OSI standards process: a deep conviction that traditional standards-setting organizations, such as the CCITT and ISO, were incapable of producing standards at the speed demanded by vendors and users. Hundreds of consortia emerged throughout all sectors of the 1990s high-tech economy. Consortia in the computer networking industries included the X/Open Consortium (created in 1984) to set standards for the Unix operating system, CableLabs (created in 1988), the Video Electronic Standards Association (created in 1989), and the Asynchronous Transfer Mode Forum (created in 1991). Success stories that emerged from these and other consortia proclaimed their potential to overcome some of the inadequacies of existing standard-setting organizations.[24]

By the mid-1990s, some of the more sober analysts in business, government, academia, and industry – those who encountered the Internet in a critical spirit rather than in a utopian tizzy – began to agree that changes in the international standards system constituted a "regime transformation." The "Internet religious war" became a common point of reference for analysts such as William Drake, who saw the technical and cultural conflict between the Internet and OSI as part of a larger debate "between two competing visions of how international standardization processes and network development should be organized and controlled."[25] Consortia did not bother with the safeguards that ANSI and ISO had adopted to ensure that standards committees would be balanced, operate according to consensus, and have representation from all interested parties. Instead, according to one insider, many consortia were "designed to be vendor marketing tools, devoted to providing support for specific vendor 'openness' claims" – more of a high-tech public relations machine than an industrial legislature. Consortia, like the self-interested vendors they served, were more concerned with market dynamics of scale and scope. Unlike their predecessors

[24] H. Landis Gabel, "Open Standards in the European Computer Industry: The Case of X/OPEN," in H. Landis Gabel, ed., *Product Standardization and Competitive Strategy* (New York: North-Holland, 1987); Steven C. Salop, "Deregulating Self-Regulated ATM Networks," *Economics of Innovation and New Technology* 1 (1990): 85–96; Garth Saloner, "Economic Issues in Computer Interface Standardization," *Economics of Innovation and New Technology* 1 (1990): 135–156. As of September 2013, 880 consortia are listed at Andrew Updegrove, "The Consortiuminfo. org Standard Setting Organization and Standards List," http://www.consortiuminfo.org/links (accessed September 25, 2013).

[25] William Drake, "The Internet Religious War," *Telecommunications Policy* 17 (1993): 643; Paul A. David and Mark Shurmer, "Formal Standards-Setting for Global Telecommunications and Information Services: Towards an Institutional Regime Transformation?" *Telecommunications Policy* 20 (1996): 789–815; Susanne K. Schmidt and Raymund Werle, *Coordinating Technology: Studies in the International Standardization of Telecommunications* (Cambridge, MA: The MIT Press, 1998), 43–84; Jonathan Coopersmith, "Creating Fax Standards: Technology Red in Tooth and Claw," *Japanese Journal for the History of Science and Technology* 11 (2010): 37–65.

in professional engineering societies and international technical societies, they preferred the rapid dissemination of a pragmatic kludge to a time-consuming pursuit of new technical knowledge.[26]

Had earlier generations of standards entrepreneurs – Charles Dudley, Paul Agnew, Charles le Maistre, and so on – been alive to witness 1990s debates over consortia, they would have seen many familiar arguments. Engineers were still (or once again) agonizing over trade-offs between moving quickly on the one hand and waiting to forge a broad consensus on the other. The requirement for the balance between producers and users, a keystone of the process for American standards in the 1920s, was rejected by the standards entrepreneurs of the 1990s who wanted to move more quickly. The disorganized, uncoordinated, and ad hoc creation of new consortia also provoked familiar legal and procedural concerns. The "patent thicket," for example, prompted dozens of lawsuits and law review articles that sought to clarify the obligations and rights of intellectual property owners who participated in the standards consortia (and in voluntary consensus standards bodies more generally). Inconsistencies across consortia and other standards bodies were especially noticeable in the rules that governed intellectual property rights in standards. Some rules specified patent licensing on "fair, reasonable, and non-discriminatory" terms, others insisted on a royalty-free licensing policy, and still others left these questions dangerously undefined – thus forcing standards engineers to consult intellectual property lawyers and add further cost and inconvenience to the process.[27] Consortia participants also were right to be nervous that antitrust statutes might be applied to their activities.[28]

[26] Carl F. Cargill, "A Five-Segment Model for Standardization," in Kahin and Abbate, eds., *Standards Policy for Information Infrastructure*, 87. See also Richard Hawkins, "The Rise of Consortia in the Information and Communication Technology Industries: Emerging Implications for Policy," *Telecommunications Policy* 23 (1999): 159–173; Martin Weiss and Carl F. Cargill, "Consortia in the Standards Development Process," *Journal of the American Society for Information Science* 43 (1992): 559–565.

[27] Mark A. Lemley, "Intellectual Property Rights and Standard Setting Organizations," *California Law Review* 90 (2002): 1889–1980; Philip J. Weiser, "The Internet, Innovation, and Intellectual Property Policy," *Columbia Law Review* 103 (2003): 534–613; Carl Shapiro, "Navigating the Patent Thicket: Cross Licenses, Patent Pools, and Standard-Setting," in Adam B. Jaffe, Josh Lerner, and Scott Stern, eds., *Innovation Policy and the Economy* (Cambridge, MA: The MIT Press, 2006); Jorge Contreras, ed., *Standards Development Patent Policy Manual* (Chicago: American Bar Association, 2007); Marc Rysman and Timothy Simcoe, "Patents and the Performance of Voluntary Standard Setting Organizations," *Management Science* 54 (2008): 1920–1934; Stanley M. Besen and Robert J. Levinson, "Standards, Intellectual Property Disclosure, and Patent Royalties after *Rambus*," *North Carolina Journal of Law & Technology* 10 (2009): 233–282.

[28] *American Society of Mechanical Engineers v. Hydrolevel Corp.*, 456 U.S. 556 (1982); David Balto, "Standard Setting in a Network Economy," Cutting Edge Antitrust Law Seminars International, February 17, 2000, http://www.ftc.gov/speeches/other/standardsetting.shtm (accessed January 12, 2012); David J. Teece and Edward R. Sherry, "Standards Setting and Antitrust," *Minnesota Law Review* 87 (2003): 1913–1994; Christopher L. Sagers, "Antitrust Immunity and Standard Setting Organizations: A Case Study in the Public-Private Distinction,"

Although some problematic aspects of consortia standardization were new, the structural elements were familiar: jurisdictional competition from outside organizations, legal and regulatory uncertainty, cultural and technical incompatibilities among different technological communities, widespread skepticism toward centralized authority, and difficulties in forging and sustaining a meaningful consensus around technical standards.

The ad hoc and competitive organization of consortia led to the production of hasty, overlapping, conflicting, and confusing standards. The rise of consortia at the end of the twentieth century thus re-created the same coordination problems that had haunted standards engineers since the beginning of the century. The elite engineers of the early twentieth century created an umbrella organization to bring industrial standardization into some logistical and procedural coherence; but the mature versions of those umbrella organizations – the formal national and international voluntary consensus standards bodies – were the target of the consortia critique in the late 1980s and 1990s. Standards consortia had stormed onto the scene by rejecting and subverting accepted procedures for balance, openness, and due process. Because these mechanisms had been developed over time to assure critics of the fairness and public accountability of the voluntary consensus standards process, their rejection amounted to a crisis of legitimacy for technical standardization.[29]

Standards engineers, entrepreneurs, and consultants[30] pursued two strategies to overcome the uncertainties of standards consortia. The first was a lobbying and public relations campaign to persuade Congress and American presidential administrations to endorse the legitimacy and legality of this organizational innovation. The National Cooperative Research Act of 1984, along with subsequent legislation passed under the Reagan, Bush, Clinton, and Bush administrations that protected standards consortia from prosecution under American antitrust law, may be seen as indicators of the success of these lobbying efforts.

Cardozo Law Review 25 (2004): 1393–1427; Louis Galambos, "The Monopoly Enigma, the Reagan Administration's Antitrust Experiment, and the Global Economy," in Kenneth Lipartito and David Sicilia, eds., *Constructing Corporate America: History, Politics, Culture* (New York: Oxford University Press, 2004).

[29] Raymund Werle and Eric J. Iversen, "Promoting Legitimacy in Technical Standardization," *Science, Technology & Innovation Studies* 2 (2006): 19–39.

[30] Since the mid-twentieth century, standards engineers have claimed to be a distinct profession and, in some cases, a distinct academic discipline as well. The Standards Engineering Society was founded in 1947; by the 1990s, dozens – if not hundreds – of standards professionals worked full time as organizers of, contributors to, or consultants between a great variety of standard-setting organizations and consortia. See Lal C. Verman, *Standardization: A New Discipline?* (Hamden, CT: Archon Books, 1973); Henk De Vries, "Standardization – A New Discipline?" in *Proceedings of 2nd IEEE Conference on Standardization and Innovation in Information Technology* (Boulder, CO: International Center for Standards Research, 1999); Sherrie Bolin, ed., *The Standards Edge: Future Generations* (Ann Arbor, MI: Sheridan Books, 2005); Kai Jakobs, ed., *Standardization Research in Information Technology: New Perspectives* (Hershey, PA: IGI Global, 2007).

The new laws also validated the voluntary consensus standards process and protected from antitrust prosecution any standards-setting organization that followed the time-honored principles of openness, balance, transparency, consensus, and due process.[31] In the process, the American federal government consolidated decades – if not centuries – of policy abstinence that removed it from a significant direct role in setting standards. At the same time, it continued to play a critical indirect role to assure that the standards process would be viable in the hands of the private sector. It also continued to make decisive contributions in specific standardization projects by emphasizing the power of federal procurement, which often positioned federal agencies as the largest consumers of standards and gave them a strong voice in ostensibly private negotiations.[32]

The American developments inspired acts of policy entrepreneurship in the European Union, World Intellectual Property Organization, and World Trade Organization that helped to make the world safe for private regimes of standard setting. European regulators linked standardization to their visions of a fractured Europe rife with national champion firms – through a series of regulations that enacted the recommendations of policy proposals. Prominent examples included the 1987 green paper on a common European market for telecommunications equipment and services and a 1994 report titled *Europe and the Global Information Society* that bluntly urged the European Union to "put its faith in market mechanisms as the motive power to carry us into the Information Age."[33] The Europeans were in step with the spirit of the 1997 WTO Agreement on Basic Telecommunications Services,

[31] Carl F. Cargill, "Consortia Standards: Towards a Re-Definition of a Voluntary Consensus Standards Organization," Subcommittee on Environment, Technology, and Standards, Committee on Science, United States House of Representatives, June 28, 2001. Relevant legislation includes the National Cooperative Research Act of 1984; the National Cooperative Research and Production Act of 1993; the National Technology Transfer and Advancement Act of 1995; and the Standards Development Organization Advancement Act of 2004. Federal rules that govern the inclusion of standards in federal procurement practices are defined in "Federal Participation in the Development and Use of Voluntary Consensus Standards and in Conformity Assessment Activities," Office of Management and Budget (OMB) Circular A-119, Revised February 10, 1998.

[32] D. Linda Garcia, "Standard Setting in the United States: Private and Public Sector Roles," *Journal of the American Society of Information Science* 43 (1992): 531–537; Steven Oksala, Anthony Rutkowski, Michael Spring, and Jon O'Donnell, "The Structure of IT Standardization," *StandardView* 4 (1996): 9–22; Stephen Oksala, "The Changing Standards World: Government Did It, Even though They Didn't Mean To," *Standards Engineering: The Journal of the Standards Engineering Society* 52 (2000); Andrew Updegrove, "A Work in Progress: Government Support for Standard Setting in the United States, 1980–2004," *Consortium Standards Bulletin* 4 (2005): 1–8.

[33] Council of Ministers, *Council Recommendation of 12 November 1984 Concerning the Implementation of Harmonization in the Field of Telecommunications*, 84/549/EEC (Brussels: EEC, November 12, 1984); Council of Ministers, *Council Decision of 22 December 1986 on Standardization in the Field of Information Technology and Telecommunications*, 87/95/EEC (Brussels: EEC, December 22, 1986); Commission of the European Communities, *Towards a*

which aimed to increase competition by reducing barriers to entry, facilitate the global interconnection of communication networks, and create independent regulatory authorities to replace the old monopoly PTT administrations. Standards fit perfectly into late-twentieth century trends of globalization that opened up global markets to a neoliberal order that celebrated competition, entrepreneurship, innovation, and the global circulation of capital.[34] Along the way, the standardization process was most effective when it was steered toward short-term market pressures and familiar personal networks. This market-oriented process, however, did not perform well when it was used to address enduring questions of scientific truth or engineering quality – problems that consortia members assumed that other professionals would address.[35]

The Continuities of Critique

The implicit critique – sometimes made explicit – of these American and European standards-related legal strategies echoed a normative claim made by Paul Gough Agnew in the 1920s: market-oriented voluntary consensus standards were, in institutional terms, far superior to an unordered world of de facto standards or, worse, a totalitarian world of de jure standards. To lend further support to their claims of the economic and ideological legitimacy and superiority of private regimes of standardization, advocates of the consortia and voluntary consensus standardization processes pursued a second, discursive strategy: They invented a new category, *open standards*.

It seemed fitting that a variety of overlapping definitions of open standards – rather than a single cohesive definition – came to the fore in the late 1990s and the first decade of the twenty-first century.[36] Most agreed that creators,

Dynamic European Economy: Green Paper on the Development of the Common Market for Telecommunications Services and Equipment, COM (87) 290 (Brussels: CEC, June 30, 1987); High-Level Group on the Information Society, *Europe and the Global Information Society: Recommendations to the European Council* (Brussels: European Council, May 26, 1994); Commission of the European Communities, *Modernising ICT Standardisation – The Way Forward*, COM (2009) 324 (Brussels: CEC, 2009).

34 Peter F. Cowhey and Jonathan D. Aronson, *Transforming Global Information and Communication Markets: The Political Economy of Innovation* (Cambridge, MA: The MIT Press, 2009).

35 See the powerful critiques of Internet research as "groupthink" and mere "craft" work (as opposed to true scientific research) in John Day, *Patterns in Network Architecture: A Return to Fundamentals* (Upper Saddle River, NJ: Prentice-Hall PTR, 2007), 351–383. For the broader sociological issues involved, see Lee Smolin, *The Trouble with Physics: The Rise of String Theory, the Fall of a Science, and What Comes Next* (New York: Houghton Mifflin Harcourt, 2006).

36 See for example Carl Cargill, "Evolution and Revolution in Open Systems," *StandardView* 2 (1994): 3–13; Jonathan Band, "Competing Definitions of 'Openness' in the NII," in Kahin and Abbate, eds., *Standards Policy for Information Infrastructure*; Ken Krechmer, "The Principles of Open Standards," *Standards Engineering* 50 (November/December 1998): 1–6; Elliott Maxwell "Open Standards, Open Source, and Open Innovation: Harnessing the Benefits of Openness," *Innovations: Technology, Governance, Globalization* 1 (2006): 119–176; Joel West, "Seeking

implementers, and end users needed to have a voice in the development of open standards, and that more openness indicated a larger scope of inclusion for end users and the general public. There emerged further agreement around the paramount importance of well-defined *procedures* to guarantee public participation in the production of standards and liberal terms of *access* to allow public use of standardized technologies.

The open standards of the twenty-first century, however defined, rested on the core principles of voluntary consensus standardization invented in the late nineteenth century in Britain and the United States. These principles were codified in 1928 by the ASA, which placed the "consensus principle" – "A national standard implies a consensus of those substantially concerned with its scope and provisions" – at the beginning of its overhauled Rules of Procedure. Consensus was more than an intellectual or scientific judgment; it was also a strategy that enabled the ASA to extend its jurisdiction within a crowded and contested organizational field. It was an attempt to create a flexible system of alliances – neither market nor hierarchy – that linked producers and consumers in a transparent and rational way. Rather than trying to create a strong centralized organization, Agnew and his colleagues in the ASA devised the consensus principle as a way to settle a jurisdictional dispute – the "states' righters versus federalists" controversy that raged throughout the ASA's first decade.[37]

The tension between centralized and decentralized control also was the perpetual preoccupation for the builders of information and communication networks. These architects and engineers experimented with hybrid forms of control that borrowed from the guiding principles of representative and participatory democracies, elite republics, benevolent dictatorships, and technically oriented meritocracies. In the twenty-first century, the trade-offs between centralized and decentralized control – or, to put it another way, between closed and open systems – continue to occupy the engineers, executives, and policy makers who oversee network design and governance. Their challenge is to reconcile, as best they can, autocratic and democratic modes of design. Despite the rhetorical appeal of openness, closed systems and tamper-proof black boxes have retained popular appeal among system designers and users alike, as

---

Open Infrastructure: Contrasting Open Standards, Open Source, and Open Innovation," *First Monday* 12 (June 2007), http://firstmonday.org/ojs/index.php/fm/article/view/1913/1795 (accessed September 16, 2013); Andrew Updegrove, "ICT Standards Today: A System Under Stress," in DeNardis, ed., *Opening Standards*; Brad Biddle, "Definition of 'Open Standards,'" March 8, 2010, http://standardslaw.org/2010/03/open-standards-definitions (accessed January 12, 2012); Joseph M. Reagle, Jr., "The Puzzle of Openness," in *Good Faith Collaboration: The Culture of Wikipedia* (Cambridge, MA: The MIT Press, 2010); OpenStand, "The Modern Standards Paradigm – Five Key Principles," http://open-stand.org/principles (accessed September 5, 2012).

[37] American Standards Association, "Procedure," *American Standards Year Book* (New York: American Standards Association, 1929), 84; P. G. Agnew, "Twenty Years of Standardization," *Industrial Standardization* (1938): 229–237.

demonstrated by the continued popularity of closed systems such as the iPad and Microsoft Office applications over open source alternatives.[38]

The qualities of twenty-first century open standards, just like the 1920s consensus standards, need to be understood as the consequences of critiques. There is continuity in the pattern of standards engineers who sought to build what they believed to be a better society – one that embodied their own moral, economic, political, and technical preferences. Standards engineers such as Charles Dudley, Paul Agnew, Bancroft Gherardi, Isaac Auerbach, Louis Pouzin, Charles Bachman, and Vinton Cerf believed their organizational innovations would expose the shortcomings of existing power structures. When they took advantage of opportunities to reimagine the future and build it in a different way, the new power structures and new information networks they created registered the enduring tension between centralized and decentralized control.

The ideology of open standards in the twenty-first century thus embodies competing visions of technology and political power in a global capitalist economy. It came into being neither as a triumph of the centralizing forces of monopoly capitalism nor as a revolution of the countercultural or libertarian critics of the capitalist order. Rather, the ideological origins of digital networks lie primarily in critiques that emerged within the center of American society through conflicts among technical professionals who sought to impose their visions of order and control. The ideals of openness fit equally as comfortably in the spirit of entrepreneurial capitalism as they do in the liberatory impulse of the hacker ethic: witness the prominence of the term "open" in the marketing campaigns of IBM, American Express, free software collectives, and advocates of transparency in government. All of these groups traffic in the discourse of openness, and all have legitimate claims for doing so.[39] The rhetorical success of "openness" depends on its ability to capture widely shared values that privilege individual autonomy, reject coercion from industrial monopolists, promote private cooperation over state coercion, and celebrate the liberating and empowering potential of information and communication technologies. The ideology of openness expresses, above all else, full support for the neoliberal order that helped to produce the Internet and that the Internet, in turn, continues to reinforce on a global scale.

---

[38] See for example Zittrain, *The Future of the Internet*. A Web search for "open vs. closed" will yield a great number of results that discuss (with varying levels of lucidity) the trade-offs between centralized and decentralized control.

[39] Of the 880 consortia in the library of consortia at www.consortiuminfo.org, 70 begin with the word "open." Andrew Updegrove, "Standard Setting Organizations and Standard List," http://www.consortiuminfo.org/links/linksall.php (accessed September 25, 2013). See more generally David E. Nye, "Shaping Communication Networks: Telegraph, Telephone, Computer," *Social Research* 64 (1997): 1067–1091; Richard Barbrook and Andy Cameron, "The Californian Ideology," *Science as Culture* 6 (1996): 44–72; Paulina Borsook, *Cyberselfish: A Critical Romp through the Terribly Libertarian Culture of High Tech* (New York: PublicAffairs, 2001); Pekka Himanen, *The Hacker Ethic and the Spirit of the Information Age* (New York: Random House, 2001); Fred Turner, *From Counterculture to Cyberculture: Stewart Brand, the Whole Earth Network, and the Rise of Digital Utopianism* (Chicago: University of Chicago Press, 2006).

Bibliography

Manuscript Collections

Alex McKenzie Collection of Computer Networking Development Records (CBI 123), Charles Babbage Institute, University of Minnesota, Minneapolis.

American Engineering Standards Committee Minutes, American National Standards Institute, New York, New York.

AT&T Archives and History Center, Warren, New Jersey.

AT&T Archives and History Center, San Antonio, Texas.

Charles W. Bachman Papers (CBI 125), Charles Babbage Institute, University of Minnesota, Minneapolis.

Commerce Papers Series, The Papers of Herbert Hoover, Herbert Hoover Presidential Library, West Branch, Iowa.

Herbert S. Bright Papers (CBI 42), Charles Babbage Institute, University of Minnesota, Minneapolis.

Honeywell, Inc., X3.2 Standards Subcommittee Records (CBI 67), Charles Babbage Institute, University of Minnesota, Minneapolis.

Isaac L. Auerbach Papers (CBI 52), Charles Babbage Institute, University of Minnesota, Minneapolis.

Oral History Interviews

Bachman, Charles. Oral history interview by Andrew L. Russell, April 9, 2011, IEEE Computer Society History Committee, available from http://www.computer.org/comphistory/pubs/2012-03-russell.pdf.

Baran, Paul. Oral history interview by David Hochfelder, October 24, 1999, IEEE History Center, Rutgers University, New Brunswick, New Jersey.

Baran, Paul. Oral history interview by Judy E. O'Neill, March 5, 1990, Menlo Park, California. Charles Babbage Institute, University of Minnesota, Minneapolis.

Cerf, Vinton. Oral history interview by Judy E. O'Neill, April 24, 1990, Reston, Virginia. Charles Babbage Institute, University of Minnesota, Minneapolis.

Crocker, Steve. Oral history interview by Judy E. O'Neill, October 24, 1991, Glenwood, Maryland. Charles Babbage Institute, University of Minnesota, Minneapolis.

Danthine, Andre. Oral history interview by Andrew L. Russell, April 6, 2012, Liege, Belgium. Charles Babbage Institute, University of Minnesota, Minneapolis.

Day, John. Oral history interview by James Pelkey, July 11, 1988. Courtesy of James Pelkey.

Déspres, Rémi. Oral history interview by Valérie Schafer, May 16, 2012, Paris, France. Charles Babbage Institute, University of Minnesota, Minneapolis.

Gien, Michel. Oral history interview by Andrew L. Russell, April 3, 2012, Paris, France. Charles Babbage Institute, University of Minnesota, Minneapolis.

Grangé, Jean-Louis. Oral history interview by Andrew L. Russell, April 3, 2012, Paris, France. Charles Babbage Institute, University of Minnesota, Minneapolis.

Kahn, Robert. Oral history interview by Judy E. O'Neill, April 24, 1990, Reston, Virginia. Charles Babbage Institute, University of Minnesota, Minneapolis.

Le Lann, Gérard. Oral history interview by Andrew L. Russell, April 3, 2012, Paris, France. Charles Babbage Institute, University of Minnesota, Minneapolis.

Levilion, Marc E. Oral history interview by Andrew L. Russell, April 2, 2012, Paris, France. Charles Babbage Institute, University of Minnesota, Minneapolis.

Licklider, J. C. R. Oral history interview by William Aspray and Arthur L. Norberg, October 28, 1988, Cambridge, Massachusetts. Charles Babbage Institute, University of Minnesota, Minneapolis.

Mills, David A. Oral history interview by Andrew L. Russell, February 26, 2004, Newark, Delaware. Charles Babbage Institute, University of Minnesota, Minneapolis.

Naffah, Najah. Oral history interview by Andrew L. Russell, April 2, 2012, Paris, France. Charles Babbage Institute, University of Minnesota, Minneapolis.

Pouzin, Louis. Oral history interview by Andrew L. Russell, April 2, 2012, Paris, France. Charles Babbage Institute, University of Minnesota, Minneapolis.

Roberts, Lawrence G. Oral history interview by Arthur L. Norberg, April 4, 1989, San Francisco, California. Charles Babbage Institute, University of Minnesota, Minneapolis.

Ruina, Jack. Oral history interview by William Aspray, 20 April 1989, Cambridge, Massachusetts. Charles Babbage Institute, University of Minnesota, Minneapolis.

Strassburg, Bernard. Oral history interview by James Pelkey, May 3, 1988, Washington, DC.

Taylor, Robert. Oral history interview by William Aspray, February 28, 1989, San Francisco, California. Charles Babbage Institute, University of Minnesota, Minneapolis.

Zimmermann, Hubert. Oral history interview by James Pelkey, May 25, 1988.

Zimmermann, Hubert. Oral history interview with Mariann Unterluggauer, July 14, 2005.

Selected References

Abbate, Janet. *Inventing the Internet*. Cambridge, MA: The MIT Press, 1999.

Abbate, Janet. "Government, Business, and the Making of the Internet." *Business History Review* 75 (2001): 147–176.

Abbate, Janet. "Privatizing the Internet: Competing Visions and Chaotic Events, 1987–1995." *IEEE Annals of the History of Computing* 32 (2010): 10–22.

Agnew, Paul Gough. "How Business Is Policing Itself." *The Nation's Business* (December 1925), 41–43.

Agnew, Paul Gough. "A Step Toward Industrial Self-Government." *The New Republic* (March 17, 1926), 92–95.

Agnew, Paul Gough. "Twenty Years of Standardization." *Industrial Standardization* (December 1938): 229–237.

Akera, Atsushi. "Voluntarism and the Fruits of Collaboration: The IBM User Group, Share." *Technology and Culture* 42 (2001): 710–736.

Aldrich, Mark. *Safety First: Technology, Labor, and Business in the Building of American Work Safety, 1870–1939*. Baltimore: The Johns Hopkins University Press, 1997.

Band, Jonathan. "Competing Definitions of 'Openness' in the NII." In Brian Kahin and Janet Abbate, eds. *Standards Policy for Information Infrastructure*. Cambridge, MA: The MIT Press, 1995.

Bartky, Ian R. *Selling the True Time: Nineteenth-Century Timekeeping in America*. Stanford, CA: Stanford University Press, 2000.

Botein, Stephen. "'Meer Mechanics' and an Open Press: The Business and Political Strategies of Colonial American Printers." *Perspectives in American History* 9 (1975): 127–225.

Bowker, Geoffrey and Susan Leigh Star. *Sorting Things Out: Classification and Its Consequences*. Cambridge, MA: The MIT Press, 1999.

Bradner, Scott. "The Internet Engineering Task Force." In Chris DiBona, Sam Ockman, and Mark Stone, eds. *Open Sources: Voices from the Open Source Revolution*. Cambridge, MA: O'Reilly, 1999.

Busch, Lawrence. *Standards: Recipes for Reality*. Cambridge, MA: The MIT Press, 2011.

Büthe, Tim. "Engineering Uncontestedness? The Origin and Institutional Development of the International Electrotechnical Commission (IEC)." *Business & Politics* 12 (2010).

Campbell-Kelly, Martin and Daniel D. Garcia-Swartz. "The History of the Internet: The Missing Narratives." *Journal of Information Technology* 28 (2013): 18–33.

Cargill, Carl. *Open Systems Standardization: A Business Approach*. Upper Saddle River, NJ: Prentice-Hall PTR, 1997.

Castells, Manuel. *The Rise of the Network Society*. Cambridge, MA: Blackwell Publishers, 1996.

Ceruzzi, Paul E. *A History of Modern Computing*. Cambridge, MA: The MIT Press, 1998.

Chandler, Jr., Alfred D. *The Visible Hand: The Managerial Revolution in American Business*. Cambridge, MA: The Belknap Press of Harvard University Press, 1977.

Chandler, Jr., Alfred D. and James W. Cortada, eds. *A Nation Transformed by Information: How Information Has Shaped the United States from Colonial Times to the Present*. New York: Oxford University Press, 2000.

Chesbrough, Henry. *Open Innovation: The New Imperative for Creating and Profiting from Technology*. Boston: Harvard Business Review Press, 2003.

Chiet, Ross E. *Setting Safety Standards: Regulation in the Public and Private Sectors*. Berkeley: University of California Press, 1990.

Chumas, Sophie J., ed. *Directory of United States Standardization Activities*. Washington, DC: U. S. Department of Commerce, National Bureau of Standards, 1975.

Cochrane, Rexmond. *Measures for Progress: A History of the National Bureau of Standards*. Washington, DC: Department of Commerce, 1966.

Computer Science and Telecommunications Board. *Global Networks and Local Values: A Comparative Look at Germany and the United States.* Washington, DC: National Academies Press, 2001.

Cowhey, Peter F. and Jonathan D. Aronson. *Transforming Global Information and Communication Markets: The Political Economy of Innovation.* Cambridge, MA: The MIT Press, 2009.

Crocker, Dave. "Making Standards the IETF Way." *StandardView* 1 (1993): 48–56.

Cronon, William. *Nature's Metropolis: Chicago and the Great West.* New York: W. W. Norton, 1991.

Darnton, Robert. "An Early Information Society: News and Media in Eighteenth-Century Paris." *American Historical Review* 105 (2000): 1–35.

David, Paul A. "Clio and the Economics of QWERTY." *American Economic Review* 75 (1985): 332–337.

David, Paul A. "Standardization Policies for Network Technologies: The Flux between Freedom and Order Revisited." In Richard Hawkins, Robin Mansell, and Jim Skea, eds. *Standards, Innovation and Competitiveness.* Aldershot: Edward Elgar, 1995.

David, Paul A. "Understanding the Emergence of 'Open Science' Institutions: Functionalist Economics in Historical Context." *Industrial and Corporate Change* 13 (2004): 571–589.

David, Paul A. and Shane Greenstein. "The Economics of Compatibility Standards: An Introduction to Recent Research." *Economics of Innovation and New Technology* 1 (1990): 3–41.

David, Paul A. and Mark Shurmer. "Formal Standards-Setting for Global Telecommunications and Information Services: Towards an Institutional Regime Transformation?" *Telecommunications Policy* 20 (1996): 789–815.

David, Paul A. and W. Edward Steinmueller. "Economics of Compatibility Standards and Competition in Telecommunication Networks." *Information Economics and Policy* 6 (1994): 217–241.

Day, John. *Patterns in Network Architecture: A Return to Fundamentals.* Upper Saddle River, NJ: Prentice-Hall PTR, 2007.

DeNardis, Laura. *Protocol Politics: The Globalization of Internet Governance.* Cambridge, MA: The MIT Press, 2009.

Déspres, Rémi. "X.25 Virtual Circuits – Transpac in France – Pre-Internet Data Networking." *IEEE Communications Magazine* (November 2010), 40–46.

DiMaggio, Paul J. and Walter W. Powell. "The Iron Cage Revisited: Institutional Isomorphism and Collective Rationality in Organizational Fields." *American Sociological Review* 48 (1983): 147–160.

Downey, Gregory J. "Virtual Webs, Physical Technologies, and Hidden Workers: The Spaces of Labor in Information Internetworks." *Technology and Culture* 42 (2001): 209–235.

Downey, Gregory J. *Telegraph Messenger Boys: Labor, Technology, and Geography, 1850–1950.* New York: Routledge, 2002.

Drake, William J. "The Internet Religious War." *Telecommunications Policy* 17 (1993): 643–649.

Ensmenger, Nathan. *The Computer Boys Take Over: Computers, Programmers, and the Politics of Technical Expertise.* Cambridge, MA: The MIT Press, 2010.

Ensmenger, Nathan. "The Digital Construction of Technology." *Technology & Culture* 53 (2012): 753–776.

Flichy, Patrice. *The Internet Imaginaire.* Cambridge, MA: The MIT Press, 2007.

Foucault, Michel. "What Is Critique?" In David Ingram, ed. *The Political.* Malden, MA: Blackwell, 1978.

Friedman, Thomas L. *The World Is Flat: A Brief History of the Twenty-First Century.* New York: Farrar, Straus and Giroux, 2005.

Froomkin, A. Michael. "Habermas@Discourse.Net: Toward a Critical Theory of Cyberspace." *Harvard Law Review* 116 (2003): 749–873.

Galambos, Louis. "Technology, Political Economy, and Professionalization: Central Themes of the Organizational Synthesis." *Business History Review* 57 (1983): 471–493.

Galambos, Louis. "Theodore N. Vail and the Role of Innovation in the Modern Bell System." *Business History Review* 66 (1992): 95–126.

Galambos, Louis. "Recasting the Organizational Synthesis: Structure and Process in the Twentieth and Twenty-First Centuries." *Business History Review* 79 (2005): 1–37.

Galambos, Louis. *The Creative Society – And the Price Americans Paid for It.* New York: Cambridge University Press, 2012.

Garcia, D. Linda. "Standard Setting in the United States: Public and Private Sector Roles." *Journal of the American Society for Information Science* 43 (1992): 531–537.

Garcia, D. Linda. "Standards for Standard Setting: Contesting the Organizational Field." In Sherrie Bolin, ed. *The Standards Edge: Dynamic Tension.* Ann Arbor, MI: Sheridan Press, 2004.

Garnet, Robert W. *The Telephone Enterprise: The Evolution of the Bell System's Horizontal Structure, 1876–1909.* Baltimore: The Johns Hopkins University Press, 1985.

Goldsmith, Jack and Tim Wu. *Who Controls the Internet? Illusions of a Borderless World.* New York: Oxford University Press, 2006.

Gooday, Graeme J. N. *The Morals of Measurement: Accuracy, Irony, and Trust in Late Victorian Electrical Practice.* New York: Cambridge University Press, 2004.

Grindley, Peter, David C. Mowery, and Brian Silverman. "SEMATECH and Collaborative Research: Lessons in the Design of High-Technology Consortia." *Journal of Policy Analysis and Management* 13 (1994): 723–758.

Hafner, Katie and Matthew Lyon. *Where Wizards Stay Up Late: The Origins of the Internet.* New York: Simon & Schuster, 1996.

Haigh, Thomas. "Software in the 1960s as Concept, Service, and Product." *IEEE Annals of the History of Computing* 24 (2002): 5–13.

Handley, Mark. "Why the Internet Only Just Works." *BT Technology Journal* 24 (2006): 119–129.

Hart, David M. "Herbert Hoover's Last Laugh: The Enduring Significance of the 'Associative State.'" *Journal of Policy History* 10 (1998): 419–444.

Hart, David M. "Antitrust and Technological Innovation in the U. S.: Ideas, Institutions, Decisions, and Outcomes, 1890–2000." *Research Policy* 30 (2001): 923–936.

Hawkins, Richard. "The Rise of Consortia in the Information and Communication Technology Industries: Emerging Implications for Policy." *Telecommunications Policy* 23 (1999): 159–173.

Hawley, Ellis W. "Herbert Hoover, the Commerce Secretariat, and the Vision of an 'Associative State,' 1921–1928." *Journal of American History* 61 (1974): 116–140.

Heide, Lars. "The Danish Welding Institute and Force Technology, 1940–2005: Technical Standardization and the Shaping of Business." *Enterprises et Histoire* 51 (2008): 57–68.

Hoke, Donald R. *Ingenious Yankees: The Rise of the American System of Manufactures in the Private Sector*. New York: Columbia University Press, 1990.

Houldsworth, Jack. "Standards for Open Network Operation." *Computer Communications* 1 (1978): 5–12.

Hounshell, David A. *From the American System to Mass Production, 1800–1932*. Baltimore: The Johns Hopkins University Press, 1984.

Hughes, Thomas P. "The Evolution of Large Technological Systems." In Wiebe E. Bijker, Thomas P. Hughes, and Trevor Pinch, eds. *The Social Construction of Technological Systems: New Directions in the Sociology and History of Technology*. Cambridge, MA: The MIT Press, 1989.

Jesiek, Brent K. *Between Discipline and Profession: A History of Persistent Instability in the Field of Computer Engineering, circa 1951–2006*. PhD dissertation, Virginia Polytechnic Institute and State University, 2006.

John, Richard R. "Theodore N. Vail and the Civic Origins of Universal Service." *Business and Economic History* 28 (1999): 71–81.

John, Richard R. "Recasting the Information Infrastructure for the Industrial Age." In Alfred D. Chandler, Jr. and James W. Cortada, eds. *A Nation Transformed by Information: How Information Has Shaped the United States from Colonial Times to the Present*. New York: Oxford University Press, 2000.

John, Richard R. *Network Nation: Inventing American Telecommunications*. Boston: Harvard University Press, 2010.

Kahin, Brian and Janet Abbate, eds. *Standards Policy for Information Infrastructure*. Cambridge, MA: The MIT Press, 1995.

Kahn, Robert E. "The Role of the Government in the Evolution of the Internet." *Communications of the ACM* 37 (1994): 15–19.

Kasson, John F. *Civilizing the Machine: Technology and Republican Values in America, 1776–1900*. New York: Hill and Wang, 1999.

Katz, Michael L. and Carl Shapiro. "Systems Competition and Network Effects." *The Journal of Economic Perspectives* 8 (1994): 93–115.

Kelty, Christopher M. *Two Bits: The Cultural Significance of Free Software*. Durham, NC: Duke University Press, 2008.

Kindleberger, Charles P. "Standards as Public, Collective and Private Goods." *Kyklos* 36 (1983): 377–396.

Krechmer, Ken. "Open Standards Requirements." *International Journal of IT Standards and Standardization Research* 4 (2006): 43–61.

Krislov, Samuel. *How Nations Choose Product Standards and Standards Change Nations*. Pittsburgh, PA: University of Pittsburgh Press, 1997.

Lamoreaux, Naomi R., Daniel M. G. Raff, and Peter Temin. "Beyond Markets and Hierarchies: Toward a New Synthesis of American Business History." *American Historical Review* 108 (2003): 404–433.

Langlois, Richard N. "External Economies and Economic Progress: The Case of the Microcomputer Industry." *Business History Review* 66 (1992): 1–50.

Langlois, Richard N. "Modularity in Technology and Organization." *Journal of Economic Behavior & Organization* 49 (2002): 19–37.

Layton, Edwin T. *The Revolt of the Engineers: Social Responsibility and the American Engineering Profession.* Cleveland, OH: Press of Case Western Reserve University, 1971.

Lee, Michelle K. and Mavis K. Lee. "High Technology Consortia: A Panacea for America's Competitiveness Problems?" *Berkeley Technology Law Journal* 6 (1992): 335–372.

Lemley, Mark A. "Intellectual Property Rights and Standard-Setting Organizations." *California Law Review* 90 (2002): 1889–1980.

Lessig, Lawrence. *Code and Other Laws of Cyberspace.* New York: Basic Books, 1999.

Libicki, Martin. *Standards: The Rough Road to the Common Byte.* Cambridge, MA: Harvard Program on Information Resources Policy, 1994.

Lipartito, Kenneth. *The Bell System and Regional Business: The Telephone in the South, 1877–1920.* Baltimore: The Johns Hopkins University Press, 1989.

Lipartito, Kenneth. "When Women Were Switches: Technology, Work, and Gender in the Telephone Industry, 1890–1920." *American Historical Review* 99 (1994): 1074–1111.

Lipartito, Kenneth. "'Cutthroat' Competition, Corporate Strategy, and the Growth of Network Industries." *Research on Technological Innovation, Management and Policy* 6 (1997): 1–53.

Lukasik, Stephen J. "Why the Arpanet Was Built." *IEEE Annals of the History of Computing* 33 (2011): 4–21.

MacDougall, Robert. "Long Lines: AT&T's Long-Distance Network as an Organizational and Political Strategy." *Business History Review* 80 (2006): 297–327.

Malamud, Carl. *Exploring the Internet: A Technical Travelogue.* Englewood Cliffs, NJ: PTR Prentice Hall, 1992.

Mattli, Walter and Tim Büthe. "Setting International Standards: Technological Rationality or Primacy of Power?" *World Politics* 56 (2003): 1–42.

Maxwell, Elliott. "Open Standards, Open Source, and Open Innovation: Harnessing the Benefits of Openness." *Innovations: Technology, Governance, Globalization* 1 (2006): 119–176.

McKenzie, Alexander. "INWG and the Conception of the Internet: An Eyewitness Account." *IEEE Annals of the History of Computing* 33 (2011): 66–71.

McMahon, A. Michal. *The Making of a Profession: A Century of Electrical Engineering in America.* New York: Institute of Electrical and Electronics Engineers, 1984.

Metcalfe, J. S. and Ian Miles. "Standards, Selection and Variety: An Evolutionary Approach." *Information Economics and Policy* 6 (1994): 243–268.

Morozov, Evgeny. *The Net Delusion: The Dark Side of Internet Freedom.* New York: Public Affairs, 2011.

Moskowitz, Marina. *Standard of Living: The Measure of the Middle Class in Modern America.* Baltimore: The Johns Hopkins University Press, 2004.

Mueller, Milton. *Universal Service: Competition, Interconnection, and Monopoly in the Making of the American Telephone System.* Cambridge, MA: The MIT Press and American Enterprise Institute, 1997.

Mumford, Lewis. "Authoritarian and Democratic Technics." *Technology & Culture* 5 (1964): 1–8.

Murphy, Craig N. and JoAnne Yates. *The International Organization for Standardization (ISO): Global Governance through Voluntary Consensus.* New York: Routledge, 2009.

National Research Council. *The Internet's Coming of Age*. Washington, DC: National Academy Press, 2000.

Nissenbaum, Helen. "How Computer Systems Embody Values." *Computer* (March 2001), 118–120.

Noam, Eli. *Telecommunications in Europe*. New York: Oxford University Press, 1992.

Noble, David F. *America by Design: Science, Technology, and the Rise of Corporate Capitalism*. New York: Oxford University Press, 1977.

Nora, Simon and Alain Minc. *The Computerization of Society: A Report to the President of France*. Cambridge, MA: The MIT Press, 1981.

Norberg, Arthur L. and Judy E. O'Neill. *Transforming Computer Technology: Information Processing for the Pentagon, 1962–1986*. Baltimore: The Johns Hopkins University Press, 1996.

Nuechterlein, Jonathan and Philip J. Weiser. *Digital Crossroads: American Telecommunications Policy in the Internet Age*. Cambridge, MA: The MIT Press, 2013.

Nye, David E. "Shaping Communication Networks: Telegraph, Telephone, Computer." *Social Research* 64 (1997): 1067–1091.

Obama, Barack. "Transparency and Openness in Government." The White House, http://www.whitehouse.gov/open. Accessed August 19, 2010.

Olshan, Marc A. "Standards-Making Organizations and the Rationalization of American Life." *The Sociological Quarterly* 34 (1993): 319–335.

OpenStand. "The Modern Standards Paradigm – Five Key Principles." http://openstand.org/principles. Accessed September 5, 2012.

Oxman, Jason. "The FCC and the Unregulation of the Internet." Federal Communications Commission Office of Plans and Policy Working Paper No. 31, July 1999.

Padlipsky, M. A. *The Elements of Networking Style and Other Animadversions on the Art of Intercomputer Networking*. 2nd ed. Lincoln, NE: iUniverse, 2000.

Pelkey, James. "Entrepreneurial Capitalism and the Emergence of Computer Communications, 1968–1988." http://www.historyofcomputercommunications.info/. Accessed September 8, 2012.

Perry, John. *The Story of Standards*. New York: Funk & Wagnalls Company, 1955.

Piscitello, David M. and A. Lyman Chapin. *Open Systems Networking: TCP/IP and OSI*. Reading, MA: Addison-Wesley, 1993.

Puffert, Douglas J. *Tracks across Continents, Paths through History: The Economic Dynamics of Standardization in Railway Gauge*. Chicago: University of Chicago Press, 2009.

Pugh, Emerson W. *Building IBM: Shaping an Industry and Its Technology*. Cambridge, MA: The MIT Press, 1995.

Raunig, Gerald. "What Is Critique? Suspension and Recomposition in Textual and Social Machines." *Transversal* (2008), http://eipcp.net/transversal/0808/raunig/en. Accessed January 17, 2012.

Reagle, Jr., Joseph M. *Good Faith Collaboration: The Culture of Wikipedia*. Cambridge, MA: The MIT Press, 2010.

Reck, Dickson, ed. *National Standards in a Modern Economy*. New York: Harper, 1956.

Reich, Leonard S. *The Making of Industrial Research: Science and Business at GE and Bell, 1876–1926*. New York: Cambridge University Press, 1985.

Rohlfs, Jeffrey. "A Theory of Interdependent Demand for a Communications Service." *Bell Journal of Economics and Management Science* 5 (1974): 16–37.

Rosenberg, Nathan. "Technological Change in the Machine Tool Industry, 1840–1910." *The Journal of Economic History* 23 (1963): 414–443.

Rosenzweig, Roy. "Wizards, Bureaucrats, Warriors, and Hackers: Writing the History of the Internet." *American Historical Review* 103 (1998): 1530–1552.

Russell, Andrew L. "Standardization in History: A Review Essay with an Eye to the Future." In Sherrie Bolin, ed. *The Standards Edge: Future Generations*. Ann Arbor, MI: Sheridan Books, 2005.

Russell, Andrew L. "'Rough Consensus and Running Code' and the Internet-OSI Standards War." *IEEE Annals of the History of Computing* 28 (2006a): 48–61.

Russell, Andrew L. "Telecommunications Standards in the Second and Third Industrial Revolutions." *The Journal of the Communications Network* 5 (2006b): 100–106.

Russell, Andrew L. "Constructing Legitimacy: The W3C's Patent Policy." In Laura DeNardis, ed. *Opening Standards: The Global Politics of Interoperability*. Cambridge, MA: The MIT Press, 2011.

Russell, Andrew L. "Modularity: An Interdisciplinary History of an Ordering Concept." *Information & Culture: A Journal of History* 47 (2012): 257–287.

Rybczynski, Tony. "Commercialization of Packet Switching (1975–1985): A Canadian Perspective." *IEEE Communications Magazine* (December 2009), 26–32.

Saloner, Garth. "Economic Issues in Computer Interface Standardization." *Economics of Innovation and New Technology* 1 (1990): 135–156.

Salus, Peter H. *Casting the Net: From ARPANET to INTERNET and Beyond*. New York: Addison-Wesley Publishing Company, 1995.

Schafer, Valérie. *La France en Reseaux*. Paris: Nuvis, 2012.

Schmidt, Susanne K. and Raymund Werle. *Coordinating Technology: Studies in the International Standardization of Telecommunications*. Cambridge, MA: The MIT Press, 1998.

Schoechle, Timothy. *Standardization and Digital Enclosure: The Privatization of Standards, Knowledge, and Policy in the Age of Global Information Technology*. Hershey, PA: IGI Global, 2009.

Schumpeter, Joseph A. *Capitalism, Socialism, and Democracy*. New York: Harper & Row, 1976 [1942].

Scott, W. Richard. *Organizations: Rational, Natural, and Open Systems*. Englewood Cliffs, NJ: Prentice-Hall, 1981.

Scranton, Philip. *Endless Novelty: Specialty Production and American Industrialization, 1865–1925*. Princeton, NJ: Princeton University Press, 1997.

Shapiro, Andrew L. *The Control Revolution: How the Internet Is Putting Individuals in Charge and Changing the World We Know*. New York: PublicAffairs, 1999.

Shapiro, Carl and Hal Varian. *Information Rules: A Strategic Guide to the Network Economy*. Boston: Harvard Business School Press, 1999.

Shewart, W. A. "Nature and Origin of Standards of Quality." *Bell System Technical Journal* 37 (1958): 1–22.

Simcoe, Timothy S. "Open Standards and Intellectual Property Rights." In Henry Chesbrough, Wim Vanhaverbeke, and Joel West, eds. *Open Innovation: Researching a New Paradigm*. New York: Oxford University Press, 2006.

Simon, Craig Lyle. *Launching the DNS War: Dot-Com Privatization and the Rise of Global Internet Governance*. PhD dissertation, University of Miami, 2006.

Sinclair, Bruce. "At the Turn of a Screw: William Sellers, the Franklin Institute, and a Standard American Thread." *Technology & Culture* 10 (1969): 20–34.

Sinclair, Bruce. *A Centennial History of the American Society of Mechanical Engineers, 1880–1980*. Toronto: University of Toronto Press, 1980.

Sirbu, Marvin A. and Laurence E. Zwimpfer. "Standards Setting for Computer Communication: The Case of X.25." *IEEE Communications Magazine* 23 (1985): 35–44.

Slaton, Amy E. *Reinforced Concrete and the Modernization of American Building, 1900–1930*. Baltimore: The Johns Hopkins University Press, 2001.

Slaton, Amy and Janet Abbate. "The Hidden Lives of Standards: Technical Prescriptions and the Transformation of Work in America." In Michael Thad Allen and Gabrielle Hecht, eds. *Technologies of Power: Essays in Honor of Thomas Parke Hughes and Agatha Chipley Hughes*. Cambridge, MA: The MIT Press, 2001.

Smith, George David. *The Anatomy of a Business Strategy: Bell, Western Electric, and the Origins of the American Telephone Industry*. Baltimore: The Johns Hopkins University Press, 1985.

Stango, Victor. "The Economics of Standards Wars." *Review of Network Economics* 3 (2004): 1–19.

Suber, Peter. *Open Access*. Cambridge, MA: The MIT Press, 2012.

Tate, Jay. "National Varieties of Standardization." In Peter A. Hall and David Soskice, eds. *Varieties of Capitalism: The Institutional Foundations of Comparative Advantage*. New York: Oxford University Press, 2001.

Temin, Peter with Louis Galambos. *The Fall of the Bell System: A Study in Prices and Politics*. New York: Cambridge University Press, 1987.

Thompson, George V. "Intercompany Technical Standardization in the Early American Automobile Industry." *The Journal of Economic History* 14 (1954): 1–20.

Toth, Robert, ed. *Standards Activities of Organizations in the United States*. Gaithersburg, MD: National Institute of Standards and Technology, 1996.

Turner, Fred. *From Counterculture to Cyberculture: Stewart Brand, the Whole Earth Network, and the Rise of Digital Utopianism*. Chicago: University of Chicago Press, 2006.

United States Congress, Office of Technology Assessment. *Global Standards: Building Blocks for the Future*. Washington, DC: U.S. Government Printing Office, 1992.

Updegrove, Andrew. "A Work in Progress: Government Support for Standard Setting in the United States, 1980–2004." *Consortium Standards Bulletin* 4 (2005): 1–8.

Updegrove, Andrew. "ICT Standards Today: A System Under Stress." In Laura DeNardis, ed. *Opening Standards: The Global Politics of Interoperability*. Cambridge, MA: The MIT Press, 2011.

Usselman, Steven W. "IBM and Its Imitators: Organizational Capabilities and the Emergence of the International Computer Industry." *Business and Economic History* 22 (1993): 1–35.

Usselman, Steven W. *Regulating Railroad Innovation: Business, Technology, and Politics in America, 1840–1920*. New York: Cambridge University Press, 2002.

Usselman, Steven W. "Public Policies, Private Platforms: Antitrust and American Computing." In Richard Coopey, ed. *Information Technology Policy: An International History*. New York: Oxford University Press, 2004.

Verman, Lal C. *Standardization: A New Discipline?* Hamden, CT: Archon Books, 1973.

Vietor, Richard H. K. *Contrived Competition: Regulation and Deregulation in America.* Cambridge, MA: Belknap Press, 1994.

Vinsel, Lee Jared. "The Crusade for Credible Energy Information and Analysis in the United States, 1973–1982." *History and Technology* 28 (2012): 149–176.

Vincent, Charles and L. Jean Camp. "Looking to the Internet for Models of Governance." *Ethics and Information Technology* 6 (2004): 161–173.

von Bertalanffy, Ludwig. "An Outline of General System Theory." *British Journal for the Philosophy of Science* 1 (1950a): 139–164.

von Bertalanffy, Ludwig. "The Theory of Open Systems in Physics and Biology." *Science* New Series 111 (1950b): 23–29.

von Burg, Urs. *The Triumph of Ethernet: Technological Communities and the Battle for the LAN Standard.* Stanford, CA: Stanford University Press, 2001.

Waldrop, M. Mitchell. *The Dream Machine: J. C. R. Licklider and the Revolution That Made Computing Personal.* New York: Viking, 2001.

Wasserman, Neil H. *From Invention to Innovation: Long-Distance Telephone Transmission at the Turn of the Century.* Baltimore: The Johns Hopkins University Press, 1985.

Weber, Steven. *The Success of Open Source.* Boston: Harvard University Press, 2004.

Weiser, Philip J. "Law and Information Platforms." *Journal of Telecommunications and High Technology Law* 1 (2002), 1–35.

Weiser, Philip J. "The Internet, Innovation, and Intellectual Property Policy." *Columbia Law Review* 103 (2003): 534–613.

Weiss, Martin and Carl F. Cargill. "Consortia in the Standards Development Process." *Journal of the American Society for Information Science* 43 (1992): 559–565.

Werle, Raymund and Eric J. Iversen. "Promoting Legitimacy in Technical Standardization." *Science, Technology & Innovation Studies* 2 (2006): 19–39.

West, Joel. "Seeking Open Infrastructure: Contrasting Open Standards, Open Source, and Open Innovation." *First Monday* 12 (June 2007), http://firstmonday.org/ojs/index.php/fm/article/view/1913/1795 (accessed September 16, 2013).

Whitney, Albert W. *The Place of Standardization in Modern Life.* Washington, DC: Government Printing Office, 1924.

Wiebe, Robert H. *The Search for Order: 1877–1920.* New York: Hill and Wang, 1967.

Winner, Langdon. "Do Artifacts Have Politics?" *Daedelus* 109 (1980): 121–136.

Yates, JoAnne. *Control through Communication: The Rise of System in American Management.* Baltimore: The Johns Hopkins University Press, 1989.

Index